T0134677

Association for Women in Mathematics Series

Volume 15

Association for Women in Mathematics Series

Focusing on the groundbreaking work of women in mathematics past, present, and future, Springer's Association for Women in Mathematics Series presents the latest research and proceedings of conferences worldwide organized by the Association for Women in Mathematics (AWM). All works are peer-reviewed to meet the highest standards of scientific literature, while presenting topics at the cutting edge of pure and applied mathematics. Since its inception in 1971, The Association for Women in Mathematics has been a non-profit organization designed to help encourage women and girls to study and pursue active careers in mathematics and the mathematical sciences and to promote equal opportunity and equal treatment of women and girls in the mathematical sciences. Currently, the organization represents more than 3000 members and 200 institutions constituting a broad spectrum of the mathematical community, in the United States and around the world.

More information about this series at http://www.springer.com/series/13764

Alyson Deines • Daniela Ferrero • Erica Graham
Mee Seong Im • Carrie Manore • Candice Price
Editors

Advances in the Mathematical Sciences

AWM Research Symposium, Los Angeles, CA, April 2017

Editors
Alyson Deines
Center for Communications Res
San Diego, CA, USA

Daniela Ferrero
Department of Mathematics
Texas State University
San Marcos, TX, USA

Erica Graham
Bryn Mawr College
Ardmore, PA, USA

Mee Seong Im
United States Military Academy
New York, NY, USA

Carrie Manore
Los Alamos National Laboratory
Los Alamos, NM, USA

Candice Price
Department of Mathematics
University of San Diego
San Diego, CA, USA

ISSN 2364-5733 ISSN 2364-5741 (electronic)
Association for Women in Mathematics Series
ISBN 978-3-030-07520-0 ISBN 978-3-319-98684-5 (eBook)
https://doi.org/10.1007/978-3-319-98684-5

Mathematics Subject Classification: 05E05, 05E10, 14Q99, 16G20, 18E10, 20F65, 20G40, 34G10, 57M27, 92B25, 92C20, 92D20, 92E20, 97D40

This Springer imprint is published by the registered company Springer Nature Switzerland AG
The registered company address is: Gewerbestrasse 11, 6330 Cham, Switzerland

Preface

This volume highlights the mathematical research presented at the 2017 Association for Women in Mathematics (AWM) Research Symposium. This event, fourth in the biennial series of AWM Research Symposia, was held at the University of California Los Angeles (UCLA) on April 8–9, 2017. The objective of the AWM Research Symposia Series is to increase awareness of the mathematical achievements of women in academia, industry, and government, as well as to provide a supportive environment for female mathematicians, at all stages of their careers, to share their research. Additionally, these symposia promote research collaboration, as they facilitate the creation of new networks of women researchers and support the research collaboration networks already existing in several areas of mathematics. The symposia also include social events to enable networking among women in different career paths, or at different career stages, while promoting the discussion of prospects, visibility, and recognition.

About the 2017 AWM Research Symposium

The 2017 AWM Research Symposium was organized by Raegan Higgins, Kristin Lauter, Magnhild Lien, Ami Radunskaya, Tatiana Toro, Luminita Vese, and Carol Woodward. The Department of Mathematics at UCLA and the Institute for Pure and Applied Mathematics (IPAM) hosted the event that featured 4 plenary talks by distinguished women mathematicians (Table 1), 19 special sessions on a broad range of areas in mathematics (Table 2), a poster sessions for graduate students and recent Ph.D.s, a job panel, a reception, a banquet, and a student chapter event.

Eight of the 19 special sessions were organized by research networks supported by the AWM ADVANCE grant: Women in Numbers (WIN), Women in Math Biology (WIMB), Women in Noncommutative Algebra and Representation Theory (WINART), Women in Numerical Analysis and Scientific Computing (WINASC), Women in Computational Topology (WinCompTop), Women in Topology (WIT),

Table 1 2017 AWM Research Symposium: plenary talks

Speaker	Title
Ruth Charney	Searching for hyperbolicity
Svitlana Marboroda	The hidden landscape of localization of eigenfunctions
Linda Petzold	Inference of the functional network controlling circadian rhythm
Mariel Vazquez	Understanding DNA topology

Women in Shape (WiSh), and Algebraic Combinatorixx (ACxx). For more details about these and other Research Collaboration Networks for Women visit: https://awmadvance.org/research-networks/.

The keynote speaker at the banquet was Maria Helena Noronha, who in a very inspirational address described her journey from Brazil to Southern California, as a researcher and as a mentor. During the banquet, the second AWM Presidential Award was presented to Deanna Haunsperger, in recognition of her contribution to advance the goals of AWM through her work in the Summer Math Program (SMP) at Carleton College.

The symposium also included a full-day session titled *Wikipedia edit-a-thon*, during which participants at the 2017 AWM Research Symposium took turns writing Wikipedia entries to enhance the visibility of women in mathematics and their contributions.

About This Volume

The first chapter in this volume corresponds to the opening plenary talk at the symposium, *Searching for Hyperbolicity* by Ruth Charney. Her chapter is an excellent introduction to geometric group theory, by a renowned expert in the field, whose research contributed to the consolidation of geometric group theory as a mathematical area.

The following three chapters comprise mathematical results presented at the *WINART Special Session: Representations of Algebras*. Mee Seong Im and Angela Wu present their work on representation theory of the generalized iterated wreath product of cyclic groups (chapter "Generalized Iterated Wreath Products of Cyclic Groups and Rooted Trees Correspondence") and symmetric groups (chapter "Generalized Iterated Wreath Products of Symmetric Groups and Generalized Rooted Trees Correspondence"). Chapter "Conway–Coxeter Friezes and Mutation: A Survey" consists of a survey on Conway-Coxeter friezes and mutation, in which Karin Baur, Eleonore Faber, Sira Gratz, Khrystyna Serhiyenko, and Gordana Todorov connect Conway-Coxeter friezes, introduced in combinatorics in the 1970s, and cluster combinatorics, arising from the introduction of cluster algebras in the early 2000s.

Table 2 2017 AWM Research Symposium: special sessions

Title	Organizers
WIN—Work from Women in Numbers	Beth Malmskog, Katherine Stange
WinCompTop: Applications of Topology and Geometry	Emilie Purvine, Radmila, Sazdanovic, Shirley Yap
WIMB—From Cells to Landscapes: Modeling Health and Disease	Erica Graham, Carrie Manore
ACxx: Algebraic Combinatorics	Hélène Barcelo, Gizem Karaali
WINASC: Recent Research Development on Numerical Partial Differential Equations and Scientific Computing	Chiu-Yen Kao, Yekaterina Epshteyn
WINART: Representations of Algebras	Susan Montgomery, Maria Vega
WiSh: Shape Modeling and Applications	Asli Genctav, Kathryn Leonard
WIT—Topics in Homotopy Theory	Julie Bergner, Angelica Osorno
Women in Sage Math	Alyson Deines, Anna Haensch
Women in Government Labs	Cindy Phillips, Carol Woodward
EDGE-y Mathematics: A Tribute to Dr. Sylvia Bozeman and Dr. Rhonda Hughes	Alejandra Alvarado, Candice Price
SMPosium: A Celebration of the Summer Mathematics Program for Women	Alissa S. Crans, Pamela A. Richardson
The Many Facets of Statistics Applied, Pure and BIG	Monica Jackson, Jo Hardin
History of Mathematics	Janet Beery
Commutative Algebra	Alexandra Seceleanu, Emily Witt
Biological Oscillations Across Time Scales	Tanya Leise, Stephanie Taylor
Geometric Group Theory	Pallavi Dani, Tullia Dymarz, Talia Fernos
Recent Progress in Several Complex Variables	Purvi Gupta, Loredana Lanzani
Research in Collegiate Mathematics Education	Shandy Hauk, Pao-sheng Hsu

Chapter "Orbit Decompositions of Unipotent Elements in the Generalized Symmetric Spaces of $\mathrm{SL}_2(\mathbb{F}_q)$" features the work of Catherine Buell, Vicky Klima, Jennifer Schaefer, Carmen Wright, and Ellen Ziliak, who jointly studied the orbit decompositions of unipotent elements in the generalized symmetric spaces of $SL_2(F_q)$. Their results were presented in the special session *EDGE-y Mathematics: A Tribute to Dr. Sylvia Bozeman and Dr. Rhonda Hughes*. Chapter "A Characterization of the $U(\Omega, m)$ Sets of a Hyperelliptic Curve as Ω and m Vary" is a paper on computational algebraic geometry by Christelle Vincent, and it was part of the special session *Women in Sage*. In "A First Step Toward Higher Order Chain Rules in Abelian Functor Calculus", Christine Osborne and Amelia Tebbe take a first step toward higher-order chain rules in abelian functor calculus, by proving the second-order directional derivative chain rule using concrete computational techniques. Their work was presented in the *WIT-Topics in Homotopy Theory*.

In "DNA Topology Review" the volume transitions to mathematical biology, as Garrett Jones and Candice Price survey how topology can be applied to model biological processes, such as actions of proteins on DNA. Their work introduces essential concepts together with a description of fundamental literature, offering an excellent guide for undergraduate or graduate students, as well as for scholars, interested in learning the basics of DNA topology. Chapter "Structural Identifiability Analysis of a Labeled Oral Minimal Model for Quantifying Hepatic Insulin Resistance" features a paper presented at the special session *From Cells to Landscapes: Modeling Health and Disease.* Jacqueline Simens, Melanie Cree-Green, Bryan Bergman, Kristen Nadeau, and Cecilia Diniz Behn present a structural identifiability analysis of a labeled oral minimal model for quantifying hepatic insulin resistance. Their research contributes to the understanding of aging, trauma, and many diseases, such as obesity and type 2 diabetes.

The following two chapters correspond to the special session *Biological Oscillations Across Time Scales*. Chapter "Spike-Field Coherence and Firing Rate Profiles of CA1 Interneurons During an Associative Memory Task" features a study of the spike-field coherence and firing rate profiles of CA1 interneurons during an associative memory task, by Pamela Riviere and Lara Rangel. Their work analyzes whether inhibitory interneurons from the CA1 region of the hippocampus contain information about task dimensions in their firing rates. Following this work is a computational study of learning-induced sequence reactivation during sharp-wave ripple activity, as observed during sleep states, by Paola Malerba, Katya Tsimring, and Maxim Bazhenov. The authors use a model of spiking neuron networks of excitatory and inhibitory neurons in the CA3 and CA1 regions of the hippocampus to study the firing behavior of neurons during sharp-wave-ripple activity.

Chapter "Learning-Induced Sequence Reactivation During Sharp-Wave Ripples: A Computational Study" corresponds to a presentation in the special session *WINASC: Recent Research Development on Numerical Partial Differential Equations and Scientific Computing*. It consists of the work by Beatrice Riviere and Xin Yang using a DG method for the simulation of CO_2 storage in a saline aquifer. Their work has important practical applications, since porous media, such as saline aquifers or oil and gas reservoirs, are a major cause of the excessive amount of carbon dioxide in the atmosphere.

In "A DG Method for the Simulation of CO_2 Storage in Saline Aquifer", Beth M. Campbell Hetrick presents a study of regularization for an ill-posed inhomogeneous Cauchy problem, extending previous results for the homogeneous problem to the inhomogeneous case. While results and numerical experiments in Hilbert space are plentiful, this chapter contains regularization results for inhomogenous ill-posed problems for true Banach space, where little is known and many exciting problems await solutions.

The 2017 AWM Research Symposium included a special session on mathematics education comprising six presentations on various theoretical perspectives on the nature of human cognition and knowledge structures. Research methods ranged from individual interview and classroom observation to national survey and in-depth study of a particular instance or case. The span of topics covered calculus, combinatorics, linear algebra, foundations of proof, application of mathematics to teaching, and development of future teachers. This volume concludes with a chapter in which Shandy Hauk, Chris Rasmussen, Nicole Engelke Infante, Elise Lockwood, Michelle Zandieh, Stacy Brown, Yvonne Lai, and Pao-sheng Hsu offer highlights of the six presentations in this session.

Acknowledgments

All chapters in this volume have been peer-reviewed. The editors thank the authors who submitted their manuscripts and the referees who carefully and thoroughly reviewed them. Their highly valuable work was essential to ensure the quality of this volume meets the high standards that characterize the AWM Springer Series.

The editors express their gratitude to the organizers of the symposium and the hosting institutions, whose dedication and hard work were central to the success of the 2017 AWM Research Symposium. They also extend their appreciation to the National Science Foundation (NSF), the National Security Agency (NSA), and the Mathematical Sciences Research Institute (MSRI) for their funding to support special sessions. The editors extend their appreciation to Microsoft for their help funding the student event, to Wolfram and Maple for sponsoring prizes for best poster contest, and to Springer, Oxford University Press, Basic Books, AMS, and *Mathematical Reviews* for presenting booths in the exhibit area.

Special thanks to Kristin Lauter, editor of the AWM Springer Series, for trusting us with the task of publishing this volume and to Jeffrey Taub and Dimana Tzvetkova, our editors at Springer, for their guidance and help while editing these proceedings.

San Diego, CA, USA — Alyson Deines
San Marcos, TX, USA — Daniela Ferrero
Ardmore, PA, USA — Erica Graham
New York, NY, USA — Mee Seong Im
Los Alamos, NM, USA — Carrie Manore
San Diego, CA, USA — Candice Price
July 2018

Contents

Searching for Hyperbolicity

Ruth Charney

Abstract This paper is an expanded version of a talk given at the AWM Research Symposium 2017. It is intended as a gentle introduction to geometric group theory with a focus on the notion of hyperbolicity, a theme that has inspired the field from its inception to current-day research. The last section includes a discussion of some current approaches to extending techniques from hyperbolic groups to more general classes of groups.

1 Introduction

This paper is an expanded version of a talk given at the AWM Research Symposium 2017. It is intended as a gentle introduction to geometric group theory for the non-expert, with a focus on the notion of hyperbolicity. Geometric group theory came into its own in the 1990s, in large part due to a seminal paper of Mikhail Gromov [14]. While the field has grown considerably since that time, hyperbolicity remains a central theme and continues to drive much current research in the field.

As the name suggests, geometric group theory provides a bridge between groups viewed as algebraic objects and geometry. Groups arise in all areas of mathematics and can be described in many different ways. Some arise purely algebraically (such as certain matrix groups), others have combinatorial descriptions (via presentations), and still others are defined topologically (such as fundamental groups of topological spaces). Geometric group theory is based on the principle that if a group acts as symmetries of some geometric object, then one can use geometry to better understand the group.

For many groups, it is easy to find such an action. The symmetric group on n-letters acts by symmetries on an n-simplex and the dihedral group of order $2n$ is the symmetry group of a regular n-gon. This is also the case for some infinite groups.

R. Charney (✉)
Brandeis University, Waltham, MA, USA
e-mail: charney@brandeis.edu

A. Deines et al. (eds.), *Advances in the Mathematical Sciences*, Association for Women in Mathematics Series 15, https://doi.org/10.1007/978-3-319-98684-5_1

The free abelian group $\mathbb{Z}^n$ acts by translation on $\mathbb{R}^n$ (preserving the Euclidean metric), and the free group on n-generators acts by translation on a regular tree of valence $2n$ with edges of length one.

So the first question one might ask is, when can one find such an action? Given an abstract group G, can we always realize G as a group of symmetries of some geometric object? As we will see below, the answer is yes. However, some geometric objects are more useful than others for this purpose. In the early 1900s, Max Dehn was interested in groups arising as fundamental groups of hyperbolic surfaces. These groups act by isometries on the hyperbolic plane $\mathbb{H}^2$. Dehn used the geometry of the hyperbolic plane to prove some amazing properties of these groups. We will discuss one of these results in Sect. 3 below. Decades later, these ideas motivated Gromov, who introduced the notion of a hyperbolic metric space. He showed that the properties that Dehn deduced held more generally for any group acting nicely on such a space.

Gromov's notion of hyperbolic spaces and hyperbolic groups have been much studied since that time. Many well-known groups, such as mapping class groups and fundamental groups of surfaces with cusps, do not meet Gromov's criteria, but nonetheless display some hyperbolic behavior. In recent years, there has been much interest in capturing and using this hyperbolic behavior wherever and however it occurs.

In this paper, I will review some basic notions in geometric group theory, discuss Dehn's work and Gromov's notion of hyperbolicity, then introduce the reader to some recent developments in the search for hyperbolicity. My goal is to be comprehensible, not comprehensive. For those interested in learning more about the subject, I recommend [1] and [6] for a general introduction to geometric group theory and [7] and [13] for more detail about hyperbolic groups.

2 Geodesic Metric Spaces, Isometries and Quasi-Isometries

We begin with some basic definitions. Let X be a metric space with distance function $d : X \times X \to \mathbb{R}$. A *geodesic* in X is a distance preserving map from an interval $I \subset \mathbb{R}$ into X, that is, a map $\alpha : I \to X$ such that for all $t_1, t_2 \in I$,

$$d(\alpha(t_1), \alpha(t_2)) = |t_1 - t_2|.$$

The interval I may be finite or infinite. This definition is analogous to the notion of a geodesic in a Riemannian manifold. In particular, a geodesic between two points in X is a length-minimizing path.

A *geodesic metric space* is a metric space X in which any two points are connected by a geodesic. For such metric spaces, the distance is intrinsic to the space; the distance between any two points is equal to the minimal length of a path connecting them. Often, we also require that our metric space be *proper*, that is, closed balls in X are compact.

Example 1 Consider the unit circle S^1 in the plane. There are two natural metrics we could put on S^1. The first is the induced Euclidean metric: the distance between two points is the length of the straight line in $\mathbb{R}^2$ between them. The other is the arc length metric: the distance between two points is the length of the (shortest) circular arc between them. The first of these is not a geodesic metric (since, for example, there is no path in S^1 of length 2 connecting a pair of antipodal points) whereas the second one is geodesic.

Example 2 Suppose Γ is a connected graph. There is a natural geodesic metric on Γ obtained by identifying each edge with a copy of the unit interval $[0, 1]$ and defining the distance between any two points in Γ to be the length of the shortest path between them. This metric is proper if and only if each vertex has finite valence.

Example 3 Let M be a complete Riemannian manifold. Then the usual distance function given by minimizing path lengths is a proper, geodesic metric on M.

A map between two metric spaces $f : X \to Y$ is an *isometry* if it is bijective and preserves distances. In lay terms, an isometry of X to itself is a "symmetry" of X. These symmetries form a group under composition.

Now suppose we are given a group G. Our first goal is to find a nice metric space on which G acts as a group of symmetries. Sometimes, such an action arises naturally. For example, suppose G is the fundamental group of a Riemannian manifold M. Then passing to the universal cover $\widetilde{M}$, we get an action of G by deck transformations on $\widetilde{M}$. This action is distance preserving since it takes geodesic paths to geodesic paths.

More generally, the same works for the fundamental group of any geodesic metric space X that admits a universal cover. The universal cover $\widetilde{X}$ inherits a geodesic metric such that the projection to X is a local isometry, and the deck transformations act isometrically on $\widetilde{X}$.

Example 4 Consider the free group on two generators F_2. This group is the fundamental group of a wedge of two circles, $S^1 \vee S^1$, so it acts by isometries on the universal cover, namely the regular 4-valent tree.

In general, a group G can act by isometries on a variety of different geodesic metric spaces. Some of these actions, however, are not helpful in studying the group. For example, any group acts on a single point! To have any hope that the geometry of the space will produce information about the group, we will need some extra conditions on the action.

Definition 1 A group G is said to act *geometrically* on a metric space X if the action satisfies the following three properties.

- *isometric:* Each $g \in G$ acts as an isometry on X.
- *proper:* For all $x \in X$, there exits $r > 0$ such that $\{g \in G \mid B(x,r) \cap gB(x,r) \neq \emptyset\}$ is finite, where $B(x,r)$ denotes the ball of radius r centered at x.
- *cocompact:* There exists a compact set $K \subset X$ whose translates by G cover X (or equivalently, X/G is compact).

In particular, the fundamental group of a compact metric space X acts geometrically on the universal covering space $\widetilde{X}$. But now suppose that our group G arises purely algebraically. How can we find a metric space X on which G acts geometrically? The following is a construction that works for any finitely generated group.

Choose a finite generating set S for G. Define the *Cayley graph* for G with respect to S to be the graph $\Gamma_S(G)$ whose vertices are in one-to-one correspondence with the elements of G and for each $s \in S$, $g \in G$, there is an edge (labelled by s) connecting the vertex g to the vertex gs.

G acts on $\Gamma_S(G)$ by left multiplication on the vertices. That is, $h \in G$ maps the vertex labelled g to the vertex labelled hg. Note that h takes edges to edges, the edge connecting g to gs maps to the edge connecting hg to hgs. Thus, if we put the path-length metric on $\Gamma_S(G)$ as described above, then this action preserves the metric and is easily seen to be geometric.

Example 5 Consider the free abelian group $\mathbb{Z}^2$ with generating set $S = \{(1, 0), (0, 1)\}$. Viewing $\mathbb{Z}^2$ as the set of integer points in the plane $\mathbb{R}^2$, we can identify the Cayley graph of $\mathbb{Z}^2$ with the square grid connecting these points. Distances are measured by path lengths in this grid, so the distance from $(0, 0)$ to (n, m) is $|n| + |m|$. Note that there are, in general, many geodesic paths between any two points. See Fig. 1.

Example 6 Let F_2 be the free group with generating set $S = \{a, b\}$. View F_2 as the fundamental group of the wedge of two circles labelled a and b. Lifting these labels to the universal covering space, we get a 4-valent tree, as in Fig. 2, with every red edge labelled a and every blue edge labelled b (Metrically, you should picture every edge as having the same length.) This tree is precisely the Cayley graph $\Gamma_S(F_2)$. To

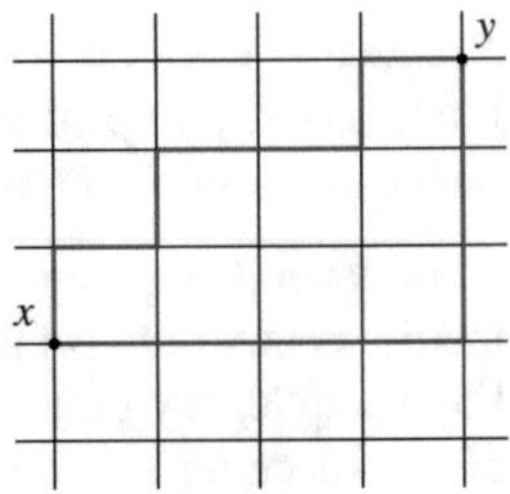

Fig. 1 Geodesics from x to y in the Cayley graph of $\mathbb{Z}^2$

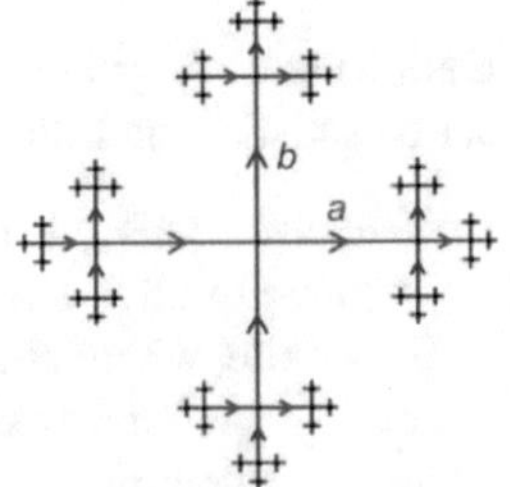

Fig. 2 Cayley graph of F_2

see this, choose a base vertex v and identify each vertex with the element of F_2 that translates v to that vertex.

Clearly, the Cayley graph depends on the choice of generating set. For example, were we to add a third generator $(1, 1)$ to our generating set for $\mathbb{Z}^2$, the Cayley graph would get additional edges which cross the grid diagonally and would hence change the distance between vertices.

This could be a cause for concern; we seek geometric properties of the Cayley graph that are intrinsic to the group, so they should not be dependent on a choice of generating set. Luckily, in the case of a finitely generated group, replacing one finite generating set by another does not distort distances too badly. This leads to a fundamental concept in geometric group theory.

Definition 2 A map $f : X \to Y$ between two metric spaces is a *quasi-isometric embedding* if there exists constants K, C such that for all $x, z \in X$

$$\frac{1}{K} d_X(x, z) - C \leq d_Y(f(x), f(z)) \leq K\, d_X(x, z) + C.$$

If in addition, every point in Y lies within C of some point in $f(X)$, then f is a *quasi-isometry*. In this case we write $X \sim_{QI} Y$.

It can be shown that any quasi-isometry has a "quasi-inverse," so the relation $X \sim_{QI} Y$ is an equivalence relation. We remark that quasi-isometries need not be continuous maps.

Example 7 Consider the inclusion of the integer grid into the plane $\mathbb{R}^2$. This is a quasi-isometry. The quasi-inverse is a discontinuous map sending the interior of each square to its boundary.

Example 8 Consider the graph in Fig. 3. Collapsing each of the triangles to a point gives a quasi-isometry of this graph onto the 3-valent tree. We call such a graph a quasi-tree.

It is an easy exercise to show that if G is a finitely generated group, then the Cayley graphs with respect to different finite generating sets are all quasi-isometric. (In fact, they are bi-Lipschitz, i.e., we can take the maps to be continuous and the constant C to be zero.) Thus, identifying G with the vertex set in its Cayley graph, we can view G itself as a metric space and this metric is well-defined up to quasi-isometry. In fact, we have the following more general statement.

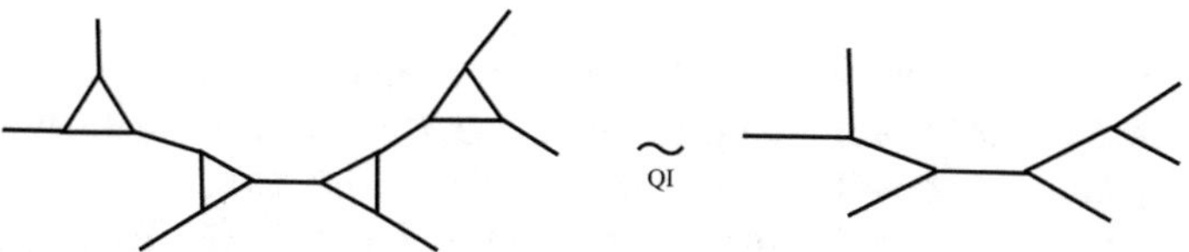

Fig. 3 A graph quasi-isometric to a tree

Proposition 1 (Milnor-Švarc Lemma) *Suppose G acts geometrically on a geodesic metric space X. Then G is finitely generated and for any choice of basepoint $x_0 \in X$, the map $G \to X$ taking $g \mapsto gx_0$ is a quasi-isometry.*

As a result, the notion of quasi-isometry is a fundamental concept in geometric group theory. The properties of a group that one can hope to glean from its action on a metric space are generally properties preserved by quasi-isometries. Moreover, since any finite index subgroup $H < G$ is quasi-isometric to G, often the strongest statement we can make regarding a property (P) is that our group G is *virtually* (P), meaning that some finite index subgroup of G satisfies property (P).

The classification of finitely generated groups up to quasi-isometry is a meta-problem in the field. It is easy to see that any two groups that are commensurable (i.e., they contain subgroups of finite index that are isomorphic) are quasi-isometric. So a related problem is the question of rigidity: for a given group G, is every group quasi-isometric to G also commensurable to G?

3 Hyperbolic Groups

Once we have our group acting geometrically on a metric space, we can ask how geometric properties of the space are reflected in algebraic or combinatorial properties of the group. The classical example of this comes from the work of Max Dehn [12]. Dehn was interested in fundamental groups of surfaces. A closed orientable surface of genus $g \geq 2$ (i.e. a torus with g holes) can be given a Riemannian metric of constant curvature -1 and its universal covering space can be identified with the hyperbolic plane $\mathbb{H}^2$. This gives a geometric action of the fundamental group on $\mathbb{H}^2$. Using geometric properties of the hyperbolic plane, Dehn proved some very strong combinatorial properties for these groups. I will describe two of his results here.

One way to describe a group is by means of a presentation. Given a set of generators S for a group G, there is a natural surjection from the free group $F(S)$ onto G. The kernel K of this map is a normal subgroup of $F(S)$. Formally, a *presentation* of G consists of a generating set S, together with a set $R \subset F(S)$ such that R generates K as a normal subgroup, or in other words, G is the quotient of $F(S)$ by the normal closure of R. The elements of R are called *relators*. We denote such a presentation for G by writing

$$G = \langle S \mid R \rangle.$$

In practice, we usually indicate R by a set of equations that hold in G. Viewing elements of $F(S)$ as "words" in the alphabet $S \cup S^{-1}$, the elements in R are words that are equal to the identity in G. However, many other words, such as products and conjugates of those in R, are also equal to the identity in G. The idea is that *all* relations among the generators that hold in G should be consequences of the ones listed in the presentation.

Here are some examples. The cyclic group of order n has presentation

$$\mathbb{Z}/n\mathbb{Z} = \langle s \mid s^n = 1 \rangle$$

while the free abelian group on two generators has presentation

$$\mathbb{Z}^2 = \langle a, b \mid ab = ba \rangle.$$

Can you recognize the following group?

$$G = \langle u, v \mid u^4 = 1, u^2 = v^3 \rangle.$$

It turns out that this group is isomorphic to the special linear group $SL(2, \mathbb{Z})$. The isomorphism is given by identifying

$$u = \begin{bmatrix} 0 & -1 \\ 1 & 0 \end{bmatrix} \qquad v = \begin{bmatrix} 0 & -1 \\ 1 & 1 \end{bmatrix}$$

Every group can be described by a presentation, though in general S and R need not be finite. Presentations can be extremely useful, and are the starting point for combinatorial group theory. On the other hand, presentations can sometimes be very mysterious and frustratingly difficult to decipher. For example, consider the following questions.

1. *The Word Problem:* Given a finite presentation $\langle S \mid R \rangle$ and a word w in $F(S)$, is there an algorithm to decide whether w represents the identity element in G?
2. *The Isomorphism Problem:* Given two finite presentations $\langle S \mid R \rangle$ and $\langle S' \mid R' \rangle$, is there an algorithm to decide whether the groups they represent are isomorphic?

It turns out that there are groups for which no such algorithms exist. In this case we say the Word Problem (or the Isomorphism Problem) is unsolvable. What Dehn showed, using the geometry of hyperbolic space, was that for fundamental groups of hyperbolic surfaces, both of these problems are solvable. Moreover, he showed that for an appropriate choice of presentation, the word problem has a particularly nice solution. Namely, any word $w \in F(S)$ that represents the identity element in G must contain more than half of a relator r, and hence can be shortened by applying the equation $r = 1$. It follows that the word problem is solvable in linear time, that is, the time it takes to decide whether $w \in F(S)$ represents the identity element in G, is linear in the length of w. An algorithm of this type is now known as a Dehn's algorithm.

Some 75 years later, Mikhail Gromov made a startling observation: the only property of the hyperbolic plane that Dehn really needed to derive his results was the fact that triangles in $\mathbb{H}^2$, no matter how far apart their vertices may be, are always "thin." And from this observation, the modern field of hyperbolic geometry (and more generally geometric group theory) was born.

What do we mean by "thin"? Let X be any geodesic metric space. A triangle $T(a, b, c)$ in X consists of three vertices $a, b, c \in X$ together with a choice of geodesics connecting them.

Definition 3 Let X be a geodesic metric space and let $\delta \geq 0$. We say a triangle $T(a, b, c)$ in X is *δ-thin* if each side of T lies in the union of the δ-neighborhoods of the other two sides. (See Fig. 4.)

One can show that in $\mathbb{H}^2$, every triangle, even those with vertices at infinity (ideal triangles) are δ-thin for $\delta = \ln(1 + \sqrt{2})$. This fact was crucial to Dehn's work.

This brings us finally to Gromov's notion of hyperbolicity.

Definition 4 A geodesic metric space X is *δ-hyperbolic* if every triangle in X is δ-thin. We say X is *hyperbolic* if it is δ-hyperbolic for some δ. A finitely generated group G is *hyperbolic* if it acts geometrically on a hyperbolic metric space.

One can show that if two spaces are quasi-isometric and one of them is hyperbolic, then so is the other (though the constant δ may change). In particular, a finitely generated group G is hyperbolic if and only if some (hence any) Cayley graph of G is hyperbolic.

Example 9 We begin with a trivial example. Any bounded metric space X is δ-hyperbolic where δ is the diameter of X, and hence any finite group is hyperbolic.

Example 10 Let X be an infinite tree. Then for any three points a, b, c in X, the triangle connecting them degenerates into a tripod and is hence 0-hyperbolic! (See Fig. 5.) Since the Cayley graph of a free group is a tree, it follows that finitely generated free groups are hyperbolic.

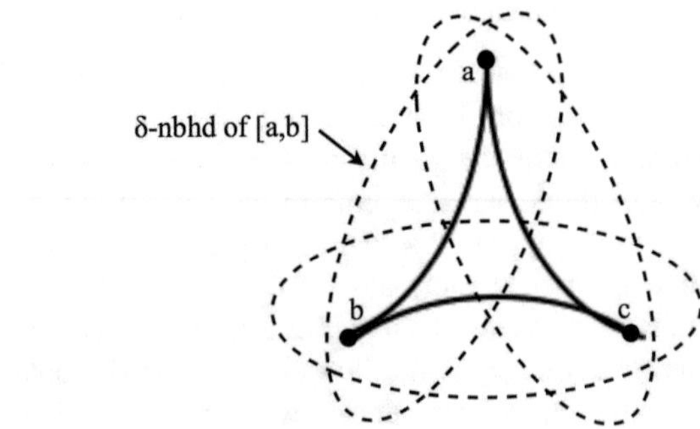

Fig. 4 A δ-thin triangle

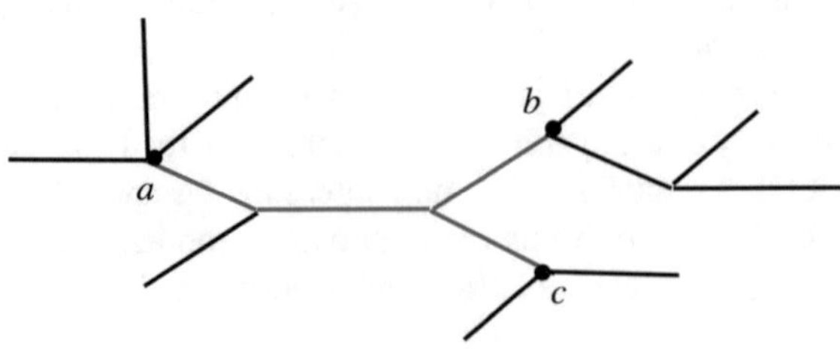

Fig. 5 A triangle in a tree is 0-thin

Example 11 Recall the presentation of $SL(2, \mathbb{Z})$ given above. The center of $SL(2, \mathbb{Z})$ is the order two subgroup generated by $u^2 = v^3$. Modding out by this subgroup gives the group $PSL(2, \mathbb{Z})$ with presentation

$$PSL(2, \mathbb{Z}) = \langle u, v \mid u^2 = v^3 = 1 \rangle.$$

The Cayley graph of $PSL(2, \mathbb{Z})$ with respect to this generating set is the quasi-tree drawn in Fig. 3 (continued out to infinity), with the edges of triangles labelled v and the remaining edges labeled u. $SL(2, \mathbb{Z})$ also acts geometrically on this quasi-tree (with the center acting trivially), so $SL(2, \mathbb{Z})$ and $PSL(2, \mathbb{Z})$ are both hyperbolic groups.

Example 12 Here is a non-example. Let $\mathbb{R}^2$ be the plane with the standard Euclidean metric. Taking larger and larger isosceles right triangles, we can see that there is no bound on "thinness." Since the Cayley graph of $\mathbb{Z}^2$ is quasi-isometric to $\mathbb{R}^2$, $\mathbb{Z}^2$ is not hyperbolic. Indeed, it is a theorem that a hyperbolic group cannot contain a copy of $\mathbb{Z}^2$.

Now suppose that we are given a hyperbolic group G. What does the geometry tell us about the group? Here is a list of some consequences of hyperbolicity. We refer the reader to [3] and [7] for proofs and additional references.

1. G has a finite presentation.
2. G has a Dehn's algorithm, hence a linear time solution to the Word Problem.
3. The Isomorphism Problem is solvable for the class of hyperbolic groups.
4. The centralizer of every element of G is virtually cyclic.
5. G has at most finitely many conjugacy classes of torsion elements.
6. For any finite set of elements $g_1, \ldots g_k$ in G, there exists $n > 0$ such that the set $\{g_1^n \ldots g_k^n\}$ generates a free subgroup of rank at most k.
7. For n sufficiently large, $H^n(G; \mathbb{Q}) = 0$.
8. If G is torsion-free, it has a finite $K(G, 1)$-space (i.e., a finite CW-complex with fundamental group G and contractible universal covering space).

The proofs of these properties are beyond the scope of this paper, but the conclusion should be clear: geometry can have strong implications for algebraic and combinatorial properties of a group.

4 Beyond Hyperbolicity

Classical hyperbolic geometry and Gromov's generalization to δ-hyperbolic spaces have provided powerful tools for studying hyperbolic groups. But this class of groups is very special. For example, any group containing a subgroup isomorphic to $\mathbb{Z}^2$ cannot be hyperbolic (see property (4) above). In recent years, there has been much interest in generalizing some of these techniques to broader classes of groups. Gromov himself introduced a notion of "non-positive curvature" for geodesic metric spaces, called CAT(0) spaces, and groups acting on these spaces have also been extensively studied. CAT(0) geometry, for example, played a major role in the recent work of Agol, Wise, and others leading to a proof of the Virtual Haken Conjecture, the last remaining piece in Thurston's program to classify three-manifolds. But this is a topic for another day.

Other approaches to generalizing the theory of hyperbolic groups involve looking at groups that act on hyperbolic spaces, but where the actions are not geometric; instead, they satisfy some weaker conditions. This includes, for example, "relatively hyperbolic groups," "acylindrically hyperbolic groups," and "hierarchically hyperbolic groups." The first of these is modeled on fundamental groups of hyperbolic manifolds with cusps; the others are inspired by mapping class groups and their actions on curve complexes. Some very nice introductions to these topics can be found in [2, 16, 17].

My own work in this area has focused on a somewhat different approach to capturing hyperbolic behavior in more general spaces and groups. Let's begin with an example.

Example 13 Let Z be the space obtained by gluing a circle and a torus together at a single point,

$$Z = S^1 \vee T^2.$$

Let $\widetilde{Z}$ be its universal cover. In $\widetilde{Z}$, the inverse image of the torus consists of infinitely many copies of the Euclidean plane (which we refer to as "flats") and emanating from each lattice point in each of these planes is a line segment which projects to the circle. The result is a tree-like configuration of planes and lines (see Fig. 6), which we will refer to as the "tree of flats." Triangles lying in a single flat can

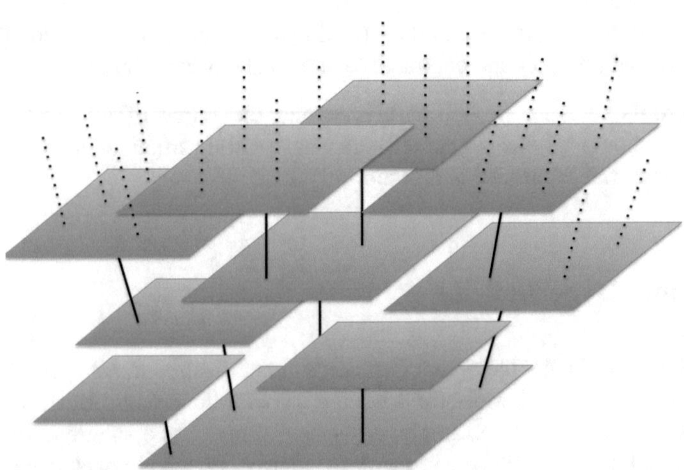

Fig. 6 Tree of flats

be arbitrarily "fat" whereas triangles whose sides lie mostly along the vertical lines are "thin." Thus, in $\widetilde{Z}$, we have hyperbolic-like directions, and non-hyperbolic directions. Intuitively, the more we travel vertically, the more hyperbolic it feels.

We are interested in identifying geodesics in a metric space X that behave like geodesics in a hyperbolic space. A good way to encode such geodesics is by means of a boundary. In general, unbounded metric spaces do not come equipped with a boundary. For example, in the hyperbolic plane (or the Euclidean plane) one can travel forever in any direction. To create a boundary for such a space, we need to add a point for each "direction to infinity." In the case of the hyperbolic or Euclidean plane, this space of directions forms a circle. Adding this circle to the plane "at infinity" compactifies the space.

It turns out that this idea generalizes nicely to any hyperbolic metric space. For a δ-hyperbolic space X, we define the boundary as follows. A *ray* in X is an isometric embedding $\alpha : [0, \infty) \to X$. As a set, the boundary of X is defined to be

$$\partial X = \{\alpha \mid \alpha : [0, \infty) \to X \text{ isaray}\}/\sim$$

where $\alpha \sim \beta$ if α and β remain bounded distance from each other. In the Euclidean plane, for example, two rays are equivalent if and only if they are parallel.

To topologize ∂X, think of two rays as representing nearby points in the boundary if they remain close to each other for a long time. More precisely, define a neighborhood $N(\alpha, R)$ of a ray α to be the set of rays β such that $\beta(t)$ lies within 2δ of $\alpha(t)$ for $0 \leq t \leq R$. As R increases, these neighborhood get smaller and smaller, and together, they form a neighborhood basis for a topology on ∂X.

For example, the boundary of the hyperbolic plane $\mathbb{H}^2$ is a circle while the boundary of an infinite tree is a Cantor set.

From the geometric group theory viewpoint, a key property of the boundary of a hyperbolic space is quasi-isometry invariance.

Theorem 1 *Let $f : X \to Y$ be a quasi-isometry between two hyperbolic metric spaces. Then f induces a homeomorphism $\partial f : \partial X \cong \partial Y$. In particular, a hyperbolic group G has a well-defined boundary, namely the boundary of a Cayley graph of G.*

These boundaries have many nice properties and applications. The boundary gives rise to a compactification of X, $\overline{X} = X \cup \partial X$, and it provides a powerful tool for studying the dynamics of groups actions, rigidity theorems, geodesic flows, etc.

In the quest to extend the techniques of hyperbolic geometry to more general spaces and groups, it is natural to ask whether analogous boundaries can be defined in more general contexts. Certainly we can consider equivalence classes of geodesic rays in any geodesic metric space X. However, if X is not hyperbolic, many things can go wrong. In some cases, it is not even clear how to define a topology on this set as the neighborhoods described above need not satisfy the requirements for a neighborhood basis.

Moreover, even when there is a nice topology on ∂X, other fundamental properties of hyperbolic boundaries can fail to hold. Consider, for example, the boundary of the Euclidean plane. This boundary is a circle and each point on the boundary can be represented by a unique ray based at the origin. As observed above, it provides a compactification of the plane. That's the good news. Here is some bad news.

- Many isometries (in particular all translations) act trivially on the boundary.
- The only pairs of points on the boundary that can be joined by a bi-infinite geodesic are pairs of antipodal points.
- A quasi-isometry of the plane to itself need not extend to a map on the boundary. For example, the map $f : \mathbb{R}^2 \to \mathbb{R}^2$ taking $re^{i\theta} \mapsto re^{i(\theta+\ln(r))}$ is a quasi-isometry which twists each ray emanating from the origin into a spiral.

In short, many of the properties of hyperbolic boundaries that permit applications to dynamics, rigidity, etc. fail to hold for this boundary.

Most significantly from the point of view of geometric group theory, quasi-isometry invariance fails miserably for non-hyperbolic boundaries. There are examples of groups that act geometrically on two CAT(0) spaces (spaces of non-positive curvature) whose boundaries are not homeomorphic [11]. Thus, we don't have a well-defined notion of a boundary for these groups.

What goes wrong is the failure of the Morse property. A quasi-isometry $f : X \to Y$ of hyperbolic spaces takes a geodesic ray in X to a quasi-geodesic ray in Y, that is, a quasi-isometric embedding of the half-line $\mathbb{R}^+ = [0, \infty)$ into Y. The Morse property guarantees that this quasi-geodesic ray lies close to some geodesic ray and hence determines a well-defined point at infinity.

Definition 5 A ray (or bi-infinite geodesic) α in X is *Morse* if there exists a function $N : \mathbb{R}^+ \times \mathbb{R}^+ \to \mathbb{R}^+$ such that for any (K, C)-quasi-geodesic β with endpoints on α, β lies in the $N(K, C)$-neighborhood of α. The function N is called a *Morse gauge* for α and we say that α is N-Morse.

If X is hyperbolic, then there exists a Morse gauge N such that every ray in X is N-Morse. This property is the key to proving quasi-isometry invariance for boundaries of hyperbolic spaces, and it plays a key role in the proofs of many other properties of hyperbolic spaces as well. If X is not hyperbolic, the Morse property may fail for some (or perhaps all) rays in X. On the other hand, many non-hyperbolic spaces contain a large number of Morse rays. Consider, for example, our "tree of flats" described above. It can be shown that a ray that spends a uniformly bounded amount of time in any flat is Morse. We view Morse rays as "hyperbolic-like" directions in X. Indeed, it can be shown that these rays share many other nice properties with rays in hyperbolic space [4, 10]. For example, if two sides of a triangle are N-Morse, then the triangle is δ-thin where δ depends only on the Morse gauge N.

This brings us to the Morse boundary. The *Morse boundary*, $\partial_M X$ can be defined for any proper geodesic metric space X. As a set, it consists of equivalence classes of Morse rays,

$$\partial_M X = \{\alpha \mid \alpha : [0, \infty) \to X \text{ isaMorseray}\}/\sim$$

where the equivalence $\sim$ is defined as before. The topology is more subtle. For a sequence of rays $\{\alpha_i\}$ to converge to α in this topology, they must not only converge pointwise, they must also be uniformly Morse, that is, there exists a Morse gauge N such that all of the α_i are N-Morse.

This boundary was first introduced for CAT(0) spaces by myself and Harold Sultan in [4] and shown to be quasi-isometry invariant. This was then generalized to arbitrary proper geodesic metric spaces by Matt Cordes in [8].

Theorem 2 *Let $f : X \to Y$ be a quasi-isometry between two proper geodesic metric spaces. Then f induces a homeomorphism $\partial f : \partial_M X \to \partial_M Y$. In particular, $\partial_M G$ is well-defined for any finitely generated group.*

Example 14

(1) If X is hyperbolic, then all rays are N-Morse for some fixed N. So in this case, $\partial_M X = \partial X$, the usual hyperbolic boundary. For example, $\partial_M \mathbb{H}^2$ is a circle.
(2) For $X = \mathbb{R}^2$ the Euclidean plane, there are no Morse rays at all, so $\partial_M X = \emptyset$.
(3) Let $\widetilde{Z}$ be the "tree of flats" from Example 13. Then the Morse geodesics in $\widetilde{Z}$ are those that spend a uniformly bounded amount of time in any flat and the maximum time spent in a flat is determined by the Morse gauge. Thus, for a given Morse gauge N, the N-Morse rays emanating from a fixed basepoint z_0 lie in a subspace quasi-isometric to a tree. It follows that these rays determine a Cantor set in the boundary and the Morse boundary of $\widetilde{Z}$ is the direct limit of these Cantor sets. Note that since $\widetilde{Z}$ is the universal cover of $S^1 \vee T^2$, the fundamental group $\mathbb{Z} * \mathbb{Z}^2 = \pi_1(S^1 \vee T^2)$ acts geometrically on $\widetilde{Z}$. Hence by the theorem above, the Morse boundary of the group $\mathbb{Z} * \mathbb{Z}^2$ is homeomorphic to the Morse boundary of $\widetilde{Z}$.

The Morse boundary was designed to capture hyperbolic-like behavior in non-hyperbolic metric spaces and to give a well-defined notion of a boundary for a finitely generated group G. When the Morse boundary is non-trivial, it provides a new tool for studying these spaces and groups. It can be used, for example, to study the dynamics of isometries [15] and to determine when two groups are quasi-isometric [5]. It can also be used to study geometric properties of subgroups $H < G$ [10]. For a survey of recent results on Morse boundaries, see [9].

Geometric group theory is a broad and growing area of mathematics. This article is intended only as a snapshot of some themes that run through the field. I invite you to investigate further!

Acknowledgements R. Charney was partially supported by NSF grant DMS-1607616.

References

1. B. Bowditch, *A Course on Geometric Group Theory*. MSJ Memoirs, vol. 16 (Mathematical Society of Japan, Tokyo, 2006)
2. B. Bowditch, Relatively hyperbolic groups. Int. J. Algebra Comput. **22**(3), 1250016 [66 pp.] (2012)
3. M. Bridson, A. Haefliger, *Metric Spaces of Non-positive Curvature*. Grundlehren der Mathematischen Wissenschaften, vol. 319 (Springer, Berlin, 1999)
4. R. Charney, H. Sultan, Contracting boundaries of CAT(0) spaces. J. Topol. **8**, 93–117 (2013)
5. R. Charney, M. Cordes, D. Murray, Quasi-mobius homeomorphisms of Morse boundaries (Jan 2018). arXiv:1801.05315
6. M. Clay, D. Margalit (eds.), *Office Hours with a Geometric Group Theorist* (Princeton University Press, Princeton, 2017)
7. M. Coornaert, T. Delzant, A. Papadopoulos, *Géométrie et théorie des groupes: Les groupes hyperboliques de Gromov*. Lecture Notes in Mathematics, vol. 1441 (Springer, Berlin, 1990)
8. M. Cordes, Morse boundaries of proper geodesic metric spaces. Groups Geom. Dyn. **11**(4), 1281–1306 (2017)
9. M. Cordes, A survey on Morse boundaries and stability, in *Beyond Hyperbolicity*. LMS Lecture Notes. arXiv:1704.07598 (to appear)
10. M. Cordes, D. Hume, Stability and the Morse boundary. J. Lond. Math. Soc. (2) **95**(3), 963–988 (2017)
11. C. Croke, B. Kleiner, Spaces with nonpositive curvature and their ideal boundaries. Topology **39**, 549–556 (2000)
12. M. Dehn, *Papers on Group Theory and Topology*. Translated from the German and with introductions and an appendix by J. Stillwell (Springer, New York, 1987)
13. E. Ghys, P. de la Harpe, *Sur les groupes hyperboliques d'aprés Mikhael Gromov (Bern, 1988)*. Progress in Mathematics, vol. 83 (Birkhäuser, Boston, 1990)
14. M. Gromov, Hyperbolic groups, in *Essays in Group Theory*. Mathematical Sciences Research Institute Publications, vol. 8 (Springer, New York, 1987), pp. 75–263
15. D. Murray, Topology and dynamics of the contracting boundary of cocompact CAT(0) spaces. Pac. J. Math. arXiv:1509.09314 (to appear)
16. D. Osin, Acylindrically hyperbolic groups. Trans. Am. Math. Soc. **368**(2), 851–888 (2016)
17. A. Sisto, What is a hierarchically hyperbolic space? (July 2017). arXiv:1707.00053

Generalized Iterated Wreath Products of Cyclic Groups and Rooted Trees Correspondence

Mee Seong Im and Angela Wu

Abstract Consider the generalized iterated wreath product $\mathbb{Z}_{r_1} \wr \mathbb{Z}_{r_2} \wr \ldots \wr \mathbb{Z}_{r_k}$ where $r_i \in \mathbb{N}$. We prove that the irreducible representations for this class of groups are indexed by a certain type of rooted trees. This provides a Bratteli diagram for the generalized iterated wreath product, a simple recursion formula for the number of irreducible representations, and a strategy to calculate the dimension of each irreducible representation. We calculate explicitly fast Fourier transforms (FFT) for this class of groups, giving the literature's fastest FFT upper bound estimate.

Keywords Iterated wreath products · Cyclic groups · Rooted trees · Irreducible representations · Fast Fourier transform · Bratteli diagrams

AMS Subject Classification Primary 20C99, 20E08; Secondary 65T50, 05E25

1 Introduction

Representations of groups appear naturally in nature, more often than groups themselves. They appear in the form of a linear representation, a permutation representation, and automorphisms of an algebra, a group, a variety or scheme, or a manifold. For example, one can study functions on the circle S^1, which could be thought of as a group under addition, which form representations of S^1. Such functions could also be thought of as periodic functions on the set $\mathbb{R}$ of real numbers, and the decomposition of the space of functions on S^1 is known as the theory of Fourier series. One can also study the additive group $\mathbb{R}$ of real numbers acting on

M. S. Im (✉)
Department of Mathematical Sciences, United States Military Academy, West Point, NY, USA

A. Wu
Department of Mathematics, University of Chicago, Chicago, IL, USA
e-mail: wu@math.uchicago.edu

A. Deines et al. (eds.), *Advances in the Mathematical Sciences*, Association for Women in Mathematics Series 15, https://doi.org/10.1007/978-3-319-98684-5_2

itself under addition. Then, one may ask how the function space of $\mathbb{R}$ decompose under the action of the group of real numbers; this is the study of Fourier transform.

A cyclic group may be thought of as the set of rotational symmetries of a regular polygon and of a generalized wreath product $\mathbb{Z}_{r_1} \wr \mathbb{Z}_{r_2} \wr \ldots \wr \mathbb{Z}_{r_k}$ as the automorphisms of a corresponding complete rooted tree generated by cyclic shifts of the children of each node. With applications to functions on rooted trees, pixel blurring (cf. [1, 6, 11–13, 19]), nonrigid molecules in molecular spectroscopy (cf. [2, 3, 18, 24]), and visual information processing (cf. [4, 16]), we generalize Orellana–Orrison–Rockmore's manuscript [20]. Denoting the iterated wreath product as $W(\mathbf{r}|_k) := \mathbb{Z}_{r_1} \wr \ldots \wr \mathbb{Z}_{r_k}$ (see Sect. 2.1), we show that the equivalence classes of irreducible representations of the iterated wreath products $W(\mathbf{r}|_k)$ are indexed by classes of labels on the vertices of the complete $\mathbf{r}|_k$-ary trees (see Sect. 2.3) of height k (Proposition 1).

Let G be a finite group and let V be a vector space over the set $\mathbb{C}$ of complex numbers. Let $GL(V)$ be the general linear group on V, and let $\rho : G \to GL(V)$ be a representation of G, i.e., ρ is a group homomorphism. We say two representations $\rho : G \to GL(V)$ and $\eta : G \to GL(W)$ are equivalent, and write $\rho \sim \eta$, if there exists a vector space isomorphism $f : V \to W$ such that $f \circ \rho(g) = \eta(g) \circ f$ for all $g \in G$. We denote by $\widehat{G}$ the set of equivalence classes of irreducible representations of G. We say that $\mathcal{R}$ is a *traversal* for G if $\mathcal{R} := \mathcal{R}_G \subset \widehat{G}$ contains one irreducible representation for each isomorphism class in $\widehat{G}$. As a basic consequence of representation theory, the equality $\sum_{\rho \in \mathcal{R}} \dim(\rho)^2 = |G|$ holds, where the sum is over all irreducible representations in $\mathcal{R}$.

We denote by $[n] := \{1, 2, \ldots, n\}$ the set of integers from 1 to n, and we denote the set of length ℓ words with letters in $[n]$ by:

$$[n]^\ell := \{x_1 x_2 \cdots x_\ell : x_i \in [n]\} .$$

Now given a subgroup $H \leq G$, we write $\mathrm{Ind}_H^G : \mathrm{Rep}(H) \to \mathrm{Rep}(G)$ to be the induction functor from the category of representations of H to the category of representations of G. That is, given a representation $\eta \in \widehat{H}$, $\eta : H \to GL(V)$, we write $\mathrm{Ind}_H^G \eta = \mathbb{C}[G] \otimes_{\mathbb{C}[H]} V$, the *induced representation* of G from η with dimension $[G : H] \cdot \dim \eta$. We also have the dual construction to induction, which is called restriction. Given a subgroup H of G, $\mathrm{Res}_H^G : \mathrm{Rep}(G) \to \mathrm{Rep}(H)$ is the restriction functor from the category of representations of G to the category of representations of H, i.e., given a representation ρ of G, we obtain the *restricted representation* $\mathrm{Res}_H^G \rho$ of H by restricting ρ to H. The induction and restriction functors are related by Frobenius reciprocity. We refer the reader to [7] for an explicit and elegant discussion on the duality of induction and restriction.

For $x = x_1 \cdots x_{r_k} \in [h]^{r_k}$, define

$$d_x := \min\{i \in \mathbb{N} : x^i = x\}, \text{ where } x^i = (x_1 \cdots x_{r_k})^i := x_{i+1} \cdots x_{r_k} x_1 \cdots x_i . \quad (1)$$

Note that $d_x | r_k$ for any $x \in [h]^{r_k}$. We write $x^G = \{x^g : g \in G\}$, the orbit of x under G. In the case $G = \mathbb{Z}_r$, then $i \in \mathbb{Z}_r$ acts on x by $i \cdot x = x^i$, cyclically rotating the letters in the word x.

We now state our first theorem, which generalizes Theorem 2.1 in [20]:

Theorem 1 *Suppose that $\mathcal{R} = \{\rho_1, \ldots, \rho_h\}$ is a traversal for the iterated wreath product $W(\mathbf{r}|_{k-1})$. Let $J \subseteq [h]^{r_k}$ denote a set of $\mathbb{Z}_{r_k}$-orbit representatives of $[h]^{r_k}$ such that $[h]^{r_k} = \bigsqcup_{x \in J} x^{\mathbb{Z}_{r_k}}$. Then, a traversal for $W(\mathbf{r}|_k)$ is given by:*

$$\mathcal{R}_{W(\mathbf{r}|_k)} = \left\{ \mathrm{Ind}_{W(\mathbf{r}|_{k-1}) \rtimes \mathbb{Z}_{d_x}}^{W(\mathbf{r}|_k)} (\rho_{x_1} \otimes \cdots \otimes \rho_{x_{r_k}} \otimes \tau) : x \in J, \tau \in \widehat{\mathbb{Z}_{d_x}} \right\}. \quad (2)$$

Now, an efficient algorithm for applying a discrete Fourier transform is called a fast Fourier transform (FFT). For a finite group G, we denote by $T(G)$ the maximum number of computations required to compute $\{\widehat{f}(\rho) : \rho \in \mathcal{R}_G\}$ over all complex-valued functions $f : G \to \mathbb{C}$ on G, where $\widehat{f}$ is the Fourier transform of f at ρ, i.e., it is the matrix

$$\widehat{f}(\rho) = \sum_{g \in G} f(g)\rho(g).$$

For the class of finite abelian groups G, the Cooley–Tukey algorithm given in [8] and [10], combined with techniques provided in [22] and [21], give an order of $O(|G| \log |G|)$ operation bound on the FFT computation time, where the O-notation is some universal constant. Rockmore in [23] provides the fastest algorithm to date in the literature for abelian group extensions:

Theorem 2 (Lemma 5 and Theorem 4, [23]) *Let $K \trianglelefteq G$ be a normal subgroup of G and assume that G/K is abelian. Let $\rho \in \widehat{G}$. Then, there exists a subgroup H with $K \leq H \leq G$ and $\widetilde{\eta} \in \widehat{H}$ such that*

- *$\eta = \widetilde{\eta}|_K$ is irreducible,*
- $\mathrm{Ind}_H^G \widetilde{\eta} = \rho$,
- *$H := \{g \in G : \rho^{(g)} \sim \rho\}$, the inertia group of ρ in G, where $\rho^{(g)}(h) = \rho(g^{-1}hg)$ for all $h \in G$, and*
- *if $G = \bigsqcup_{i \in [G:H]} s_i H$, then $\rho = \widetilde{\eta}^{(1)} \otimes \widetilde{\eta}^{(s_2)} \otimes \cdots \otimes \widetilde{\eta}^{(s_{[G:H]})}$, where $\widetilde{\eta}^{(s)}(h) = \widetilde{\eta}(s^{-1}hs)$.*

The representation $\rho^{(g)}$ is called a *conjugate representation* of ρ. By Theorem 2, we see that it suffices to take a traversal $\mathcal{R}_K$ of K in order to find a complete traversal of G. Then, we need to construct the set of extensions of η to its inertia group H_η for each representation $\eta \in \mathcal{R}_K$. Finally, we need to build up the induced representation of G from each extension. Applying [23] to give an upper bound on the number of operations needed to compute a fast Fourier transform of an iterated wreath product, we present an explicit running time of fast Fourier transforms for $W(\mathbf{r}|_k)$, thus proving a tighter upper bound estimate than Theorem 1 in [23]:

Theorem 3 *For $f : K \wr \mathbb{Z}_r \to \mathbb{C}$, we have*

$$T(K \wr \mathbb{Z}_r) = r \cdot T(K^r) + \sum_{\eta \in E} m \left(r \dim(\eta)^\alpha + \dim(\eta)^2 O \left(r \log \frac{r}{m} \right) \right)$$
$$+ r\, m \cdot \dim(\eta)^\alpha. \tag{3}$$

Using Bratteli diagrams (see Sect. 2.4), we calculate the irreducible representations iteratively (see Sect. 5), thus proving Theorem 3.

1.1 Summary of the Sections

In Sect. 2, we provide some background and notation. We begin by defining wreath products of cyclic groups in Sect. 2.1, give an introduction to Clifford theory in Sect. 2.2, give a construction of **r**-trees in Sect. 2.3, and then define Bratteli diagrams in Sect. 2.4.

In Sect. 3, we prove Theorem 1 and then give the number of irreducible representations for the iterated wreath product $\mathbb{Z}_{r_1} \wr \ldots \wr \mathbb{Z}_{r_k}$ in Theorem 5. In Sect. 4, we prove the one-to-one correspondence between equivalence classes of irreducible representations of the generalized wreath product and orbits of compatible $\mathbf{r}|_k$-labels (Proposition 1), give the number of $\mathbf{r}|_k$-trees of height k in Corollary 2, and write the dimension of an irreducible representation of an iterated wreath product of cyclic groups in terms of companion trees in Proposition 2. In Sect. 5, we prove Theorem 3, generalizing Theorem 1 in [23] by giving an explicit computation, and finally, in Sect. 6, we conclude by providing an open problem.

2 Background

2.1 Wreath Products

We refer to Section 1.1 in [20] for a beautiful exposition with illustrative examples about the construction of the wreath product $G \wr H$ of a finite group G with a subgroup H of S_n, which is summarized as follows. We define an action of H on $G^n = G \times \cdots \times G$ by if $\pi \in H$ and $a = (a_1, a_2, \ldots, a_n) \in G^n$, then $\pi \cdot a := a^\pi = (a_{\pi^{-1}(1)}, a_{\pi^{-1}(2)}, \ldots, a_{\pi^{-1}(n)})$. The wreath product $G \wr H$ is defined to be $G^n \times H$ as a set, with multiplication given by:

$$(a; \pi)(b; \sigma) = (ab^\pi; \pi\sigma). \tag{4}$$

Throughout this chapter, we will fix $\mathbf{r} = (r_1, r_2, r_3, \ldots) \in \mathbb{N}^{\omega}$, a positive integral vector. We denote by $\mathbf{r}|_k := (r_1, r_2, \ldots, r_k)$ the k-length vector found by truncating $\mathbf{r}$.

Definition 1 We define the *(generalized) k-th* $\mathbf{r}$*-cyclic wreath product* $W(\mathbf{r}|_k)$ recursively by:

$$W(\mathbf{r}|_0) = \{0\} \text{and} W(\mathbf{r}|_k) = W(\mathbf{r}|_{k-1}) \wr \mathbb{Z}_{r_k}.$$

Note that multiplication for the wreath product $W(\mathbf{r}|_k)$ is defined recursively using (4).

Example 1 We have $W(\mathbf{r}|_1) = \mathbb{Z}_{r_1}$, $W(\mathbf{r}|_2) = \mathbb{Z}_{r_1} \wr \mathbb{Z}_{r_2}$, and $W(\mathbf{r}|_k) = \mathbb{Z}_{r_1} \wr \ldots \wr \mathbb{Z}_{r_k}$.

Throughout this chapter, we will be considering the chain of groups given in Definition 1.

2.2 *Clifford Theory*

The following [9, 15], and [5] contain an extensive background on Clifford theory, which allows one to recursively construct the irreducible representations of a group. In this chapter, we will give a brief overview of the main results of Clifford theory.

Let G be a finite group and let K be a normal subgroup of G. Then, G acts on the set of inequivalent irreducible representations of K. For any irreducible representation σ of K, let $\Delta(\sigma)$ denote its orbit under this action, i.e., inequivalent conjugates of σ. Let $\mathrm{Stab}_G(\sigma)$ be the isotropy subgroup of σ under the G-action.

Theorem 4 (Clifford Theory) *Let K be a normal subgroup of G.*

1. *([7], Theorem 10) If σ be a representation of K, then*

$$\mathrm{Res}^G_K \mathrm{Ind}^G_K \sigma = [\mathrm{Stab}(\sigma) : K] \cdot \Delta(\sigma).$$

2. *([7], Theorem 14) If $\rho : G \to GL(V)$ is an irreducible representation of G, then*

$$\mathrm{Res}^G_K \rho = \frac{d_\rho}{[G : \mathrm{Stab}(\sigma)] d_\sigma} \cdot \Delta(\sigma),$$

where σ is any irreducible representation of K which appears in $\mathrm{Res}^G_K \rho$, and d_ρ is the dimension of the vector space V.

We also call d_ρ the *degree* of ρ.

2.3 Rooted Trees of a Fixed Height

We define $\mathbf{r}$-trees, a generalization of the r-trees in Section 3 of [20]. A *rooted tree* is a connected simple graph with no cycles and with a distinguished vertex called the *root*. We say a node v is in the j*-th layer* of a rooted tree if it is at distance j from the root.

Definition 2 We define the complete $\mathbf{r}$-tree, denoted by $T(\mathbf{r}|_k)$, of height k, or $\mathbf{r}|_k$-tree, recursively as follows: let $T(r_1)$ be the 1-layer tree consisting of a root node only. Let $T(\mathbf{r}|_2)$ consist of a root with r_2 children. Let $T(\mathbf{r}|_k)$ consist of a root node with r_k children, with each the root of a copy of the $(k-1)$-layer tree $T(\mathbf{r}|_{k-1})$, which yields a tree with k levels of nodes.

Example 2 The tree $T(r_1)$ is given by • and $T(r_1, r_2)$ with r_2 leaves is given by:

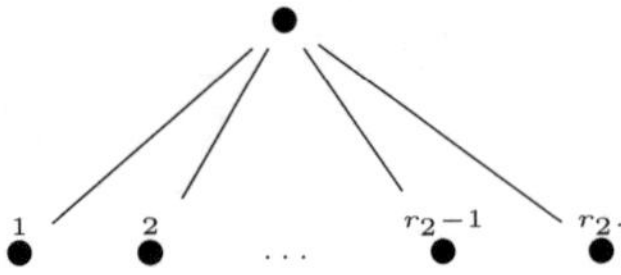

Example 3 Writing $\mathbf{r}|_3 = (r_1, r_2, r_3)$, the following is the complete tree $T(\mathbf{r}|_3)$ of height 3 with 3 levels of nodes:

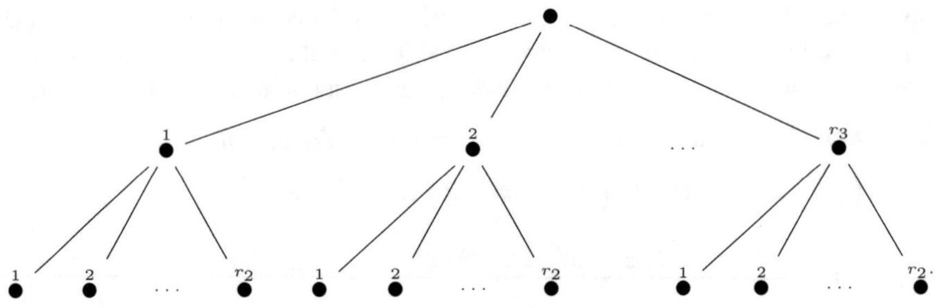

Example 4 The complete tree $T(\mathbf{r}|_4)$ of height 4 with 4 levels of nodes is

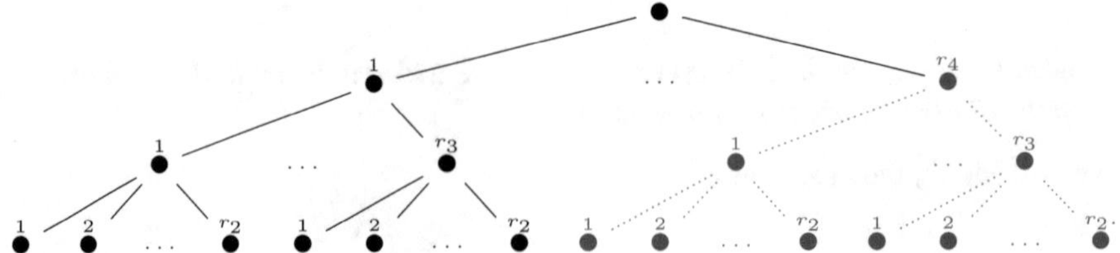

Notice that $T(\mathbf{r}|_k)$ has $\prod_{i=2}^{k} r_i$ leaves, with $\prod_{i=k-j+1}^{k} r_i$ nodes in the j-th layer. The subtree T_v of $T = T(\mathbf{r}|_k)$ is the tree rooted at v consisting of all the children and descendants of v. We call T_v a maximal subtree of T if v is a child of the root, or equivalently if v is in the second layer.

Example 5 In Example 4, the subtree indicated by dotted edges in magenta is a maximal subtree of $T(\mathbf{r}|_4)$.

Definition 3 Let $V_{T(\mathbf{r}|_k)}$ be the set of vertices of the tree $T(\mathbf{r}|_k)$. An $\mathbf{r}|_k$*-label* is a function $\phi : V_{T(\mathbf{r}|_k)} \to \mathbb{N}$ on the vertices of the tree $T(\mathbf{r}|_k)$ that assigns a natural number to each vertex. A $\mathbf{r}|_k$-label is *compatible* if it satisfies the following:

1. for $k = 1$: $\phi(\text{rootnode}) \in [r_1]$, and
2. for $k > 1$:

   a. given any child of the root v, $\phi_v := \phi|_{T_v}$ is a compatible $\mathbf{r}|_{k-1}$-label, and
   b. $\phi(\text{rootnode}) \in [d]$, where $\mathbb{Z}_d$ is the stabilizer of the action of $\mathbb{Z}_{r_k}$ on equivalence classes of $\{(\phi_v) : v \text{ is a child of the root}\}$,

where ϕ_v denotes the restriction of ϕ to the maximal subtree T_v.

We say that two compatible labels ϕ and ψ of $T(\mathbf{r}|_k)$ are *equivalent*, and write $\phi \sim \psi$, if they are in the same orbit under the action of $W(\mathbf{r}|_k)$ or, equivalently, if $\psi^{W(\mathbf{r}|_k)} = \phi^{W(\mathbf{r}|_k)}$, where $\psi^g(v) := \psi(v^g)$.

2.4 Bratteli Diagrams

We refer to Section 4.1 in [20] or to [17] for a detailed discussion on Bratteli diagrams.

A *Bratteli diagram* B is a weighted graph, which can be described by a set of vertices from a disjoint collection of sets B_m, $m \geq 0$, and edges that connect vertices in B_m to vertices in B_{m+1}. Assuming that the set B_0 contains a unique vertex, the edges are labeled by positive integer weights. The set B_m is the set of vertices at level m. If a vertex $T_1 \in B_m$ is connected to a vertex $T_2 \in B_{m+1}$, then we write $T_1 \leq T_2$.

Given a tower of subgroups $\langle 1 \rangle = G_0 \leq G_1 \leq \ldots \leq G_n$, the corresponding Bratteli diagram has vertices of set B_i labeling the irreducible representations of G_i. If ρ and η are irreducible representations of G_i and G_{i-1}, respectively, then the corresponding vertices are connected by an edge weighted by the multiplicity of η in ρ when restricted to G_{i-1}.

Example 6 The Hasse diagram of the partially ordered set with a labeling of the edges is the Bratteli diagram of the iterated wreath product of cyclic groups (see Section 4.1 in [20] for an illustrated example for the n-fold iterated wreath product of $\mathbb{Z}_r$).

3 Irreducible Representations of Iterated Wreath Products

We will now prove Theorem 1.

Proof First, we note that $\mathcal{R}^{r_k} := \{\rho_{x_1} \otimes \rho_{x_2} \otimes \cdots \otimes \rho_{x_{r_k}} : x \in [h]^{r_k}\}$ is a traversal for $W(\mathbf{r}|_{k-1})^{r_k}$. Consider the action of $\mathbb{Z}_{r_k}$ on $\mathcal{R}^{r_k}$ by its action on the indices, indexed by $[h]^{r_k}$. This is isomorphic to the action of $W(\mathbf{r}|_k) = W(\mathbf{r}|_{k-1})^{r_k} \rtimes \mathbb{Z}_{r_k}$ on $\widehat{W}(\mathbf{r}|_{k-1})^{r_k}$ by conjugation.

Fix some $\sigma_x := \rho_{x_1} \otimes \cdots \otimes \rho_{x_{r_k}} \in \mathcal{R}^{r_k}$ (which corresponds to the word $x = x_1 \cdots x_{r_k} \in [h]^{r_k}$). The stabilizer of x under the cyclic action of $\mathbb{Z}_{r_k}$ is a subgroup corresponding to:

$$\mathbb{Z}_{d_x} \cong \left(\frac{r_k}{d_x}\mathbb{Z}\right) \Big/ (r_k\mathbb{Z})$$

for some $d_x | r_k$. Notice that $W(\mathbf{r}|_{k-1})^{r_k} \trianglelefteq W(\mathbf{r}|_k)$. In the language of Clifford theory, the inertia group for σ_x is given by:

$$I = I_{\sigma_x} = W(\mathbf{r}|_{k-1})^{r_k} \rtimes \mathbb{Z}_{d_x}. \tag{5}$$

Also, notice the inclusion $W(\mathbf{r}|_{k-1})^{r_k} \leq I \leq W(\mathbf{r}|_k)$ of a chain of subgroups. For $H \leq G$ and $\tau \in \widehat{H}$, denote

$$\widehat{G}(\tau) = \left\{\theta \in \widehat{G} : \tau \leq \operatorname{Res}_H^G \theta\right\}. \tag{6}$$

In applying Clifford theory, we find that $\widehat{I}(\sigma) = \{\sigma \otimes \tau : \tau \in \widehat{\mathbb{Z}_d}\}$. More importantly,

$$\widehat{W}(\mathbf{r}|_k)(\sigma) = \left\{\operatorname{Ind}_{W(\mathbf{r}|_{k-1})^{r_k} \rtimes \mathbb{Z}_{d_\sigma}}^{W(\mathbf{r}|_k)} \sigma \otimes \tau : \tau \in \widehat{\mathbb{Z}_d}\right\} \tag{7}$$

In addition, $\widehat{W}(\mathbf{r}|_k) = \bigcup_{\sigma \in \widehat{W}(\mathbf{r}|_{k-1})^{r_k}} \widehat{W}(\mathbf{r}|_k)(\sigma)$. Also, if $\theta \in \widehat{W}(\mathbf{r}|_k)(\sigma)$ and $\theta \in \widehat{W}(\mathbf{r}|_k)(\sigma')$ and $I_\sigma^G = I_{\sigma'}^G$, then there exists $g \in W(\mathbf{r}|_k)$ such that $\sigma = \sigma'^g$. The result follows.

Corollary 1 *For a particular $\sigma = \rho_{x_1} \otimes \cdots \otimes \rho_{x_{r_k}} \in \widehat{W}(\mathbf{r}|_{k-1})^{r_k}$ and $\tau \in \widehat{\mathbb{Z}_{d_x}}$, let I be the inertia group of σ. Then, we have*

$$\operatorname{Res}_{W(\mathbf{r}|_{k-1})^{r_k}}^{W(\mathbf{r}|_k)} \left(\operatorname{Ind}_{W(\mathbf{r}|_{k-1})^{r_k} \rtimes \mathbb{Z}_{d_x}}^{W(\mathbf{r}|_k)} \sigma \otimes \tau\right) = \sigma \oplus \sigma^1 \oplus \ldots \oplus \sigma^{r/d_x}. \tag{8}$$

Proof This follows from Theorem 1 and an application of Clifford theory.

3.1 Number of Irreducible Representations

Following the exact same argument for Theorem 2.2 in [20], we have the following recursion for the number of irreducible representations for $W(\mathbf{r}|_k)$.

Theorem 5 *The number $M(\mathbf{r}|_k)$ of irreducible representations of $W(\mathbf{r}|_k)$ satisfies the recursion:*

$$M(\mathbf{r}|_k) = \frac{1}{r_k}\sum_{d|r_k} f(d)d^2 = \frac{1}{r_k}\sum_{d|c|r_k} \mu(c/d)M\left(\mathbf{r}|_{k-1}\right)^{r_k/c} d^2, \tag{9}$$

where $M(\mathbf{r}|_1) = r_1$ and $\mu(n)$ is the Euler number for a natural number $n \in \mathbb{N}$.

4 Bijection Between the Branching Diagram for Generalized Iterated Wreath Products and Rooted Trees

In this section, we will give a combinatorial structure describing the branching diagrams for the iterated wreath products of cyclic groups. In a similar spirit to Proposition 4.6 in [20], we have the following:

Proposition 1 *There exists a one-to-one correspondence between equivalence classes of irreducible representations of the wreath product $W(\mathbf{r}|_k)$ of cyclic groups and orbits of compatible $\mathbf{r}|_k$-labels.*

Proof We inductively define a map $F : \widehat{W}(\mathbf{r}|_k) \to \{\mathbf{r}|_k\text{-labels}\}$ that gives a bijection between equivalence classes of irreducible representations of $W(\mathbf{r}|_k)$ and orbits of labels. Denote by z_1 a fixed generator of $\mathbb{Z}_{r_1}$ and w_1 a fixed r_1-th root of unity. For $i = 1, \ldots, r_1$, let $\tau_i^{(1)}$ denote the irreducible representation of $\mathbb{Z}_{r_1}$ such that $\tau(z_1) = (w_1)^j$.

Recall that $T(\mathbf{r}|_1)$ consists of only a root node, and all $\mathbf{r}|_1$-labels $\phi : V_{T(\mathbf{r}|_1)} \to [r_1]$ are determined exactly by their value on the root node. Thus, $\widehat{W}(\mathbf{r}|_1) = \{\tau_j^{(1)} : j \in [r_1]\}$ is clearly in bijection with $\mathbf{r}|_1$-labels. To be precise, $F : \widehat{W}(\mathbf{r}|_1) = \widehat{\mathbb{Z}_{r_1}} \to \{\mathbf{r}|_1\text{-labels}\}$ is defined by $F(\tau_j^{(1)})(\text{root}) = j$ for all $j \in [r_1]$.

Suppose for induction that $F : \widehat{W}(\mathbf{r}|_{k-1}) \to \mathbf{r}|_{k-1}\text{-labels}$ gives a bijection of the form desired. We define $F : \widehat{W}(\mathbf{r}|_k) \to \{\mathbf{r}|_k\text{-labels}\}$ as follows.

Let $\{\rho_1, \ldots, \rho_h\}$ be an enumeration for a traversal $\mathcal{R}_{W(\mathbf{r}|_{k-1})}$ for $W(\mathbf{r}|_{k-1})$. It suffices to define F on a traversal for $\widehat{W}(\mathbf{r}|_k)$, such as $\mathcal{R}_{W(\mathbf{r}|_k)}$ defined above. Then, let $\rho_{x_1} \otimes \cdots \otimes \rho_{x_{r_k}} \otimes \tau$ be an arbitrary element of $\mathcal{R}_{W(\mathbf{r}|_k)}$. Let

$$\phi := F\left(\text{Ind}_{W(\mathbf{r}|_{k-1})\wr\mathbb{Z}_{d_x}}^{W(\mathbf{r}|_k)} \rho_{x_1} \otimes \cdots \otimes \rho_{x_{r_k}} \otimes \tau\right) : V_{T(\mathbf{r}|_k)} \to \mathbb{N}$$

be defined as follows: let $U = \{u_1, \ldots, u_{r_k}\}$ be the set of r_k children of the root node. For any $u \in U$, let ϕ_u denote the vector $\mathbf{r}|_{k-1}$ found by restricting ϕ to the maximal subtree T_u. Let $d = d_x$ be as defined in (1). Let ζ be a fixed generator of $\mathbb{Z}_d$ and ω a fixed d-th root of unity. For $i = 1, \ldots, d$, let π_i denote the irreducible representation of $\mathbb{Z}_d$ such that $\pi(\zeta) = \omega^i$.

- For any non-root node of $T(\mathbf{r}|_k)$, we let the value of ϕ be defined to satisfy $\phi_{u_i} := F(\rho_{x_i})$.
- For the root node, notice that $\tau \in \widehat{\mathbb{Z}_{d_x}}$ by definition of $\mathcal{R}_{W(\mathbf{r}|k)}$. But, d_x is exactly the integer such that $\mathbb{Z}_{d_x}$ is the stabilizer of $\mathbb{Z}_{r_k}$ on the equivalence classes of $\{\phi_u : u \in U\}$, so $\tau \in \widehat{\mathbb{Z}_d}$. Thus, $\tau = \pi_j$ for some $j \in [d_x]$. Let $\phi(\text{root}) := j$.

It follows from induction that F is a bijection from the equivalence classes of $\widehat{W}(\mathbf{r}|_k)$ to the orbits of $\mathbf{r}|_k$ labels.

Corollary 2 *The number $h_k(\mathbf{r}|_k)$ of $\mathbf{r}|_k$-trees of height k is given by the recursion:*

$$h_k(\mathbf{r}|_k) = \frac{1}{r_k} \sum_{d|c|r_k} \mu(c/d) h_{k-1}(\mathbf{r}|_{k-1})^{r_k/c} d^2,$$

where $\mu(n)$ is the Euler number of $n \in \mathbb{N}$.

4.1 Degrees of Irreducible Representations

Following the discussion in Section 4.1.1 in [20], we define for any $\mathbf{r}|_k$-label ϕ the companion tree C_ϕ.

Definition 4 Fix an $\mathbf{r}|_k$-label ϕ. Define the *companion label* C_ϕ to be the $\mathbf{r}|_k$-label $C : V_{T(\mathbf{r}|_k)} \to \mathbb{N}$ as follows: an arbitrary vertex v on the ℓ-th layer of the complete $\mathbf{r}$-tree $T(\mathbf{r}|_k)$ is

$$C(v) = \left| \left\{ \phi_u^{W(\mathbf{r}|_{k-\ell-1})} : u \text{ is a child of } v \right\} \right|.$$

Recall that $x^G = \{x^g : g \in G\}$, the orbit of x under the action of G. So, $C(v)$ is the number of orbits occupied by the $\mathbf{r}|_{k-\ell-1}$-labels of the $r_{k-\ell-1}$ maximal $\mathbf{r}|_{k-\ell-1}$-subtrees of T_v, or the number of inequivalent sublabels on maximal subtrees of T_v given by ϕ.

Example 7 The companion tree C_ϕ to Example 3 is

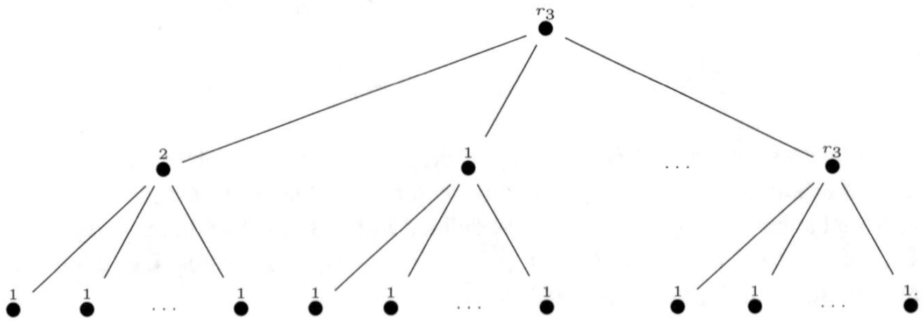

Similar to Proposition 4.3 in [20], we obtain:

Proposition 2 *Let ρ be an irreducible representation of $W(\mathbf{r}|_k)$ associated to $\mathbf{r}|_k$-tree T with companion tree C_T. Then, the dimension d_ρ of ρ is given by:*

$$d_\rho = \prod_v C(v), \tag{10}$$

the product of the value of the companion label C on all vertices.

5 Fast Fourier Transforms, Adapted Bases, and Upper Bound Estimates

We will now prove Theorem 3.

Proof Enumerate $\mathcal{R}_K = \{\eta_1, \ldots, \eta_L\}$ so that K has L inequivalent irreducible representations. Notice that

$$\mathcal{R}_{K^r} = \{\eta_{\ell_1} \otimes \cdots \otimes \eta_{\ell_r} : \ell_1, \ldots, \ell_r \in [L]\} = \{\eta_{\boldsymbol{\ell}} : \boldsymbol{\ell} \in [L]^r\}.$$

Notice that $\{(1, i) : i \in [r]\}$ is a complete set of coset representatives of K^r in $K \wr \mathbb{Z}_r = K^r \rtimes \mathbb{Z}_r$. Let $f_i : K^r \to \mathbb{C}$ be defined by $f_i(\mathbf{k}) = f(\mathbf{k}, i)$. Calculating $\{\widehat{f_i}(\eta_{\boldsymbol{\ell}}) : i \in [r], \boldsymbol{\ell} \in [L]^r\}$ requires $r \cdot T(K^r)$ computations. We will use the $\widehat{f_i}$'s to compute the Fourier transform $\widehat{f}$.

It suffices to compute $\widehat{f}$ for every induced irreducible representation of maximal extensions for all irreducible representations of K^r. So, fix some $\boldsymbol{\eta} = \eta_{\boldsymbol{\ell}} \in \mathcal{R}_{K^r}$ corresponding to some fixed $\boldsymbol{\ell} \in [L]^r$. Let $m = m_{\boldsymbol{\ell}} := \min\{i \neq 0 : \boldsymbol{\ell}^i = \boldsymbol{\ell}\}$, where $(\ell_1\ell_2 \cdots \ell_r)^i := \ell_{i+1} \cdots \ell_r \ell_1 \cdots \ell_i$. Identify $K \rtimes \mathbb{Z}/r\mathbb{Z}$ with $K \rtimes (\mathbb{Z}/m\mathbb{Z} \times m\mathbb{Z}/r\mathbb{Z})$ to relabel the element $(\mathbf{k}, i)$ by $(\mathbf{k}, i_1, i_2) = (\mathbf{k}, i \mod m, \lfloor i/m \rfloor)$.

The inertia group of $\boldsymbol{\eta}$ is given by $H = K^r \rtimes \mathbb{Z}_{r/m} = K^r \rtimes (1 \times m\mathbb{Z}/r\mathbb{Z})$. Fix some (r/m)-th root of unity ζ. Denote the irreducible representations of $H/K^r \cong \mathbb{Z}_{r/m}$ by χ_j for $j \in [r/m]$, where $\chi_j(i) = \zeta^{ij}$. Let $\widetilde{\chi} : H \to \mathbb{C}$ be given by $\widetilde{\chi}(\mathbf{k}, i_2) = \chi(i_2)$. If $\widetilde{\eta}$ is one extension of $\boldsymbol{\eta}$ to H, then the set $\{\widetilde{\chi}_j \otimes \widetilde{\eta} : j \in [r/m]\}$ gives the complete set of inequivalent extensions of $\boldsymbol{\eta}$ to H. All these extensions are irreducible.

Similarly, we relabel $\{f_i : K^r \to \mathbb{C} : i \in [r]\}$ by $\{f_{i_1,i_2} : K^r \to \mathbb{C} : i_1 \in [m], i_2 \in [r/m]\}$ where $f_{i_1,i_2}(\mathbf{k}) = f(\mathbf{k}, i_1, i_2)$. For all $i_1 \in [m]$, let $f_{i_1} : H \to \mathbb{C}$ be given by $f_{i_1}(\mathbf{k}, i_2) = f(\mathbf{k}, i_1, i_2)$. For all i_1, we have

$$\begin{aligned}
\widehat{f_{i_1}}(\widetilde{\chi} \otimes \widetilde{\eta}) &= \sum_{\mathbf{k} \in K^r, i_2 \in [r/m]} \widetilde{\chi}(i_2) \cdot \widetilde{\eta}(\mathbf{k}, i_2) \cdot f_{i_1}(\mathbf{k}, i_2) \\
&= \sum_{\mathbf{k} \in K^r, i_2 \in [r/m]} \widetilde{\chi}(i_2) \cdot \widetilde{\eta}\,((\mathbf{1}, i_2)(\mathbf{k}, 1))\, f_{i_1,i_2}(\mathbf{k})
\end{aligned}$$

$$= \sum_{i_2 \in [r/m]} \widetilde{\chi}(i_2)\widetilde{\eta}(\mathbf{1}, i_2) \sum_{\mathbf{k} \in K^r} \eta(\mathbf{k}) f_{i_1,i_2}(\mathbf{k})$$

$$= \sum_{i_2 \in [r/m]} \widetilde{\chi}(i_2)\widetilde{\eta}(\mathbf{1}, i_2) \widehat{f_{i_1,i_2}}(\eta).$$

Notice that since $\widetilde{\eta}$ is an extension of η, $\widetilde{\eta}(\mathbf{k}, 1) = \eta(\mathbf{k})$ for any $\mathbf{k} \in K^r$. Let $\alpha_{i_1}(i_2) = \widetilde{\eta}(\mathbf{1}, i_2)\widehat{f_{i_1,i_2}}(\eta)$. Since $\widehat{f_{i_1,i_2}}(\eta)$ and $\widetilde{\eta}$ are $\dim(\eta) \times \dim(\eta)$ matrices, finding $\alpha_{i_1}(i_2)$ for all i_1, i_2 requires $m \cdot r/m \cdot \dim(\eta)^\alpha$ computations, where α is the constant of matrix multiplication. Further, the (j, k)-th entry of the resulting matrix of the summation is the Fourier transform of α_{jk} at the irreducible χ. As the associated group is $\mathbb{Z}_{r/m}$, Fourier transforms require time $O(\frac{r}{m} \log \frac{r}{m})$. There are $\dim(\eta)^2$ of these functions, so computation of the final matrix for all i_1 requires

$$m \cdot \dim(\eta)^2 \cdot O\left(\frac{r}{m} \log \frac{r}{m}\right) = \dim(\eta)^2 \cdot O\left(r \log \frac{r}{m}\right)$$

computations.

Now, we are interested in computing the matrix value of the transform on the induced representation $\rho := \operatorname{Ind}_H^G \widetilde{\chi} \otimes \widetilde{\eta}$ for a specific $\widetilde{\chi} \otimes \widetilde{\eta}$. However, $\rho(\mathbf{k}, i_1, 1)$ is a block diagonal matrix with entries $(\widetilde{\chi} \otimes \widetilde{\eta})^{(j)}$ in the j-th place, where the j represents the irreducible found by conjugating by $(\mathbf{1}, 1, j)$. Using this, we obtain:

$$\widehat{f}(\rho) = \sum_{\mathbf{k}, i_1, i_2} \rho(\mathbf{k}, i_1, i_2) \cdot f_{i_1,i_2}(\mathbf{k})$$

$$= \sum_{i_2} \rho(\mathbf{1}, 1, i_2) \sum_{\mathbf{k}, i_1} \begin{pmatrix} (\widetilde{\chi} \otimes \widetilde{\eta})^{(1)}(\mathbf{k}, i_1) & \cdots & 0 \\ \vdots & \ddots & \vdots \\ 0 & \cdots & (\widetilde{\chi} \otimes \widetilde{\eta})^{(m)}(\mathbf{k}, i_1) \end{pmatrix} f_{i_1,i_2}(\mathbf{k})$$

$$= \sum_{i_2} \rho(\mathbf{1}, 1, i_2) \begin{pmatrix} \widehat{f_{i_2}}(\widetilde{\chi} \otimes \widetilde{\eta}^{(1)}) & \cdots & 0 \\ \vdots & \ddots & \vdots \\ 0 & \cdots & \widehat{f_{i_2}}(\widetilde{\chi} \otimes \widetilde{\eta}^{(m)}) \end{pmatrix}.$$

With $\dim(\eta)^\alpha \cdot m^2$ computations for each induced representation and a total of $\frac{r}{m}$ induced representations, we have $r \cdot m \dim(\eta)^\alpha$ computations for each orbit of irreducible representations of K^r under conjugation by G.

In total, we need

$$r \cdot T(K^r) + \sum_{\eta \in E} m \left(r \dim(\eta)^\alpha + \dim(\eta)^2 O\left(r \log \frac{r}{m}\right)\right) + r\, m \cdot \dim(\eta)^\alpha \quad (11)$$

computations, where E is a set of representatives for each G-orbit of irreducible representations of K^r. The size of E is given simply by the number of orbits of $[L]^r$ under action by $\mathbb{Z}_r$.

6 Conclusion and Future Direction

We have given an explicit description of a traversal for the iterated wreath product $W(\mathbf{r}|_k)$ of cyclic groups in Theorem 1, and we have determined a tighter upper bound of the FFT computation time for the iterated wreath product in Theorem 3.

In our sequel chapter [14], we examine the representation theory of generalized iterated wreath products of symmetric groups, where we give a complete description of the traversal for these families of generalized iterated wreath products, and show the existence of a bijection between equivalence classes of irreducible representations of the generalized iterated wreath product and orbits of labels on certain rooted trees.

We conclude with an open problem, which is to find adapted bases and fast Fourier transform operation bounds for chains of subgroups of iterated wreath products of more general classes of groups.

Acknowledgments The authors acknowledge Mathematics Research Communities for providing an exceptional working environment at Snowbird, Utah. They would like to thank Michael Orrison for helpful discussions, and the referees for extremely useful remarks on this manuscript. This manuscript was written during M.S.I.'s visit to the University of Chicago in 2014. She thanks their hospitality.

References

1. J.T. Astola, C. Moraga, R.S. Stanković, *Fourier Analysis on Finite Groups with Applications in Signal Processing and System Design* (Wiley, Hoboken, 2005)
2. K. Balasubramanian, Enumeration of internal rotation reactions and their reaction graphs. Theor. Chim. Acta **53**(2), 129–146 (1979)
3. K. Balasubramanian, Graph theoretical characterization of NMR groups, nonrigid nuclear spin species and the construction of symmetry adapted NMR spin functions. J. Chem. Phys. **73**(7), 3321–3337 (1980)
4. D. Borsa, T. Graepel, A. Gordon, The wreath process: a totally generative model of geometric shape based on nested symmetries (2015). Preprint. arXiv:1506.03041
5. T. Ceccherini-Silberstein, F. Scarabotti, F. Tolli, Clifford theory and applications. Functional analysis. J. Math. Sci. (N.Y.) **156**(1), 29–43 (2009)
6. W. Chang, Image processing with wreath product groups (2004), https://www.math.hmc.edu/seniorthesis/archives/2004/wchang/wchang-2004-thesis.pdf
7. A.J. Coleman, *Induced Representations with Applications to S_n and* GL(n). Lecture notes prepared by C. J. Bradley. Queen's Papers in Pure and Applied Mathematics, No. 4 (Queen's University, Kingston, 1966)
8. J.W. Cooley, J.W. Tukey, An algorithm for the machine calculation of complex Fourier series. Math. Comput. **19**, 297–301 (1965)
9. C.W. Curtis, I. Reiner, *Methods of Representation Theory. Vol. I.* Wiley Classics Library (Wiley, New York, 1990). With applications to finite groups and orders. Reprint of the 1981 original, A Wiley-Interscience Publication
10. P. Diaconis, Average running time of the fast Fourier transform. J. Algorithms **1**(2), 187–208 (1980)

11. R. Foote, G. Mirchandani, D.N. Rockmore, D. Healy, T. Olson, A wreath product group approach to signal and image processing. I. Multiresolution analysis. IEEE Trans. Signal Process. **48**(1), 102–132 (2000)
12. R.B. Holmes, Mathematical foundations of signal processing II. the role of group theory. MIT Lincoln Laboratory, Lexington. Technical report 781 (1987), pp. 1–97
13. R.B. Holmes, Signal processing on finite groups. MIT Lincoln Laboratory, Lexington. Technical report 873 (1990), pp. 1–38
14. M.S. Im, A. Wu, Generalized iterated wreath products of symmetric groups and generalized rooted trees correspondence. Adv. Math. Sci. https://arxiv.org/abs/1409.0604 (to appear)
15. G. Karpilovsky, *Clifford Theory for Group Representations*. North-Holland Mathematics Studies, vol. 156 (North-Holland Publishing Co., Amsterdam, 1989) Notas de Matemática [Mathematical Notes], 125
16. M. Leyton, *A Generative Theory of Shape*, vol. 2145 (Springer, Berlin, 2003)
17. D.K. Maslen, D.N. Rockmore, The Cooley-Tukey FFT and group theory. Not. AMS **48**(10), 1151–1160 (2001)
18. R. Milot, A.W. Kleyn, A.P.J. Jansen, Energy dissipation and scattering angle distribution analysis of the classical trajectory calculations of methane scattering from a Ni (111) surface. J. Chem. Phys. **115**(8), 3888–3894 (2001)
19. G. Mirchandani, R. Foote, D.N. Rockmore, D. Healy, T. Olson, A wreath product group approach to signal and image processing-part II: convolution, correlation, and applications. IEEE Trans. Signal Process. **48**(3), 749–767 (2000)
20. R.C. Orellana, M.E. Orrison, D.N. Rockmore, Rooted trees and iterated wreath products of cyclic groups. Adv. Appl. Math. **33**(3), 531–547 (2004)
21. L.R. Rabiner, R.W. Schafer, C.M. Rader, The chirp z-transform algorithm and its application. Bell Syst. Tech. J. **48**, 1249–1292 (1969)
22. C.M. Rader, Discrete Fourier transforms when the number of data samples is prime. Proc. IEEE **56**(6), 1107–1108 (1968)
23. D. Rockmore, Fast Fourier analysis for abelian group extensions. Adv. Appl. Math. **11**(2), 164–204 (1990)
24. M. Schnell, Understanding high-resolution spectra of nonrigid molecules using group theory. ChemPhysChem **11**(4), 758–780 (2010)

Generalized Iterated Wreath Products of Symmetric Groups and Generalized Rooted Trees Correspondence

Mee Seong Im and Angela Wu

Abstract Consider the generalized iterated wreath product $S_{r_1} \wr \ldots \wr S_{r_k}$ of symmetric groups. We give a complete description of the traversal for the generalized iterated wreath product. We also prove an existence of a bijection between the equivalence classes of ordinary irreducible representations of the generalized iterated wreath product and orbits of labels on certain rooted trees. We find a recursion for the number of these labels and the degrees of irreducible representations of the generalized iterated wreath product. Finally, we give rough upper bound estimates for fast Fourier transforms.

Keywords Iterated wreath products · Symmetric groups · Rooted trees · Irreducible representations · Fast Fourier transform · Bratteli diagrams

AMS Subject Classification Primary 20C30, 20E08; Secondary 65T50, 05E18, 05E10

1 Introduction

The representation theory of the symmetric group is remarkably prevalent in combinatorics; one can explicitly parametrize the irreducible representations of the symmetric group using Young diagrams, leading us to the study of the interaction of these diagrams, an examination of the decomposition of tensor products of Young diagrams, and an investigation of the dimension of the irreducible representation associated to a Young diagram. Wreath products of symmetric groups arise as the automorphism group of regular rooted trees (see Theorem 2.1.6 or Theorem 2.1.15

M. S. Im (✉)
Department of Mathematical Sciences, United States Military Academy, West Point, NY, USA

A. Wu
Department of Mathematics, University of Chicago, Chicago, IL, USA
e-mail: wu@math.uchicago.edu

A. Deines et al. (eds.), *Advances in the Mathematical Sciences*, Association for Women in Mathematics Series 15, https://doi.org/10.1007/978-3-319-98684-5_3

in [7]), with applications ranging from functions on rooted trees (see Sect. 2.5), pixel blurring (cf. [1, 8, 15, 18, 19, 31]), symmetries of nonrigid molecules in molecular spectroscopy (cf. [2, 3, 30, 34]), and visual information processing (cf. [4, 27]) to choosing subcommittees from sets of committees and voting (cf. [12, 14, 26]). With motivation from [20] and [32], we consider generalized iterated wreath product $W(\mathbf{r}|_k) := S_{r_1} \wr \ldots \wr S_{r_k}$ of symmetric groups, where S_{r_i} is the symmetric group on r_i letters, and study its representation theory.

Throughout this chapter, let G be a finite group, and let V be a vector space over the complex numbers $\mathbb{C}$. Let $GL(V)$ be the general linear group on V, and let $\rho : G \to GL(V)$ be a representation of G, i.e., ρ is a group homomorphism. We say two representations $\rho : G \to GL(V)$ and $\eta : G \to GL(W)$ are equivalent, and write $\rho \sim \eta$, if there exists a vector space isomorphism $f : V \to W$ such that $f \circ \rho(g) = \eta(g) \circ f$ for all $g \in G$. We denote by $\widehat{G}$ the set of irreducible representations of G. We say that $\mathcal{R}$ is a *traversal* for G if $\mathcal{R} := \mathcal{R}_G \subset \widehat{G}$ contains exactly one irreducible representation for each isomorphism class in $\widehat{G}$. Thus, a traversal consists of a complete list of pairwise inequivalent irreducible representations of G. As a basic consequence of representation theory, the equality $\sum_{\rho \in \mathcal{R}} \dim(\rho)^2 = |G|$ holds, where the sum is over all irreducible representations in $\mathcal{R}$.

Let $[n] := \{1, 2, \ldots, n\}$, the set of integers from 1 to n, and let

$$[n]^\ell := \{x_1 x_2 \cdots x_\ell : x_i \in [n]\},$$

the set of length ℓ words with letters in $[n]$.

Now given a subgroup $H \leq G$, we write $\mathrm{Ind}_H^G : \mathrm{Rep}(H) \to \mathrm{Rep}(G)$ to be the induction functor from the category of representations of H to the category of representations of G. That is, given a representation $\eta \in \widehat{H}$ of subgroup $H \subseteq G$, where $\eta : H \to GL(V)$, we write $\mathrm{Ind}_H^G \eta = \mathbb{C}[G] \otimes_{\mathbb{C}[H]} V$, the *induced representation* of G from η with dimension $[G : H] \cdot \dim \eta$. There also exists the dual construction to induction called restriction. Given a subgroup H of G, Res_H^G : $\mathrm{Rep}(G) \to \mathrm{Rep}(H)$ is the restriction functor from the category of representations of G to the category of representations of H, i.e., given a representation ρ of G, we obtain the *restricted representation* $\mathrm{Res}_H^G \rho$ of H by restricting ρ to H. The induction and restriction functors are related by Frobenius reciprocity. We refer the reader to [10] for a detailed and elegant discussion on the duality of the induction and restriction functors.

We say that $\boldsymbol{\alpha} = (\alpha_1, \ldots, \alpha_h)$ is a *partition* of a natural number $n > 0$, and write $\boldsymbol{\alpha} \vdash n$, if every α_i satisfies the following:

1. for each i, $\alpha_i \in \mathbb{N} = \{1, 2, 3, \ldots\}$,
2. $\alpha_1 \geq \alpha_2 \geq \ldots \geq \alpha_h$, and
3. $\sum_{i=1}^h \alpha_i = n$.

If $\alpha \vdash n$, then we will denote the length of α by $|\boldsymbol{\alpha}| = h$. We will write $\alpha \models_h n$ to denote that $\alpha = (\alpha_1, \ldots, \alpha_h) \in \left(\mathbb{Z}_{\geq 0}\right)^h$ is a *weak composition* of the natural number n with h parts so that each $\alpha_i \geq 0$ is an integer and $\sum_{i=1}^{h} \alpha_i = n$.

For $\boldsymbol{\alpha} \models_h n$, we define

$$S_{\boldsymbol{\alpha}} := S_{\alpha_1} \times S_{\alpha_2} \times \cdots \times S_{\alpha_h} \leq S_n,$$

the permutation subgroup acting on the set $[n]$ by the full action on h disjoint orbits of size α_j. We also define $S_0 = \{1\}$.

For a group G, we will now discuss inner and outer tensor products associated to describing the irreducible representations of $G \wr S_n$. If ρ and η are representations of G, then their inner tensor product $\rho \otimes \eta$ is again a representation of G defined by $(\rho \otimes \eta)(g) = \rho(g) \otimes \eta(g)$, where $g \in G$. If ρ is a representation of G and η is a representation of a group H, then their outer tensor product $\rho \boxtimes \eta$ is a representation of $G \times H$ defined by $(\rho \boxtimes \eta)(g, h) = \rho(g) \otimes \eta(h)$, where $g \in G$ and $h \in H$.

The irreducible representations of the base group $G^n = G \times \cdots \times G$ are n-fold outer tensor products of irreducible representations of G. If $\mathcal{R} = \{\rho_1, \ldots, \rho_h\}$ is a traversal for G and $\alpha \models_h n$, then $\rho_1^{\boxtimes \alpha_1} \boxtimes \cdots \boxtimes \rho_h^{\boxtimes \alpha_h}$ is an irreducible representation of G^n, which can be extended to an irreducible representation $(\rho_1^{\boxtimes \alpha_1} \boxtimes \cdots \boxtimes \rho_h^{\boxtimes \alpha_h})'$ of the inertia group $G \wr S_\alpha$. On the other hand, if $\sigma \in \mathcal{R}_{S_\alpha}$, then composing σ with the projection of $G \wr S_\alpha$ onto S_α gives an irreducible representation σ' of $G \wr S_\alpha$. Now, the inner tensor product of $(\rho_1^{\boxtimes \alpha_1} \boxtimes \cdots \boxtimes \rho_h^{\boxtimes \alpha_h})'$ and σ' is an irreducible representation of $G \wr S_\alpha$, and the induced representation $\mathrm{Ind}_{G \wr S_\alpha}^{G \wr S_n}((\rho_1^{\boxtimes \alpha_1} \boxtimes \cdots \boxtimes \rho_h^{\boxtimes \alpha_h})' \otimes \sigma')$ is an irreducible representation of $G \wr S_n$. With these remarks, we give an explicit description of the traversal of $W(\mathbf{r}|_k)$:

Theorem 1 *For $N > 0$, let $\mathcal{R}_G = \{\rho_1, \ldots, \rho_h\}$ be a traversal for a group $G \leq S_N$. Let $\boldsymbol{\alpha} \models_h n$. Then, the irreducible representations given by:*

$$\left\{\mathrm{Ind}_{G \wr S_\alpha}^{G \wr S_n}((\rho_1^{\boxtimes \alpha_1} \boxtimes \cdots \boxtimes \rho_h^{\boxtimes \alpha_h})' \otimes \sigma') : \alpha \models_h n, \sigma \in \mathcal{R}_{S_\alpha}\right\} \tag{1}$$

form a traversal for $G \wr S_n$. In particular, if $\mathcal{R}_{W(\mathbf{r}|_{k-1})} = \{\rho_1, \ldots, \rho_h\}$ is a traversal for the wreath product $W(\mathbf{r}|_{k-1})$, then a traversal for $W(\mathbf{r}|_k)$ is

$$\mathcal{R}_{W(\mathbf{r}|_k)} = \left\{\mathrm{Ind}_{W(\mathbf{r}|_{k-1}) \wr S_\alpha}^{W(\mathbf{r}|_k)}((\rho_1^{\boxtimes \alpha_1} \boxtimes \cdots \boxtimes \rho_h^{\boxtimes \alpha_h})' \otimes \sigma') : \alpha \models_h r_k, \sigma \in \mathcal{R}_{S_\alpha}\right\}, \tag{2}$$

where S_α is a subgroup of S_{r_k}.

Note that we write $\rho^\alpha := \rho_1^{\boxtimes \alpha_1} \boxtimes \cdots \boxtimes \rho_h^{\boxtimes \alpha_h}$, where $\rho_1, \ldots, \rho_h$ are traversals of a group G, and we define $\rho^0 := 1$.

We also find a recursion for the number of equivalence classes of ordinary irreducible representations of the generalized iterated wreath products in Corollary 2, and their dimensions are given in Proposition 1.

A *rooted tree* is a connected simple graph with no cycles, and with a distinguished vertex, which is called the *root*. We refer to Sect. 2.4 for a further discussion on rooted trees. We recall the following theorem:

Theorem 2 (Theorem 2.1.15, [7]) *Let $\mathcal{T}(\mathbf{r}|_k)$ be a complete* $\mathbf{r}$*-tree of height k. We have*

$$\mathrm{Aut}(\mathcal{T}(\mathbf{r}|_k)) \cong W(\mathbf{r}|_k).$$

We find a bijection between equivalence classes of ordinary irreducible representations of the generalized iterated wreath product $W(\mathbf{r}|_k)$ and the orbits of families of labels on certain complete trees, thus connecting to the geometric construction in Theorem 2:

Theorem 3 *There is a bijection between equivalence classes $\widehat{W}(\mathbf{r}|_k)$ of ordinary irreducible representations of the iterated wreath product of symmetric groups and $W(\mathbf{r}|_k)$-orbits of rooted trees $\mathcal{T}(\mathbf{r}|_k)$.*

1.1 Summary of the Sections

We begin Sect. 2 with some background. We give a summary of the representation theory of symmetric groups in Sect. 2.1, give the construction of iterated wreath products in Sect. 2.2, and discuss Clifford theory in Sect. 2.3. We give the construction of rooted trees in Sect. 2.4, and Bratteli diagrams in Sect. 2.5. We conclude the background section by reviewing adapted bases and fast Fourier transforms in Sect. 2.6. We prove Theorem 1 in Sect. 3, and we prove Theorem 3 in Sect. 4. We also give the dimension of an irreducible representation of the iterated wreath product in Proposition 1. In Sect. 5, we give coarse upper bound estimates for fast Fourier transforms for $G \wr S_n$ (and thus for the generalized iterated wreath product $W(\mathbf{r}|_k)$ in Corollary 3), and in Sect. 6, we discuss some open problems.

2 Background

We will begin by giving some necessary background.

2.1 *Representations of the Symmetric Group*

We refer to [6, 16, 21, 24, 25] for an extensive background on representations of symmetric groups. In this section, we will give a brief summary of the representation theory of symmetric groups.

The symmetric group S_n has order $n!$ whose conjugacy classes are labeled by partitions of n. Thus, the number of inequivalent irreducible representations over $\mathbb{C}$ is equal to the number of partitions of n. One may also parametrize irreducible representations by the same set that parametrizes conjugacy classes for S_n, which is by partitions of n, or, equivalently, the more commonly used of the so-called Young diagrams of size n (see Example 6).

2.2 *Wreath Products*

We refer to Chapter 2 in [7] for a beautiful exposition on the construction of the wreath product $G \wr H$ of a finite group G with a subgroup $H \leq S_n$, which is summarized as follows. Define an action of H on $G^n = G \times \cdots \times G$ by if $\pi \in H$ and $a = (a_1, a_2, \ldots, a_n) \in G^n$, then $\pi \cdot a := a^\pi = (a_{\pi^{-1}(1)}, a_{\pi^{-1}(2)}, \ldots, a_{\pi^{-1}(n)})$. The wreath product $G \wr H$ is defined to be $G^n \times H$ as a set, with multiplication given by:

$$(a; \pi)(b; \sigma) = (ab^\pi; \pi\sigma). \tag{3}$$

Throughout this paper, we will fix $\mathbf{r} = (r_1, r_2, r_3, \ldots) \in \mathbb{N}^\omega$, a positive integral vector. We denote by $\mathbf{r}|_k := (r_1, r_2, \ldots, r_k)$, the length k vector found by truncating $\mathbf{r}$.

Let H be a finite group. Let $H^X := \{f : X \to H\}$, a set of all maps from X to H, which is a group under pointwise multiplication: $(f \circ f')(x) = f(x)f'(x)$ for all $x \in X$. Now for H acting on a set X and G acting on a set Y,

$$(g, h)^{-1}(x, y) = (h^{-1}g^{-1}, h^{-1})(x, y) = (h^{-1}x, g(x)^{-1}y)$$

for all $(g, h) \in G \wr H$ and $x \in X$ and $y \in Y$.

Definition 1 Let S_{r_i} be a symmetric group acting on a finite set of order r_i for every $1 \leq i \leq k$. Assume that S_{r_i} acts on a finite set X_i, where $1 \leq i \leq k-1$. Set $V_{k+1} = \{\varnothing\}$ and, for $i = 1, 2, \ldots, k$, let

$$V_i = X_i \times X_{i+1} \times \cdots \times X_{k-1} \times X_k.$$

The *generalized iterated wreath product* $W(\mathbf{r}|_k) := S_{r_1} \wr \ldots \wr S_{r_k}$ of symmetric groups consists of all k-tuples $(g_1, \ldots, g_k)$, where $g_k \in S_{r_k}$, and $g_i : V_{i+1} \to S_{r_i}$, $1 \le i < k$, with the multiplication law and action on V_1 recursively defined in the following way:

$$(g_i, \ldots, g_k)(g_i', \ldots, g_k') =$$

$$\left(g_i \cdot (g_{i+1}, \ldots, g_{k-1}, g_k)g_i', (g_{i+1}, \ldots, g_{k-1}, g_k)(g_{i+1}', \ldots, g_{k-1}', g_k')\right),$$

where

$$\left((g_{i+1}, g_{i+2}, \ldots, g_{k-1}, g_k)g_i'\right)(x_{i+1}, \ldots, x_{k-1}, x_k) =$$

$$g_i'\left((g_{i+1}, \ldots, g_{k-1}, g_k)^{-1}(x_{i+1}, \ldots, x_{k-1}, x_k)\right),$$

and by:

$$\begin{aligned}(g_{i+1}, g_{i+2}, \ldots, g_{k-1}, g_k)(x_{i+1}, \ldots, x_{k-1}, x_k) = \\ \Big((g_{i+2}, \ldots, g_{k-1}, g_k)(x_{i+2}, \ldots, x_{k-1}, x_k), \\ g_{i+1}(g_{i+2}, \ldots, g_{k-1}, g_k)(x_{i+2}, \ldots, x_{k-1}, x_k)x_{i+1}\Big)\end{aligned} \quad (4)$$

for all $x_j \in X_j$, $g_j \in S_{r_j}^{V_{j+1}}$, $i \le j \le k$, and $i = 1, 2, \ldots, k$.

Remark 1 The generalized k-th $\mathbf{r}$-symmetric wreath product $W(\mathbf{r}|_k)$ could also be defined recursively by:

$$W(\mathbf{r}|_0) = \{1\} \text{and} W(\mathbf{r}|_k) = W(\mathbf{r}|_{k-1}) \wr S_{r_k},$$

where the multiplication for the wreath product $W(\mathbf{r}|_k)$ is defined recursively using (3).

Example 1 Note that $W(\mathbf{r}|_1) = S_{r_1}$, $W(\mathbf{r}|_2) = S_{r_1} \wr S_{r_2}$, and $W(\mathbf{r}|_k) = S_{r_1} \wr S_{r_2} \wr \ldots \wr S_{r_k}$.

Throughout this paper, we will be considering the chain of groups given in Remark 1.

2.3 Clifford Theory

The following references [5, 13, 22], and [7] contain an extensive background on Clifford theory, which allow one to use recursion to construct the irreducible representations of a group. In this chapter, we will give a brief overview of the main results of Clifford theory and the little-group method.

Let G be a finite group and let $N \triangleleft G$ be a normal subgroup of G. For two representations σ, ρ, we write $\sigma \prec \rho$ if σ is a *subrepresentation* of ρ. We say that $\widetilde{\sigma} \in \widehat{G}$ is an *extension* of $\sigma \in \widehat{N}$ if $\operatorname{Res}^G_N \widetilde{\sigma} = \sigma$.

Definition 2 Fix $\theta \in \widehat{N}$ and $g \in G$.

1. We define

$$\widehat{G}(\theta) := \left\{\rho \in \widehat{G} : \theta \prec \operatorname{Res}^G_N \rho\right\}. \tag{5}$$

2. The g-conjugate $\sigma^g \in \widehat{N}$ of σ is defined as $\sigma^g(h) := \sigma(ghg^{-1})$ for any $h \in N$.
3. The inertia group of σ in G is given by $I_G(\sigma) := \{g \in G : \sigma^g \sim \sigma\}$.

Now, the finite group G also acts on the set of inequivalent irreducible representations of N. For any irreducible representation σ of N, let $\Delta(\sigma)$ denote its orbit under this action, i.e., inequivalent conjugates of σ. Let $\operatorname{Stab}_G(\sigma)$ be the isotropy subgroup of σ under the G-action.

Theorem 4 (Clifford Theory) *Let N be a normal subgroup of G.*

1. *([10], Theorem 10) If σ be a representation of N, then*

$$\operatorname{Res}^G_N \operatorname{Ind}^G_N \sigma = [\operatorname{Stab}(\sigma) : N] \cdot \Delta(\sigma).$$

2. *([10], Theorem 14) If $\rho : G \to GL(V)$ is an irreducible representation of G, then*

$$\operatorname{Res}^G_N \rho = \frac{d_\rho}{[G : \operatorname{Stab}(\sigma)]d_\sigma} \cdot \Delta(\sigma),$$

where σ is any irreducible representation of N appearing in $\operatorname{Res}^G_N \rho$, and d_ρ is the dimension of the vector space V.

We also call d_ρ the *degree* of the representation ρ. Next, the little-group method provided below is motivated by Chapter 5 Section 1 in [35].

Theorem 5 (Little-Group Method) *Suppose that any $\sigma \in \widehat{N}$ has an extension $\widetilde{\sigma}$ to its inertia group $I_G(\sigma)$. Let Σ be a set of orbit representatives of the irreducible representations of N under action of G, where $g \in G$ acts on $\sigma \in \widehat{N}$ by σ^g. Then, a traversal of G is given by:*

$$\left\{ \mathrm{Ind}_{I_G(\sigma)}^G (\widetilde{\sigma} \otimes \bar{\psi}) : \sigma \in \Sigma, \psi \in \mathcal{R}_{I_G(\sigma)/N} \right\},$$

where $\bar{\psi}$ *is the representation on* $I_G(\sigma)$ *given by* $\bar{\psi}(g) = \psi(\mathrm{proj}(g))$, *where* proj : $I_G(\sigma) \to I_G(\sigma)/N$ *is a canonical projection map.*

Lemma 1 *Suppose that* $\{\rho_1, \ldots, \rho_h\}$ *is a traversal for* G. *Then,* $\{\rho(\eta) : \eta \in [h]^n\}$ *is a traversal for* G^n, *where* $\rho(\eta) := \rho_{\eta_1} \boxtimes \cdots \boxtimes \rho_{\eta_n}$ *and* $[h] = \{1, 2, \ldots, h\}$.

Definition 3 The permutation action of S_n on G^n by permuting the factors of G^n induces an action of S_n on $\widehat{G^n}$ by:

$$(\rho(\eta))^\sigma (g_1, \ldots, g_n) = \rho(\eta)(g_{\sigma^{-1}(1)}, \ldots, g_{\sigma^{-1}(n)}).$$

It follows from Definition 3 that $\rho(\eta^\sigma) \sim \rho(\eta)$.

Lemma 2 *For any* $\eta, \mu \in [h]^n$, $\rho(\eta) \sim \rho(\mu)$ *if and only if*

$$\left|\{j \in [n] : \eta_j = \ell\}\right| = \left|\{j \in [n] : \mu_j = \ell\}\right|$$

for any $\ell \in [h]$. *Thus, the set* $\{\rho^\alpha : \alpha \models_h n\}$ *forms a complete set of representatives for the orbits of* $\widehat{G^n}$ *under action by* S_n.

2.4 Rooted Trees of a Fixed Height

Let $\mathbf{r} = (r_1, r_2, r_3, \ldots) \in \mathbb{Z}_{\geq 0}^{\mathbb{N}}$. In this section, we will give the construction of $\mathbf{r}$-rooted trees, generalizing the r-trees discussed in Section 3 of [32]. A *rooted tree* is a connected simple graph with no cycles and with a distinguished vertex, which we call a *root*. We say a node v, i.e., a vertex, is in the j*-th layer* of a rooted tree if it is at distance j from the root. The *branching factor* of a vertex is its number of children, and a *leaf* is a vertex with branching factor zero.

Definition 4 We define the complete $\mathbf{r}$-tree $\mathcal{T}(\mathbf{r}|_k)$ of height k, or $\mathbf{r}|_k$-tree, recursively as follows. Let $\mathcal{T}(r_1)$ be the tree consisting of a root node only. Let $\mathcal{T}(\mathbf{r}|_k)$ consist of a root node with r_k children, with each the vertex in the first layer a copy of the $k-1$-level tree $\mathcal{T}(\mathbf{r}|_{k-1})$, which yields a tree with k levels of nodes.

We will also denote the complete $\mathbf{r}$-tree by $\mathbf{r}|_k$-tree.

Example 2 The tree $\mathcal{T}(r_1)$ is given by • and $\mathcal{T}(r_1, r_2)$ with r_2 leaves is given by:

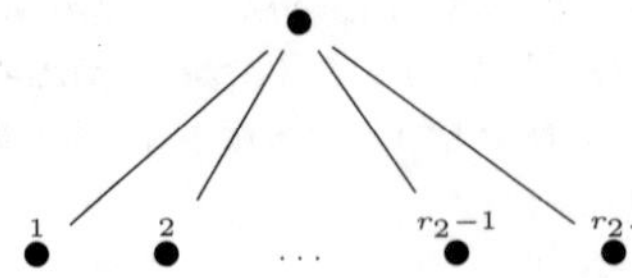

Example 3 Writing $\mathbf{r}|_3 = (r_1, r_2, r_3)$, $\mathbf{r}|_3$-tree of height 3 with 3 levels of nodes is given by:

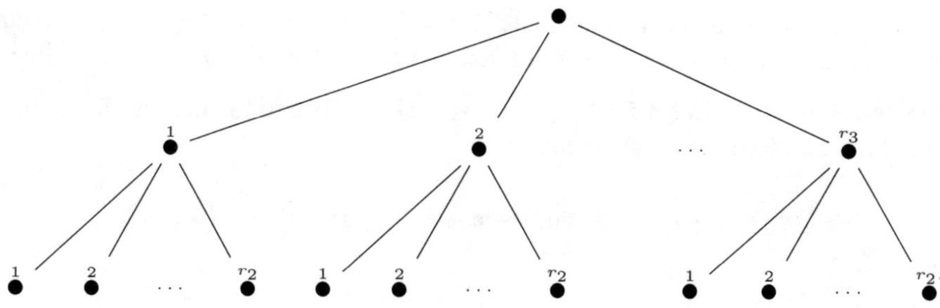

Example 4 The complete tree $\mathcal{T}(\mathbf{r}|_4)$ of height 4 with 4 levels of nodes is given by:

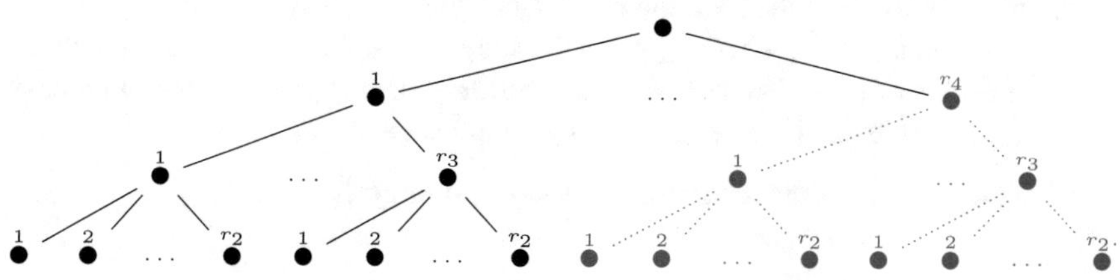

Notice that $\mathcal{T}(\mathbf{r}|_k)$ has $\prod_{i=2}^{k} r_i$ leaves, with $\prod_{i=k-j+1}^{k} r_i$ nodes in the j-th layer. The subtree $\mathcal{T}_v$ of $\mathcal{T} = \mathcal{T}(\mathbf{r}|_k)$ is the tree rooted at v consisting of all the children and descendants of v. We call $\mathcal{T}_v$ a *maximal subtree* of $\mathcal{T}$ if v is a child of the root, or equivalently if v is in the first layer. Let $\deg(v)$ denote the number of leaves of the subtree $\mathcal{T}_v$.

Example 5 In Example 4, the subtree indicated by dotted edges in magenta is a maximal subtree of $\mathcal{T}(\mathbf{r}|_4)$.

We define

$$\widehat{S}_* := \bigsqcup_{n \in \mathbb{N}} \bigsqcup_{\alpha \vdash n} \widehat{S}_\alpha. \tag{6}$$

Definition 5 An $\mathbf{r}|_k$*-label* is a function $\phi : V_{\mathcal{T}(\mathbf{r}|_k)} \to \widehat{S}_*$ on the vertices $V_{\mathcal{T}(\mathbf{r}|_k)}$ of the tree $\mathcal{T}(\mathbf{r}|_k)$ satisfying

$$\phi(v) \in \bigsqcup_{\alpha \vdash \deg(v)} \widehat{S}_\alpha.$$

We say that two labels ϕ and ψ on $\mathcal{T}(\mathbf{r}|_k)$ are *equivalent*, and write $\phi \sim \psi$, if there exists $\sigma \in \mathrm{Aut}(\mathcal{T}(\mathbf{r}|_k))$ such that $\phi^\sigma = \psi$, where $\phi^\sigma(v) := \phi(v^\sigma)$, which is defined as the right action of $\sigma \in W(\mathbf{r}|_k)$ on v.

In other words, two compatible labels ϕ and ψ of $\mathcal{T}(\mathbf{r}|_k)$ are equivalent if they are in the same orbit under the $W(\mathbf{r}|_k)$-action, or equivalently, if $\phi^{W(\mathbf{r}|_k)} = \psi^{W(\mathbf{r}|_k)}$.

Definition 6 An $\mathbf{r}|_k$-label $\phi : V_{\mathcal{T}(\mathbf{r}|_k)} \to \widehat{S_*}$ is *valid* if it satisfies all of the following recursive conditions. We denote by:

$$\mathcal{T}(\mathbf{r}|_k) := \{\phi : \phi \mathrm{isavalid} \mathbf{r}|_k\mathrm{-label}\} \quad \text{and} \quad \mathcal{T} = \bigsqcup_k \mathcal{T}(\mathbf{r}|_k).$$

1. Given an $\mathbf{r}|_1$-label $\phi : V_{\mathcal{T}(\mathbf{r}|_1)} = \{\mathrm{rootnode}\} \to \widehat{S_*}$, where we require $\phi \in \widehat{S_{r_1}}$.
2. If $k > 1$, given an $\mathbf{r}|_k$-label $\phi : V_{\mathcal{T}(\mathbf{r}|_k)} \to \widehat{S_*}$, we require
    a. for any child v of the root, the $\mathbf{r}|_{k-1}$-label $\phi|_{\mathcal{T}_v}$ is in $\mathcal{T}(\mathbf{r}_{k-1})$, and
    b. $\phi(\mathrm{rootnode}) \in \widehat{S_\alpha}$, where S_α gives the stabilizer of the action by S_{r_k} on $\mathbf{r}|_{k-1}$-sublabels of ϕ, so that $\alpha \vdash r_k$ is the partition of $[r_k]$ given by the number of $\mathbf{r}|_{k-1}$-sublabels of ϕ in each nonempty equivalence class,

    where $\phi|_{\mathcal{T}_v}$ denotes the restriction of ϕ to the subtree $\mathcal{T}_v$.

2.5 *Bratteli Diagrams*

We refer to Section 4.1 in [32] or to [29] for a detailed discussion on Bratteli diagrams.

A *Bratteli diagram* B is a weighted graph, which may be described by a set of vertices from a disjoint collection of sets B_k, $k \geq 0$, and edges that connect vertices in B_k to vertices in B_{k+1}. Assume that the set B_0 contains a unique vertex, and that the edges are labeled by positive integer weights. In the case the multiplicity is 1, we omit the labels. The set B_k is the set of vertices at level k. If a vertex $v_k \in B_k$ is connected to a vertex $v_{k+1} \in B_{k+1}$, then we write $v_k \leq v_{k+1}$.

Given a tower of subgroups $\langle 1 \rangle = G_0 \leq G_1 \leq \ldots \leq G_n$, the corresponding Bratteli diagram has vertices of set B_i labeling the irreducible representations of G_i. If ρ and η are irreducible representations of G_i and G_{i-1}, respectively, then the corresponding vertices are connected by an edge weighted by the multiplicity of η in ρ when restricted to the group G_{i-1}.

Example 6 The Young lattice is an example of a Bratteli diagram, where the vertices represent Young diagrams, or partitions, and the edge joining a partition of k to a partition of $k+1$ has weight 1. For the symmetric group S_4 for the sequence $S_1 < S_2 < S_3 < S_4$ of subgroups, where S_i permutes only the symbols $1, \ldots, i$, the Bratteli diagram has the form:

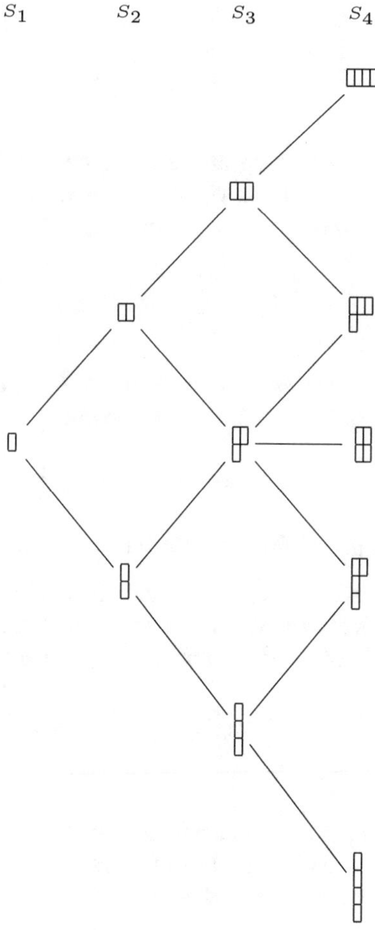

The distinct edges in the Bratteli diagram viewed as directed from level $i-1$ to level i may be viewed as mutually orthogonal S_{i-1}-equivariant morphisms $\mathbb{C}[S_{i-1}] \to \mathbb{C}[S_i]$; the paths from the root to a leaf give a natural indexing of Gelfand–Tsetlin bases for the towers of subgroups (cf. [17, 36]). These bases correspond to those matrix representations which are block diagonal with irreducible blocks at each step (with equivalent irreducibles being equal) when restricted through the tower of subgroups.

2.6 Adapted Bases and Fast Fourier Transforms

Classical discrete Fourier transform (DFT) and fast Fourier transform (FFT)-based approaches come from the use of commutative groups. We expand the original work by Holmes (cf. [19]) and Karpovsky–Trachtenberg (cf. [23]) who merged DFT and FFT for signal processing to noncommutative groups.

Let $\mathcal{R}$ be a set of traversals of $\widehat{G}$. We recall some foundational background from [33].

Definition 7 Let G be a finite group, and let $L(G)$ be the $|G|$-dimensional complex vector space of functions defined on G. If ρ is a matrix representation of G, then the *Fourier transform* $\widehat{f}(\rho)$ *of* f *at* ρ is the matrix sum:

$$\widehat{f}(\rho) = \sum_{g \in G} f(g)\rho(g).$$

Definition 8 The *discrete Fourier transform* $DFT_{\mathcal{R}}(f)$ with respect to a traversal $\mathcal{R} \subseteq \widehat{G}$ is the collection of individual Fourier transforms:

$$DFT_{\mathcal{R}}(f) = \left\{\widehat{f}(\rho) : \rho \in \mathcal{R}\right\}.$$

The following notion of adapted bases is fundamental in the FFT algorithm.

Definition 9 Let H be a subgroup of a group G and let $\mathcal{S} = \{\eta_1, \ldots, \eta_l\}$ and $\mathcal{R} = \{\rho_1, \ldots, \rho_h\}$ be sets of matrix representations for H and G, respectively. Then, the pair $(G, \mathcal{R})$ is $(H, \mathcal{S})$*-adapted* if for all $1 \le i \le h$ and $y \in H$,

$$\rho_i(y) = \eta_{i_1}(y) \oplus \ldots \oplus \eta_{i_m}(y)$$

for some $\eta_{i_j} \in \mathcal{S}$.

Let $T(G, \mathcal{R})$ be the computational time to compute discrete Fourier transform for an arbitrary function f with respect to a traversal $\mathcal{R}$. Let $T(G)$ be the minimum of $T(G, \mathcal{R})$ over all $\mathcal{R}$. We now cite a theorem:

Theorem 6 (Theorem 3.1, [9]) *The Fourier transform for the symmetric group* S_n *may be evaluated in no more than* $\left(\frac{5}{12}n^3 + \frac{1}{2}n^2 - \frac{11}{12}n\right) n!$ *arithmetic operations.*

3 Irreducible Representations of Iterated Wreath Products

In this section, we will prove Theorem 1.

Proof Let $\{\rho_1, \ldots, \rho_h\}$ be a traversal for G. For $\mathbf{i} \in [h]^n$, denote

$$\rho(\mathbf{i}) := \rho_{i_1} \boxtimes \rho_{i_2} \boxtimes \cdots \boxtimes \rho_{i_n}.$$

Let $\rho(\mathbf{i}) \in \mathcal{R}_{G^n}$ be fixed. Let $\sigma \in S_n$. The action of σ on $\rho(\mathbf{i})$ is given by:

$$(\rho(\mathbf{i}))^{\sigma}(g_1, \ldots, g_n) = (\rho_{i_1} \boxtimes \cdots \boxtimes \rho_{i_n})(g_{\sigma^{-1}(1)}, \ldots, g_{\sigma^{-1}(n)})$$

by Definition 3. Since $(\rho(\mathbf{i}))^{\sigma} \sim \rho(\mathbf{i})$ if and only if $i_{\ell} = i_{\sigma(\ell)}$ for every $\ell \in [n]$ by Lemma 2, the inertia group of $\rho(\mathbf{i})$ is

$$S_{A_1(\mathbf{i})} \times \cdots \times S_{A_h(\mathbf{i})} \cong S_{\alpha},$$

where

$$A_{\ell}(\mathbf{i}) = \{j \in [n] : i_j = \ell\} \subseteq [n]$$

and $\alpha_{\ell} = |A_{\ell}|$.

So for $\rho^{\alpha} \in \mathcal{R}_{G^n}$ and $I = I_{G \wr S_n}(\rho^{\alpha})$, the irreducible representation ρ^{α} has an extension to I by the little-group method (an application of Clifford theory) to the structure of irreducible representations of $G \wr S_n = G^n \rtimes S_n$. We thus find that a traversal for $G \wr S_n$ is precisely

$$\left\{\mathrm{Ind}_{G \wr S_{\alpha}}^{G \wr S_n}((\rho_1^{\boxtimes \alpha_1} \boxtimes \cdots \boxtimes \rho_h^{\boxtimes \alpha_h})' \otimes \sigma') : \alpha \models_h n, \sigma \in \mathcal{R}_{S_{\alpha}}\right\}.$$

The second statement of the theorem follows as a special case of the main statement.

Let $N(G)$ be the number of non-isomorphic irreducible representations of a finite group G. If we denote by $\mathcal{P}(n)$ the number of partitions of the integer n (so that $\mathcal{P}(n) = |\{\alpha : \alpha \vdash n\}|$), where $\mathcal{P}(0) := 1$, then $N(S_n) = \mathcal{P}(n)$.

Corollary 1 (Lemma 4.2.9, [21]) *Suppose that $N(G) = h$. Then, the number of non-isomorphic irreducible representations of $G \wr S_n$ is given by:*

$$N(G \wr S_n) = \sum_{\alpha \models_h n} \prod_{i \in [h]} \mathcal{P}(\alpha_i). \tag{7}$$

Proof This follows from Theorem 1 and an application of Clifford theory.

Let $N(\mathbf{r}|_k)$ denote the number of equivalence classes of ordinary irreducible representations for the wreath product $W(\mathbf{r}|_k) = W(\mathbf{r}|_{k-1}) \wr S_{r_k}$. Define $P(n, h) := \sum_{\alpha \models_h n} \prod_{i=1}^{h} \mathcal{P}(\alpha_i)$.

Corollary 2 *It follows that for $\alpha \vdash n$,*

$$N(G \wr S_{\alpha}) = \prod_{j=1}^{|\alpha|} \sum_{\beta \models_h \alpha_j} \prod_{i \in [h]} \mathcal{P}(\beta_i). \tag{8}$$

For $h := N(\mathbf{r}|_{k-1})$, we find that h satisfies the following recursion:

$$N(\mathbf{r}|_k) = P\left(r_k, N(\mathbf{r}|_{k-1})\right) = \sum_{\alpha \models_h r_k} \prod_{i \in [h]} \mathcal{P}(\alpha_i). \tag{9}$$

Proof This follows from Corollary 1.

4 Branching Diagram and r-Label Correspondence

We find a combinatorial structure describing the branching diagrams for iterated wreath products of symmetric groups by proving Theorem 3.

Proof It suffices to define a map on a traversal of $\widehat{W}(\mathbf{r}|_k)$, which is given in (2). We will define a map $F : \mathcal{R}_{W(\mathbf{r}|_k)} \to \mathcal{T}(\mathbf{r}|_k)$ recursively, and it suffices to prove that each orbit of $\mathcal{T}(\mathbf{r}|_k)$ under action by $W(\mathbf{r}|_k)$ has exactly one pre-image under F.

Let $k = 1$. For any $\rho \in \widehat{W}(\mathbf{r}|_k) = \widehat{S}_{r_1}$, we define the $\mathbf{r}|_1$-label as

$$F(\rho) : V_{\mathcal{T}(\mathbf{r}|_1)} = \{\text{root}\} \to \widehat{S}_{r_1}, \qquad \text{where } F(\rho)(\text{root}) := \rho.$$

This is clearly a bijection as desired.

Now, let $k > 1$. By the inductive hypothesis, $F : \mathcal{R}_{W(\mathbf{r}|_{k-1})} \to \mathcal{T}(\mathbf{r}|_{k-1})$ has exactly one pre-image per orbit of $\mathcal{T}(\mathbf{r}|_k)$. Suppose that a traversal for $W(\mathbf{r}|_{k-1})$ is given by the set $\{\rho_1, \ldots, \rho_h\}$. We need to define F on $\mathcal{R}_{W(\mathbf{r}|_k)}$, and show that orbits have exactly one pre-image as desired.

Pick an arbitrary element of $\rho_1^{\alpha_1} \otimes \cdots \otimes \rho_h^{\alpha_h} \otimes \sigma$ of $\mathcal{R}_{W(\mathbf{r}|_k)}$. Denote its image under F by:

$$\phi := F(\rho_1^{\alpha_1} \otimes \cdots \otimes \rho_h^{\alpha_h} \otimes \sigma) : V_{\mathcal{T}(\mathbf{r}|_k)} \to S_{r_k}.$$

Let $U \subset V_{\mathcal{T}(\mathbf{r}|_k)}$ be the r_k children of the root. Assign an ordering to $U = \{u_1, \ldots, u_{r_k}\}$. Then, partition the set U as:

$$U = U_1 \sqcup \ldots \sqcup U_h,$$

where each U_i satisfies $|U_i| = \alpha_i$ while preserving the ordering. For each $u_i \in U$, define the value of ϕ on all nodes in subtree $\mathcal{T}_{u_i}$ to satisfy $\phi|_{\mathcal{T}_{u_i}} := F(\rho_{j^i})$, where j^i satisfies $U_{j^i} \ni u_i$ and where $\phi|_{\mathcal{T}_{u_i}}$ denotes the restriction of ϕ to the subtree $\mathcal{T}_{u_i} \subseteq \mathcal{T}$. It remains to define the value of ϕ on the root node. We let $\phi(\text{root}) = \sigma$.

Notice that $\phi|_{\mathcal{T}_{u_i}} \in \mathcal{T}(\mathbf{r}|_{k-1})$ by definition and induction. Since σ is in the stabilizer of the S_{r_k}-action on ρ^α, which is exactly S_α, we see that ϕ is a compatible label for $\mathcal{T}(\mathbf{r}|_k)$. Thus, F is well-defined, and each orbit of $\mathcal{T}(\mathbf{r}|_k)$ has exactly one pre-image.

4.1 Degrees of Irreducible Representations

Following the discussion in Section 4.1.1 in [32], we define for any $\mathbf{r}|_k$-tree $\mathcal{T}$ the companion tree $C_{\mathcal{T}}$.

Definition 10 Fix $\mathcal{T}(\mathbf{r}|_k)$ and $\mathbf{r}|_k$-label ϕ. Let the *companion label* $C_\phi : V_{\mathcal{T}(\mathbf{r}|_k)} \rightarrow \mathbb{N}$ be defined by:

$$C_\phi(v) = \begin{cases} \dim(\phi(v)) & \text{if } v \text{ is a leaf of the tree } \mathcal{T}(\mathbf{r}|_k), \\ |S_{r_i}/S_\alpha| = \binom{r_i}{\alpha} & \text{otherwise, where} |v \text{ is in the } (k-i) \text{ -th layer of} \\ & \mathcal{T} \text{ and } \phi(v) \in S_\alpha. \end{cases}$$

Similar to Proposition 4.3 in [32], we obtain the following:

Proposition 1 *If ρ is an irreducible representation of $W(\mathbf{r}|_k)$ associated to $\mathbf{r}|_k$-label ϕ, then the dimension d_ρ of ρ is given by:*

$$d_\rho = \prod_v C_\phi(v), \tag{10}$$

the product of the value of the companion label C_ϕ on all vertices.

5 Fast Fourier Transforms, Adapted Bases, and Upper Bound Estimates

We use the FFT estimates derived in [9] and [33] to state a coarse, overall upper bound on the running time of FFT for the wreath product $W(\mathbf{r}|_k)$.

Theorem 7 (Theorem 3, [33]) *We have*

$$T(G \wr S_n) \leq nT(G) \cdot |G \wr S_{n-1}| + nT(G \wr S_{n-1}) \cdot |G| + n^3 2^{|\mathcal{R}_G|} |G \wr S_n|. \tag{11}$$

The separation of variables approach has been one of the primarily components that is responsible for the fastest known algorithms for almost all classes of finite groups, including symmetric groups [28] and their wreath products [11].

Corollary 3 *Let $T(\mathbf{r}|_k)$ be the computation time for the wreath product $W(\mathbf{r}|_k)$. Then,*

$$\begin{aligned} T(\mathbf{r}|_k) \leq r_k \prod_{i=1}^{k-1}(r_i!)\Bigg((r_{k-1}!)\left(T(\mathbf{r}|_{k-1}) + r_k^3 2^{|\mathcal{R}_{W(\mathbf{r}|_{k-1})}|}\right) \\ + T(W(\mathbf{r}|_{k-1}) \wr S_{r_{k-1}})\Bigg). \end{aligned} \tag{12}$$

Proof This result is a consequence of Theorem 7.

6 Conclusion and Open Problems

As a sequel to [20], we have given an explicit description of a traversal for the iterated wreath product $W(\mathbf{r}|_k)$ and we have shown the existence of a bijection between equivalence classes of ordinary irreducible representations of the generalized iterated wreath products and $W(\mathbf{r}|_k)$-orbits of complete rooted trees. We have also stated a recursion for the number of equivalence classes of ordinary irreducible representations of the iterated wreath product and have given the dimension of an irreducible representation of $W(\mathbf{r}|_k)$.

We conclude by giving several open problems. One problem is to find a tighter fast Fourier transform (FFT) bound for chains of subgroups of $W(\mathbf{r}|_k)$ than the upper bound stated in Theorem 7. Another problem is to study the representation theory of, and find adapted bases and FFT operation bounds for chains of subgroups for, iterated wreath products of more general class of groups.

Acknowledgments The authors acknowledge Mathematics Research Communities for providing an exceptional working environment at Snowbird, Utah. They would like to thank Michael Orrison for helpful discussions, and the referees for immensely valuable comments. This paper was written during M.S.I.'s visit to the University of Chicago in 2014. She thanks their hospitality.

References

1. J.T. Astola, C. Moraga, R.S. Stanković, *Fourier Analysis on Finite Groups with Applications in Signal Processing and System Design* (Wiley, Hoboken, 2005)
2. K. Balasubramanian, Enumeration of internal rotation reactions and their reaction graphs. Theor. Chim. Acta **53**(2), 129–146 (1979)
3. K. Balasubramanian, Graph theoretical characterization of NMR groups, nonrigid nuclear spin species and the construction of symmetry adapted NMR spin functions. J. Chem. Phys. **73**(7), 3321–3337 (1980)
4. D. Borsa, T. Graepel, A. Gordon, The wreath process: a totally generative model of geometric shape based on nested symmetries (2015). Preprint. arXiv:1506.03041
5. T. Ceccherini-Silberstein, F. Scarabotti, F. Tolli, Clifford theory and applications. Functional analysis. J. Math. Sci. (N.Y.) **156**(1), 29–43 (2009)
6. T. Ceccherini-Silberstein, F. Scarabotti, F. Tolli, *Representation Theory of the Symmetric Groups*. Cambridge Studies in Advanced Mathematics, vol. 121 (Cambridge University Press, Cambridge, 2010). The Okounkov-Vershik approach, character formulas, and partition algebras
7. T. Ceccherini-Silberstein, F. Scarabotti, F. Tolli, *Representation Theory and Harmonic Analysis of Wreath Products of Finite Groups*. London Mathematical Society Lecture Note Series, vol. 410 (Cambridge University Press, Cambridge, 2014)
8. W. Chang, Image processing with wreath product groups (2004), https://www.math.hmc.edu/seniorthesis/archives/2004/wchang/wchang-2004-thesis.pdf
9. M. Clausen, U. Baum, Fast Fourier transforms for symmetric groups: theory and implementation. Math. Comput. **61**(204), 833–847 (1993)
10. A.J. Coleman, *Induced Representations with Applications to S_n and* GL(n). Lecture notes prepared by C.J. Bradley. Queen's Papers in Pure and Applied Mathematics, No. 4 (Queen's University, Kingston, 1966)

11. J.W. Cooley, J.W. Tukey, An algorithm for the machine calculation of complex Fourier series. Math. Comput. **19**, 297–301 (1965)
12. K.-D. Crisman, M.E. Orrison, Representation theory of the symmetric group in voting theory and game theory, in *Algebraic and Geometric Methods in Discrete Mathematics*. Contemporary Mathematics, vol. 685 (American Mathematical Society, Providence, 2017), pp. 97–115
13. C.W. Curtis, I. Reiner, *Methods of Representation Theory. Vol. I*. Wiley Classics Library (Wiley, New York, 1990). With applications to finite groups and orders. Reprint of the 1981 original. A Wiley-Interscience Publication
14. Z. Daugherty, A.K. Eustis, G. Minton, M.E. Orrison, Voting, the symmetric group, and representation theory. Am. Math. Mon. **116**(8), 667–687 (2009)
15. R. Foote, G. Mirchandani, D.N. Rockmore, D. Healy, T. Olson, A wreath product group approach to signal and image processing. I. Multiresolution analysis. IEEE Trans. Signal Process. **48**(1), 102–132 (2000)
16. W. Fulton, J. Harris, *Representation Theory*. Graduate Texts in Mathematics, vol. 129 (Springer, New York, 1991). A first course, Readings in Mathematics
17. T. Geetha, A. Prasad, Comparison of Gelfand-Tsetlin bases for alternating and symmetric groups (2017). Preprint. arXiv:1606.04424
18. R.B. Holmes, Mathematical foundations of signal processing II. The role of group theory. MIT Lincoln Laboratory, Lexington. Technical report 781 (1987), pp. 1–97
19. R.B. Holmes, Signal processing on finite groups. MIT Lincoln Laboratory, Lexington. Technical report 873 (1990), pp. 1–38
20. M.S. Im, A. Wu, Generalized iterated wreath products of cyclic groups and rooted trees correspondence. Adv. Math. Sci., https://arxiv.org/abs/1409.0603 (to appear)
21. G. James, A. Kerber, *The Representation Theory of the Symmetric Group*. Encyclopedia of Mathematics and Its Applications, vol. 16 (Addison-Wesley Publishing Co., Reading, 1981). With a foreword by P.M. Cohn, With an introduction by Gilbert de B. Robinson
22. G. Karpilovsky, *Clifford Theory for Group Representations*. North-Holland Mathematics Studies, vol. 156 (North-Holland Publishing Co., Amsterdam, 1989). Notas de Matemática [Mathematical Notes], 125
23. M.G. Karpovsky, E.A. Trachtenberg, Fourier transform over finite groups for error detection and error correction in computation channels. Inf. Control **40**(3), 335–358 (1979)
24. A. Kerber, *Representations of Permutation Groups. I*. Lecture Notes in Mathematics, vol. 240 (Springer, Berlin, 1971)
25. A. Kleshchev, Representation theory of symmetric groups and related Hecke algebras. Bull. Am. Math. Soc. **47**(3), 419–481 (2010)
26. S. Lee, Understanding voting for committees using wreath products (2010), https://www.math.hmc.edu/seniorthesis/archives/2010/slee/slee-2010-thesis.pdf
27. M. Leyton, *A Generative Theory of Shape*, vol. 2145 (Springer, Berlin, 2003)
28. D.K. Maslen, The efficient computation of Fourier transforms on the symmetric group. Math. Comput. **67**(223), 1121–1147 (1998)
29. D.K. Maslen, D.N. Rockmore, The Cooley-Tukey FFT and group theory. Not. AMS **48**(10), 1151–1160 (2001)
30. R. Milot, A.W. Kleyn, A.P.J. Jansen, Energy dissipation and scattering angle distribution analysis of the classical trajectory calculations of methane scattering from a Ni (111) surface. J. Chem. Phys. **115**(8), 3888–3894 (2001)
31. G. Mirchandani, R. Foote, D.N. Rockmore, D. Healy, T. Olson, A wreath product group approach to signal and image processing-part II: convolution, correlation, and applications. IEEE Trans. Signal Process. **48**(3), 749–767 (2000)
32. R.C. Orellana, M.E. Orrison, D.N. Rockmore, Rooted trees and iterated wreath products of cyclic groups. Adv. Appl. Math. **33**(3), 531–547 (2004)
33. D.N. Rockmore, Fast Fourier transforms for wreath products. Appl. Comput. Harmon. Anal. **2**(3), 279–292 (1995)

34. M. Schnell, Understanding high-resolution spectra of nonrigid molecules using group theory. ChemPhysChem **11**(4), 758–780 (2010)
35. B. Simon, *Representations of Finite and Compact Groups*. Graduate Studies in Mathematics, vol. 10 (American Mathematical Society, Providence, 1996)
36. A.M. Vershik, A.Y. Okounkov, A new approach to the representation theory of the symmetric groups. II. Zapiski Nauchnykh Seminarov POMI **307**, 57–98 (2004)

Conway–Coxeter Friezes and Mutation: A Survey

Karin Baur, Eleonore Faber, Sira Gratz, Khrystyna Serhiyenko, and Gordana Todorov

Abstract In this survey chapter, we explain the intricate links between Conway–Coxeter friezes and cluster combinatorics. More precisely, we provide a formula, relying solely on the shape of the frieze, describing how each individual entry in the frieze changes under cluster mutation. Moreover, we provide a combinatorial formula for the number of submodules of a string module, and with that a simple way to compute the frieze associated to a fixed cluster-tilting object in a cluster category of Dynkin type A in the sense of Caldero and Chapoton.

Keywords AR quiver · Cluster category · Cluster mutation · Cluster-tilted algebra · Frieze pattern · Caldero–Chapoton map · String module

---

K. Baur
Institut für Mathematik und Wissenschaftliches Rechnen, Universität Graz, NAWI Graz, Graz, Austria
e-mail: baurk@uni-graz.at

E. Faber (✉)
School of Mathematics, University of Leeds, Leeds, UK
e-mail: e.m.faber@leeds.ac.uk

S. Gratz
School of Mathematics and Statistics, University of Glasgow, Glasgow, UK
e-mail: Sira.Gratz@glasgow.ac.uk

K. Serhiyenko
Department of Mathematics, University of California at Berkeley, Berkeley, CA, USA
e-mail: khrystyna.serhiyenko@berkeley.edu

G. Todorov
Department of Mathematics, Northeastern University, Boston, MA, USA
e-mail: g.todorov@neu.edu

A. Deines et al. (eds.), *Advances in the Mathematical Sciences*, Association for Women in Mathematics Series 15, https://doi.org/10.1007/978-3-319-98684-5_4

1 Introduction

Cluster algebras were introduced by Fomin and Zelevinsky in [13]. A key motivation was to provide an algebraic framework for phenomena observed in the study of dual canonical bases for quantized enveloping algebras and in total positivity for reductive groups.

Cluster categories were introduced in 2005 [4, 7], to give a categorical interpretation of cluster algebras. The following table shows the beautiful interplay and correspondences between cluster algebras and cluster categories in type A. Note that the correspondences between the first and second column hold more generally, not only in type A: Caldero and Chapoton [6] have provided a formal link between cluster categories and cluster algebras by introducing what is now most commonly known as the *Caldero Chapoton map* (short: CC-map) or *cluster character*. Fixing a cluster-tilting object (which takes on the role of the initial cluster), it associates to each indecomposable in the cluster category a unique cluster variable in the associated cluster algebra, sending the indecomposable summands of the cluster-tilting object to the initial cluster.

Cluster algebra	$\leftarrow$	Cluster category	Polygon
cluster variables	CC-map	indecomposable objects	diagonals
clusters		cluster-tilting objects	triangulations
mutations		mutations	flip

In the 1970s, Coxeter and Conway first studied frieze patterns of numbers ([9] and [8]). When these numbers are positive integers, they showed that the frieze patterns arise from triangulations of polygons. Thus, we can extend this table by a further column:

...	Polygon	Frieze
	diagonals	integers
	triangulations	sequences of 1's
	flip	??

Here, the last entry is missing: the meaning of mutation or flip on the level of frieze patterns is not known until now. The purpose of this survey chapter is to show how to complete the picture of cluster combinatorics in the context of friezes. It is based on the paper [3] where more background on cluster categories can be found and where the proofs are included. More precisely, we determine how mutation of a cluster affects the associated frieze, thus effectively introducing the notion of a mutation of friezes that is compatible with mutation in the associated cluster algebra. This provides a useful new tool to study cluster combinatorics of Dynkin type A.

In order to deal with the mutations for friezes, we will use cluster categories and generalized cluster categories as introduced by Buan et al. [4] for hereditary algebras and by Amiot [1] more generally. In both cases, cluster categories are triangulated categories in which the combinatorics of cluster algebras receives a categorical interpretation: cluster variables correspond to rigid indecomposable objects and clusters correspond to cluster-tilting objects. One of the essential features in the definition of cluster algebras is the process of mutation, which replaces one element of the cluster by another unique element such that a new cluster is created. The corresponding categorical mutation replaces an indecomposable summand of a cluster-tilting object by another unique indecomposable object using approximations in the triangulated categories; this process creates another cluster-tilting object which corresponds to the mutated cluster.

We now explain the different players appearing in the table above.

1.1 Frieze Patterns

The notion of friezes was introduced by Coxeter [10]; it was Gauss's *pentagramma mirificum* which was the original inspiration. We recall that a *frieze* is a grid of positive integers consisting of a finite number of infinite rows: the top and bottom rows are infinite rows of 0s and the second to top and second to bottom are infinite rows of 1s as one can see in the following diagram:

$$
\begin{array}{ccccccccccc}
\cdots & & 0 & & 0 & & 0 & & 0 & & \cdots \\
 & 1 & & 1 & & 1 & & 1 & & 1 & \\
\cdots & & m_{-1,-1} & & m_{00} & & m_{11} & & m_{22} & & \cdots \\
 & m_{-2,-1} & & m_{-1,0} & & m_{01} & & m_{12} & & m_{23} & \\
\cdots & & \cdots & & \cdots & & \cdots & & \cdots & & \cdots \\
 & 1 & & 1 & & 1 & & 1 & & 1 & \\
\cdots & & 0 & & 0 & & 0 & & 0 & & \cdots
\end{array}
$$

The entries of the frieze satisfy the *frieze rule*: for every set of adjacent numbers arranged in a *diamond*

$$
\begin{array}{ccc}
 & b & \\
a & & d \\
 & c &
\end{array}
$$

we have

$$ad - bc = 1.$$

The sequence of integers in the first nontrivial row, $(m_{ii})_{i\in\mathbb{Z}}$, is called a *quiddity sequence*. This sequence completely determines the frieze. Each frieze is also periodic, since it is invariant under glide reflection. The order of the frieze is defined to be the number of rows minus one. It follows that each frieze of order n is n-periodic.

Among the famous results about friezes is the bijection between the friezes of order n and triangulations of a convex n-gon, which was proved by Conway and Coxeter in [9] and [8]. This was used to set the first link with cluster combinatorics using [7] and [6] by Caldero and Chapoton. Recently, frieze patterns have been generalized in several directions and found applications in various areas of mathematics; for an overview, see [17].

1.2 Cluster Algebras

Fomin and Zelevinsky introduced the notion of cluster algebras in [13]. Cluster algebras are commutative algebras generated by *cluster variables*; cluster variables are obtained from an *initial cluster* (of variables) by replacing one element at a time according to a prescribed rule, where the rule is given either by a skew-symmetric (or more generally skew-symmetrizable) matrix or, equivalently, by a quiver with no loops nor 2-cycles. The process of replacing one element of a cluster by another unique element in order to obtain another cluster, together with the prescribed change of the quiver, is called *mutation*. Finite sequences of iterated mutations create new clusters and new cluster variables; all cluster variables are obtained in such a way.

The process of such mutations may never stop; however, if the quiver is of Dynkin type, then by a theorem of Fomin and Zelevinsky, this process stops and one obtains a finite number of cluster variables [14]. Among those cluster algebras, the best behaved and understood are the cluster algebras of type A. The clusters of the cluster algebra of type A_{n-3} are in bijection with the triangulations of a convex n-gon, for $n \geq 3$. This is exactly what is employed in this work in order to relate and use cluster categories, via triangulations of an n-gon, so that we can describe the mutations of friezes of order n. Since we will also be dealing with quivers Q' which are mutation equivalent to the quivers of type A_n and may have nontrivial potential, we need to consider generalized cluster categories $\mathscr{C}_{(Q',W)}$, which are shown to be triangle equivalent to $\mathscr{C}_Q$ [1].

1.3 Cluster Categories

Let Q be an acyclic quiver with n vertices, over an algebraically closed field. We consider the category $\operatorname{mod} kQ$ of (finitely generated) modules over kQ, or, equivalently, the category $\operatorname{rep}Q$ of representations of the quiver Q. The bounded derived category $D^b(kQ)$ can be viewed as $\cup_{i\in\mathbb{Z}} \operatorname{mod}(kQ)[i]$, with connecting morphisms.

As an example, consider the quiver

$$Q: \qquad 1 \longleftarrow 2 \longleftarrow 3$$

The module category of the path algebra kQ has six indecomposable objects up to isomorphisms, with irreducible maps between them as follows:

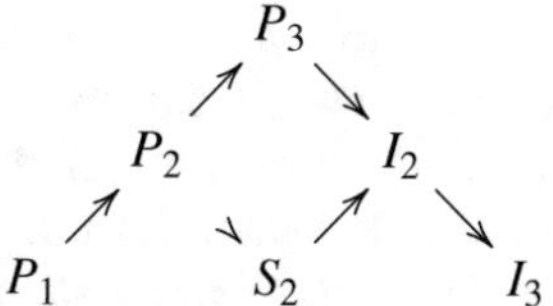

The modules P_i are indecomposable projective, the I_i are indecomposable injectives and the S_i are the simple modules, with $I_1 = P_3$, $S_1 = P_1$ and $S_3 = I_3$. The bounded derived category then looks as follows (the arrows indicate the connecting morphisms):

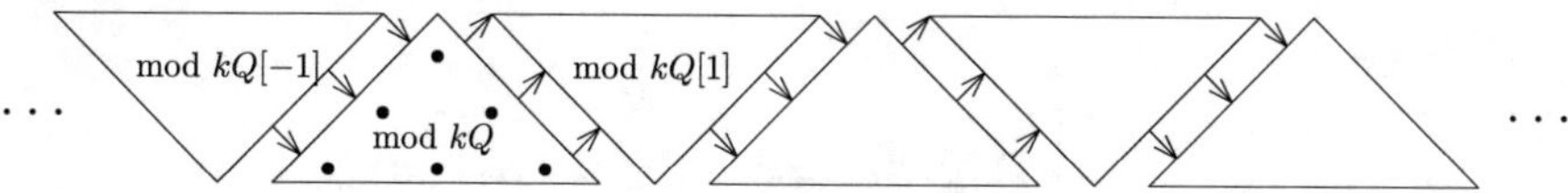

Let Q be a Dynkin quiver of type A. Let $\mathscr{C}$ be the associate cluster category, which by definition is $\mathscr{C} = \mathscr{C}_Q = D^b(kQ)/\tau^{-1}[1]$ where $D^b(kQ)$ is the bounded derived category of the path algebra kQ with the suspension functor [1] and the Auslander–Reiten functor τ. In this case, the specialized CC map gives a direct connection between the Auslander–Reiten quiver of the cluster category $\mathscr{C}$ with a fixed cluster-tilting object T, and the associated frieze $F(T)$ in the following way: recall that each vertex of the Auslander–Reiten quiver corresponds to an isomorphism class of an indecomposable object in the cluster category. When the specialized CC map is applied to a representative of each isomorphism class and the vertex is labeled by that value, one only needs to complete those rows by the rows of 1s and 0s at the top and bottom in order to obtain a frieze, cf. [6, Proposition 5.2].

2 From Cluster Categories to Frieze Patterns

Let $\mathscr{C}$ be a cluster category, let T be a cluster-tilting object and let $B_T = \mathrm{End}_{\mathscr{C}}(T)$ be the endomorphism algebra, which is also called a *cluster-tilted algebra*. The module category $\mathrm{mod}(B_T)$ is shown to be equivalent to the quotient category $\mathscr{C}/\mathrm{add}(T[1])$ of the cluster category. This result by Buan, Marsh and Reiten is used in a very essential way: each indecomposable object in $\mathscr{C}$, which is not isomorphic to a summand of $T[1]$, corresponds to an indecomposable B_T-module, preserving the structure of the corresponding Auslander–Reiten quivers; at the same time, the indecomposable summands of $T[1]$ correspond to the suspensions of the indecomposable projective B_T-modules in the generalized cluster category of the algebra B_T.

When $\mathscr{C}$ is the cluster category associated to a Dynkin quiver of type A, for each cluster-tilting object T, the associated specialized CC-map sends each indecomposable summand of $T[1]$ to 1 and each indecomposable B_T-module M to the number of its submodules, as we explain now. In the actual Caldero–Chapoton formula for cluster variable x_M in terms of the initial cluster variables, the coefficients are given as the Euler–Poincaré characteristics of the Grassmannians of submodules of the module M. In this expression, the sum is being taken over the dimension vectors of the submodules of M. However in this setup, since all indecomposable B_T-modules are string modules that have dimension at most one at every vertex, all the Grassmannians are just points. The *specialized Caldero–Chapoton map* is the map we get from postcomposing the CC-map associated to T with the specialization of the initial cluster variables to one. It will be denoted by ρ_T throughout the paper. It is given by the following formula:

$$\rho_T(M) = \begin{cases} 1 & \text{if } M = T_i[1]\,, \\ \sum_{\underline{e}} \chi(\mathrm{Gr}_{\underline{e}}(M)) & \text{if } M \text{ is a } B_T\text{-module.} \end{cases}$$

Here, $\mathrm{Gr}_{\underline{e}}(M)$ is the Grassmannian of submodules of M with dimension vector $\underline{e}$. Hence, the sum is equal to the number of submodules and the values of the specialized CC-map are positive integers. The values of the specialized CC-map are now entered in the AR quiver of the cluster category $\mathscr{C}$ at the places of the corresponding indecomposable objects. The image of this generalized CC-map only needs to be completed with the rows of 1s and 0s above and below in order to obtain the *frieze associated to the cluster-tilting object* T, denoted by $F(T)$.

Since the generalized CC-map for cluster categories of Dynkin type A is given in terms of the number of submodules of B_T-modules, the first goal of the paper is to give a formula for the number of submodules. This is determined by the following result, hence providing a combinatorial formula for the number of submodules of any given indecomposable B_T-module. Its proof can be found in [3, Section 4]. We recall that each B_T-module is a string module and hence has a description in terms of the lengths of the individual legs. Let $(k_1, \ldots, k_m)$ denote these lengths, cf. Fig. 1. We further denote by $s(M)$ the number of submodules of a B_T-module M.

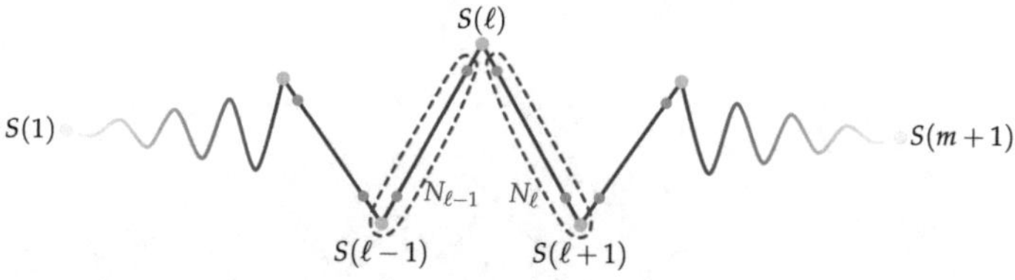

Fig. 1 A string module M of shape $(k_1, \ldots, k_m)$ with legs N_l

In order to state the formula for the number of submodules, we say that a subset I of $\{1, \ldots, m\}$ is *admissible* if, assuming that $I = \{i_1, \ldots, i_l\}$ is ordered by $i_1 < \ldots < i_l$, any two consecutive numbers i_j, i_{j+1} are either of parity even–odd or odd–even.

Theorem 2.1 *Let M be an indecomposable B_T-module, of shape $(k_1, \ldots, k_m)$. Then, the number of submodules of M is given as:*

$$s(M) = 1 + \sum_{j=0}^{m} \sum_{|I|=m-j} \prod_{i \in I} k_i ,$$

where the second sum runs over all admissible subsets I of $\{1, \ldots, m\}$.

Using the position of the module in the AR quiver and the information about the positions of the indecomposable projective B_T-modules, the procedure for finding the numerical invariants $(k_1, \ldots, k_m)$ of the module is given in [3, Section 2]. This purely combinatorial way of computing the numbers of submodules is the basis for computing the associated friezes and, eventually, mutations of friezes.

Remark 2.2 Let $\mathscr{C}$ be the cluster category associated to a Dynkin quiver of type A and let T be a cluster-tilting object in $\mathscr{C}$. Then for each indecomposable B_T-module M, we have $\rho_T(M) = s(M)$. Theorem 2.1 thus gives us a combinatorial way to compute the specialized Caldero–Chapoton map.

We end this section by giving an example illustrating the frieze pattern obtained through the specialized CC-map.

Example 2.3 We now illustrate several notions on the example of the cluster category $\mathscr{C}_{A_{11}}$: the Auslander–Reiten quiver of $\mathscr{C}_{A_{11}}$, a cluster-tilting object T, the cluster-tilted algebra B_T and the Auslander–Reiten quiver of the generalized cluster category of B_T where the modules are given by their composition factors. The Auslander–Reiten quiver of $\mathscr{C}_{A_{11}}$ is the quotient of the Auslander–Reiten quiver of $D^b(kA_{11})$ by the action of $\tau^{-1}[1]$, a fundamental domain which is depicted in black below. We pick the cluster-tilting object $T = \bigoplus_{i=1}^{11} T_i$ whose indecomposable summands are marked with circles:

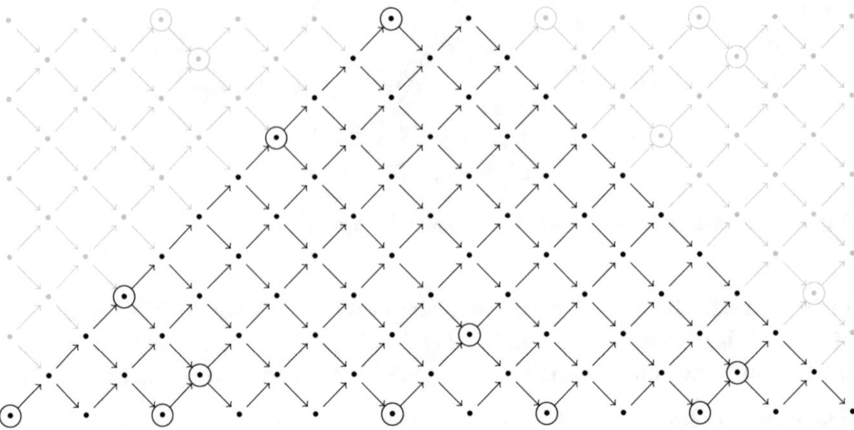

Consider the cluster-tilted algebra $B_T = \mathrm{End}_{\mathscr{C}_{A_{11}}}(T)$. Then, $B_T = kQ/I$, where Q is the quiver

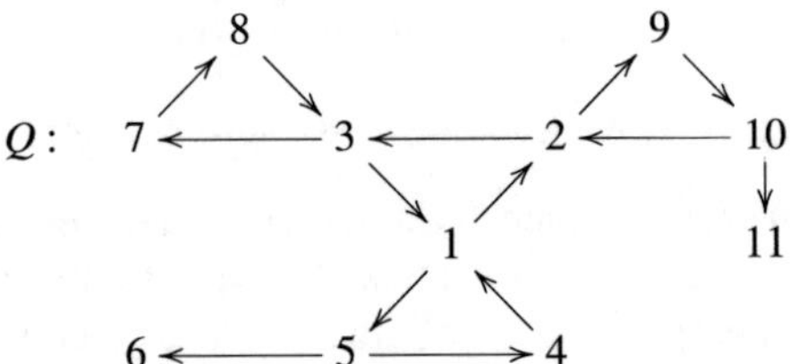

and I is the ideal generated by the directed paths of length 2 which are part of the same 3-cycle. We refer the reader to [5] for a detailed description of cluster-tilted algebras of Dynkin type A.

We can view $\mathrm{mod}(B_T)$ as a subcategory of $\mathscr{C}_{A_{11}}$ and label the indecomposable objects in $\mathscr{C}_{A_{11}}$ by modules and shifts of projective modules, respectively:

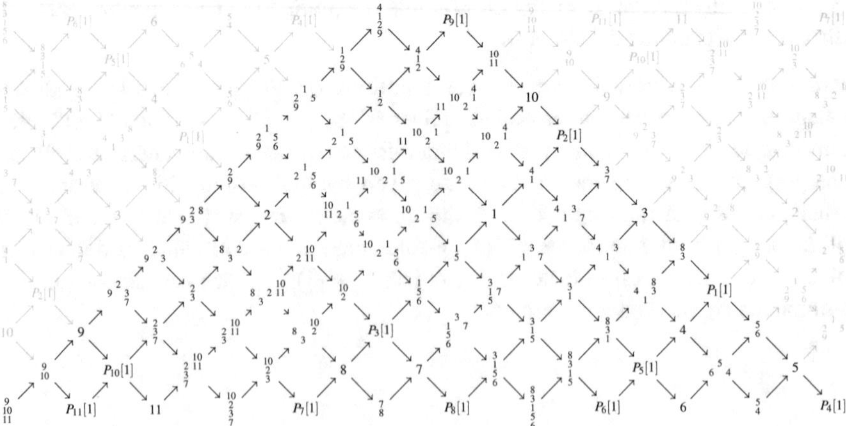

In this picture, the direct summands of T correspond to the indecomposable projective modules P_i of B_T (which lie on the right of the $P_i[1]$ in the picture). The specialized CC-map replaces each vertex labeled by a module, by the number of its submodules and the shifts of projectives by 1s. Adding in the first two and last two rows of 0s and 1s gives rise to the associated frieze $F(T)$:

0		0		0		0		0		0		0		0		0		0		0		0	
	1		1		1		1		1		1		1		1		1		1		1		1
6		1		2		3		1		5		1		4		1		2		5		1	
	5		1		5		2		4		4		3		3		1		9		4		2
4		4		2		3		7		3		11		2		2		4		7		7	
	3		7		1		10		5		8		7		1		7		3		12		3
5		5		3		3		7		13		5		3		3		5		5		5	
	8		2		8		2		18		8		2		8		2		8		2		18
3		3		5		5		5		11		3		5		5		3		3		7	
	1		7		3		12		3		4		7		3		7		1		10		5
2		2		4		7		7		1		9		4		4		2		3		7	
	3		1		9		4		2		2		5		5		1		5		2		4
4		1		2		5		1		3		1		6		1		2		3		1	
	1		1		1		1		1		1		1		1		1		1		1		1
0		0		0		0		0		0		0		0		0		0		0		0	

3 Description of the Regions in the Frieze

The Quiver of a Triangulation

Let $\mathcal{T}$ be a triangulation of an $(n+3)$-gon, and let the diagonals be labeled by $1, 2, \ldots, n$. We recall that the quiver $Q_{\mathcal{T}}$ of the triangulation $\mathcal{T}$ is defined as follows: the vertices of $Q_{\mathcal{T}}$ are the labels $\{1, 2, \ldots, n\}$. There is an arrow $i \to j$ in case the diagonals share an endpoint and the diagonal i can be rotated clockwise to diagonal j (without passing through another diagonal incident with the common vertex). This is illustrated in Example 3.2 and Fig. 5 below.

Let $B = B_{\mathcal{T}}$ be the path algebra of $Q_{\mathcal{T}}$ modulo the relations arising from triangles in $Q_{\mathcal{T}}$: whenever α, β are two successive arrows in an oriented triangle in $Q_{\mathcal{T}}$, their composition is 0. Let P_x be the indecomposable projective B-module associated to the vertex x and S_x its simple top. Let

$$T = \oplus_{x \in \mathcal{T}} P_x.$$

We considered T as an object of the generalized cluster category $\mathscr{C} = \mathscr{C}_B$. Then, T is a cluster-tilting object in $\mathscr{C}$ and $B \cong \mathrm{End}_{\mathscr{C}}(T)$. Hence, B is a cluster-tilted algebra, called the cluster-tilted algebra associated to the triangulation $\mathscr{T}$. We can extend this to an object in the Frobenius category $\mathscr{C}_f$ by adding the $n+3$ projective-injective summands associated to the boundary segments $[12], [23], \ldots, [n+3, 1]$ of the polygon, with irreducible maps between the objects corresponding to diagonals/edges as follows: $[i-1, i+1] \to [i, i+1]$, $[i, i+1] \to [i, i+2]$ [2, 11, 16]. We denote the projective-injective associated to $[i, i+1]$ by Q_{x_i}. Let

$$T_f = (\oplus_{x \in \mathscr{T}} P_x) \oplus \left(Q_{x_1} \oplus \cdots \oplus Q_{x_{n+3}}\right)$$

This is a cluster-tilting object of $\mathscr{C}_f$ in the sense of [15, Section 3]. Given a B-module M, by abuse of notation, we denote the corresponding objects in $\mathscr{C}$ and $\mathscr{C}_f$ by M, that is $\mathrm{Hom}_{\mathscr{C}}(T, M) = M$. In other words, an indecomposable object of $\mathscr{C}_f$ is either an indecomposable B-module or Q_{x_i} for some $i \in \{1, \ldots, n+3\}$ or of the form $P_x[1]$ for some $x \in \mathscr{T}$.

The *frieze* $F(\mathscr{T})$ *of the triangulation* $\mathscr{T}$ is the frieze pattern $F(T)$ for T the cluster-tilting object associated to $\mathscr{T}$.

3.1 Diagonal Defines a Quadrilateral

Let a be a diagonal in the triangulation, $a \in \{1, 2, \ldots, n\}$. This diagonal uniquely defines a quadrilateral formed by diagonals or boundary segments. Label them b, c, d, e as in Fig. 2.

3.2 Diagonal Defines Two Rays

Consider the entry 1 of the frieze corresponding to a. There are two rays passing through it. We go along these rays forwards and backwards until we reach the first entry 1. As the frieze has two rows of ones bounding it, we will always reach an entry

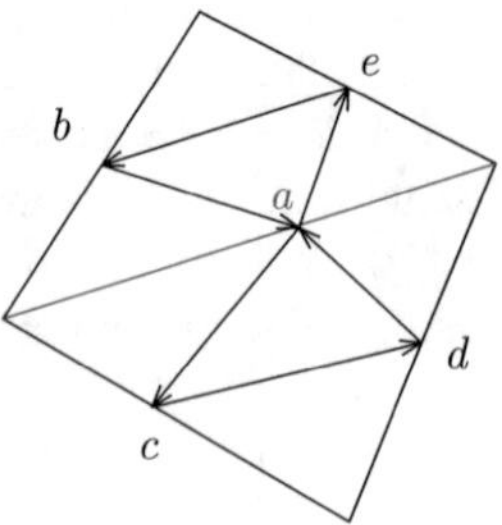

Fig. 2 Triangulation around a

1 in each of these four directions. Going forwards and upwards: the first occurrence of 1 corresponds to the diagonal b. Down and forwards: diagonal d. Backwards down from the entry corresponding to 1: diagonal c and backwards up: diagonal e. If we compare with the coordinate system for friezes of Sect. 1.1, the two rays through the object corresponding to the diagonal $a = [kl]$ are the entries $m_{i,l}$ (with i varying) and $m_{k,j}$ (with j varying).

In the frieze or in the AR quiver, we give the four segments between the entry 1 corresponding to a and the entries corresponding to b, c, d and e names (see Fig. 4 for a larger example containing these paths). Whereas a is always a diagonal, b, c, d, e may be boundary segments. If b is a diagonal, the ray through $P_a[1]$ goes through $P_b[1]$, and if b is a boundary segment, say $b = [i, i+1]$ (with $a = [ij]$) this ray goes through Q_{x_i}. By abuse of notation, it will be more convenient to write this projective-injective as $P_b[1]$ or as $P_{x_i}[1]$ (if we want to emphasize that it is an object of the Frobenius category $\mathscr{C}_f$ that does not live in $\mathscr{C}$).

Let $\mathfrak{e}$ and $\mathfrak{c}$ denote the unique sectional paths in $\mathscr{C}_f$ starting at $P_a[1]$ and ending at $P_b[1]$ and $P_d[1]$, respectively, but not containing $P_b[1]$ or $P_d[1]$. Similarly, let $\mathfrak{b}$ and $\mathfrak{d}$ denote the sectional paths in $\mathscr{C}_f$ starting at $P_e[1]$ and $P_c[1]$, respectively, and ending at $P_a[1]$, not containing $P_e[1]$, $P_c[1]$, see Fig. 4.

Note that b and d are opposite sides of the quadrilateral determined by a. In particular, the corresponding diagonals do not share endpoints. In other words, $P_b[1]$ and $P_d[1]$ do not lie on a common ray in the AR quiver. So by the combinatorics of $\mathscr{C}_f$, there exist two distinct sectional paths starting at $P_b[1]$, $P_d[1]$. These sectional paths both go through S_a. Let $\mathfrak{c}^a$, $\mathfrak{e}^a$ denote these paths starting at $P_b[1]$ and at $P_d[1]$, up to S_a, but not including $P_b[1]$, $P_d[1]$, respectively. Observe that the composition of $\mathfrak{e}$ with $\mathfrak{c}^a$ and the composition of $\mathfrak{c}$ with $\mathfrak{e}^a$ are not sectional, see Fig. 4. Similarly, let $\mathfrak{d}_a$, $\mathfrak{b}_a$ denote the two distinct sectional paths starting at S_a and ending at $P_e[1]$, $P_c[1]$, respectively, but not including $P_e[1]$, $P_c[1]$. Note that the composition of $\mathfrak{c}^a$ with $\mathfrak{b}_a$ and the composition of $\mathfrak{e}^a$ with $\mathfrak{d}_a$ are not sectional.

3.3 Diagonal Defines Subsets of Indecomposables

For x a diagonal in the triangulation $\mathscr{T}$ and P_x the corresponding projective indecomposable, we write $\mathscr{X}$ for the set of indecomposable B-modules having a non-zero homomorphism from P_x into them, $\mathscr{X} = \{M \in \operatorname{ind} B \mid \operatorname{Hom}(P_x, M) \neq 0\}$. Given a B-module M, its *support* is the full subquiver $\operatorname{supp}(M)$ of $Q_{\mathscr{T}}$ generated by all vertices x of $Q_{\mathscr{T}}$ such that $M \in \mathscr{X}$. It is well known that the support of an indecomposable module is connected.

If x is a boundary segment, we set $\mathscr{X}$ to be the empty set (there is no projective indecomposable associated to x, so there are no indecomposables reached).

We use the notation above to describe the regions in the frieze. Thus, if x, y are diagonals or boundary segments, we write $\mathscr{X} \cap \mathscr{Y}$ for the indecomposable objects in $\mathscr{C}$ that have x and y in their support.

Remark 3.1 Let M be an indecomposable B-module in $\mathscr{X} \cap \mathscr{Y}$ such that there exists a (unique) arrow $\alpha : x \to y$ in the quiver. It follows that the right action of the element $\alpha \in B$ on M is non-zero, that is $M\alpha \neq 0$.

By the remark above, we have the following equalities. Note that none of the modules below are supported at a, because the same remark would imply that such modules are supported on the entire 3-cycle in $Q_{\mathscr{T}}$ containing a. However, this is impossible as the composition of any two arrows in a 3-cycle is zero in B. We have

$$\mathscr{B} \cap \mathscr{E} = \{M \in \text{ind}\, B \mid M \text{ is supported on } e \to b\}$$

$$\mathscr{C} \cap \mathscr{D} = \{M \in \text{ind}\, B \mid M \text{ is supported on } c \to d\}$$

Moreover, since the support of an indecomposable B-module forms a connected subquiver of Q, we also have the following equalities:

$$\mathscr{B} \cap \mathscr{C} = \{M \in \text{ind}\, B \mid M \text{ is supported on } b \to a \to c\}$$

$$\mathscr{D} \cap \mathscr{E} = \{M \in \text{ind}\, B \mid M \text{ is supported on } d \to a \to e\}$$

$$\mathscr{B} \cap \mathscr{D} = \{M \in \text{ind}\, B \mid M \text{ is supported on } b \to a \leftarrow d\}$$

$$\mathscr{C} \cap \mathscr{E} = \{M \in \text{ind}\, B \mid M \text{ is supported on } c \leftarrow a \to e\}$$

Finally, using similar reasoning it is easy to see that the sets described above are disjoint. Next, we describe modules lying on sectional paths defined in Sect. 3.2. First, consider sectional paths starting or ending in $P_a[1]$, then we claim that

$$\mathfrak{i} = \{M \in \text{ind}\, B \mid i \in \text{supp}(M) \subset Q_i\} \cup \{P_a[1]\}$$

for all $i \in \{b, c, d, e\}$, for Q_i the subquiver of Q containing i, as in Fig. 3. We show that the claim holds for $i = b$, but similar arguments can be used to justify the remaining cases. Note that it suffices to show that a module $M \in \mathfrak{b}$ is supported on b but it is not supported on e or a. By construction, the sectional path $\mathfrak{b}$ starts at $P_e[1]$, so $0 = \text{Hom}(\tau^{-1} P_e[1], M) = \text{Hom}(P_e, M)$. On the other hand, $\mathfrak{b}$ ends at $P_a[1]$, so $0 = \text{Hom}(M, \tau P_a[1]) = \text{Hom}(M, I_a)$, where I_a is the injective B-module at a. This shows that M is not supported at e or a. Finally, we can see from Fig. 4 that

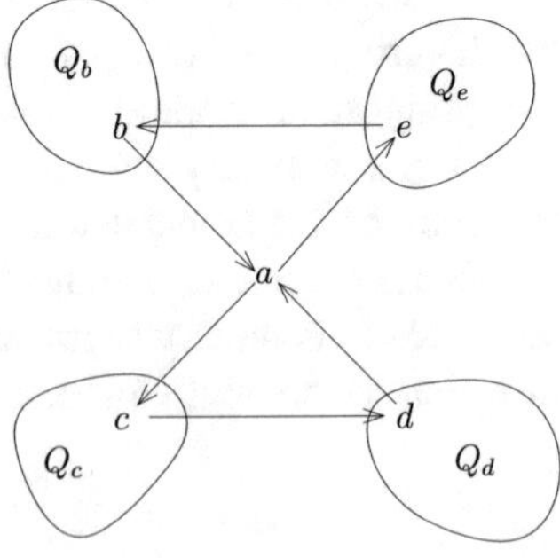

Fig. 3 Regions in quiver

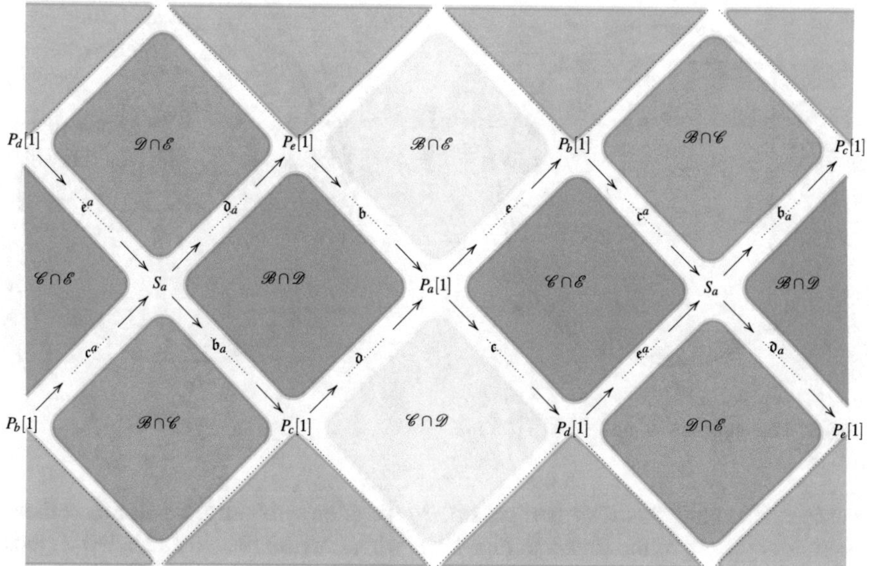

Fig. 4 Regions in the AR quiver determined by $P_a[1]$

M has a non-zero morphism into $\tau P_b[1] = I_b$, provided that b is not a boundary segment. However, if b is a boundary segment, then $\mathfrak{b} \cap \text{Ob}(\text{mod}\, B) = \emptyset$ and we have $\mathfrak{b} = \{P_a[1]\}$. Conversely, it also follows from Fig. 4 that every module M supported on b and some other vertices of Q_b lies on $\mathfrak{b}$. This shows the claim.

Now, consider sectional paths starting or ending in S_a. Using similar arguments as above, we see that

$$\mathfrak{i}^a = \{M \in \text{ind}\, B \mid a \in \text{supp}(M) \subset Q_i^a\}$$

for $i \in \{c, e\}$ and

$$\mathfrak{i}_a = \{M \in \text{ind}\, B \mid a \in \text{supp}(M) \subset Q_i^a\}$$

for $i \in \{b, d\}$, where Q_i^a is the full subquiver of Q on vertices of Q_i and the vertex a.

Finally, we define $\mathscr{F}$ to be the set of indecomposable objects of $\mathscr{C}_f$ that do not belong to

$$\mathscr{A} \cup \mathscr{B} \cup \mathscr{C} \cup \mathscr{D} \cup \mathscr{E} \cup \{P_a[1]\}.$$

The region $\mathscr{F}$ is a succession of wings in the AR quiver of $\mathscr{C}_f$, with peaks at the $P_x[1]$ for $x \in \{b, c, d, e\}$. That is, in the AR quiver of $\mathscr{C}_f$ consider two neighboured copies of $P_a[1]$ with the four vertices $P_b[1]$, $P_c[1]$, $P_d[1]$, $P_e[1]$. Then, the indecomposables of $\mathscr{F}$ are the vertices in the triangular regions below these four

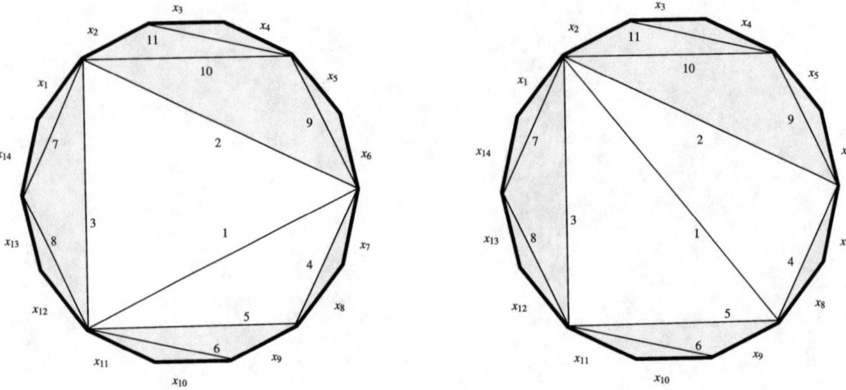

Fig. 5 Triangulations $\mathscr{T}$ and $\mathscr{T}' = \mu_1(\mathscr{T})$

vertices, including them (as their peaks). By the glide symmetry, we also have these regions at the top of the frieze. In Fig. 4, the wings are the shaded unlabeled regions at the boundary. It corresponds to the diagonals inside and bounding the shaded regions in Fig. 5. We will see in the next section that objects in $\mathscr{F}$ do not change under mutation of T_f at $P_a[1]$.

Example 3.2 We consider the triangulation $\mathscr{T}$ of a 14-gon, see the left hand of Fig. 5 and the triangulation $\mathscr{T}' = \mu_1(\mathscr{T})$ obtained by flipping diagonal 1.

The quivers of $\mathscr{T}$ and of $\mathscr{T}'$ are given below. Note that the quiver Q is the same as in Example 2.3.

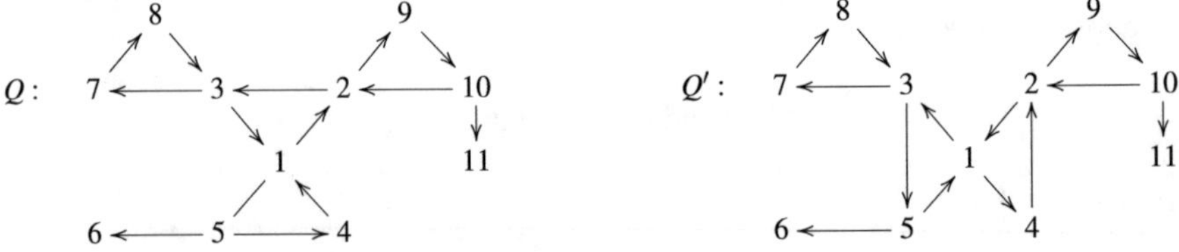

Figure 6 shows the Auslander–Reiten quiver of the cluster category $\mathscr{C}_f$ for Q. In Fig. 7 (Sect. 4.2), the frieze patterns of T and of T' are given.

4 Mutating Friezes

Assume now that our cluster-tilting object T in $\mathscr{C}$ is of the form $T = \bigoplus_{i=1}^{n} T_i$, where the T_i are mutually non-isomorphic indecomposable objects. Mutating T at T_i for some $1 \leq i \leq n$ yields a new cluster-tilting object $T' = T/T_i \oplus T_i'$, to which we can associate a new frieze $F(T')$. In terms of the frieze, we can think of this

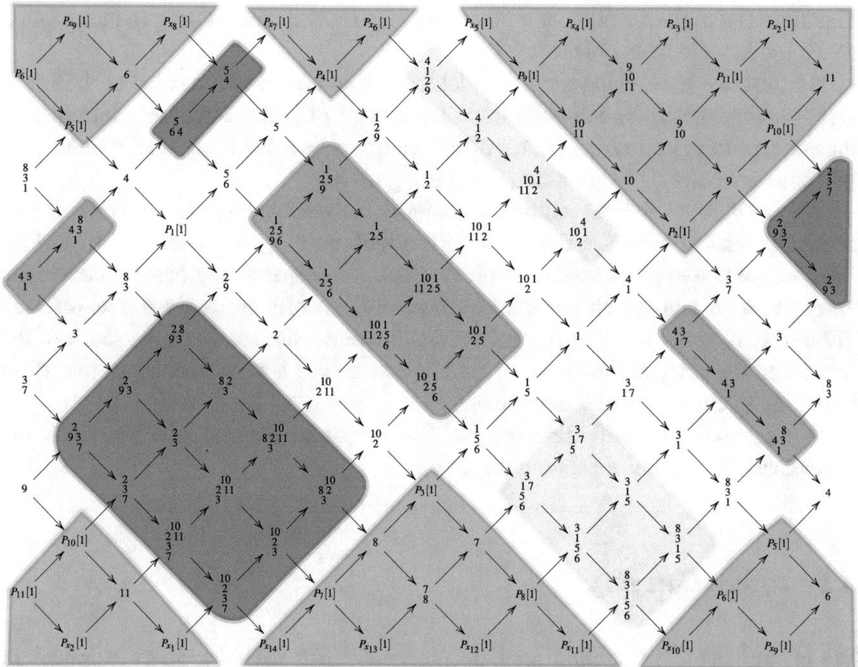

Fig. 6 AR quiver of the category $\mathscr{C}_f$ arising from Q

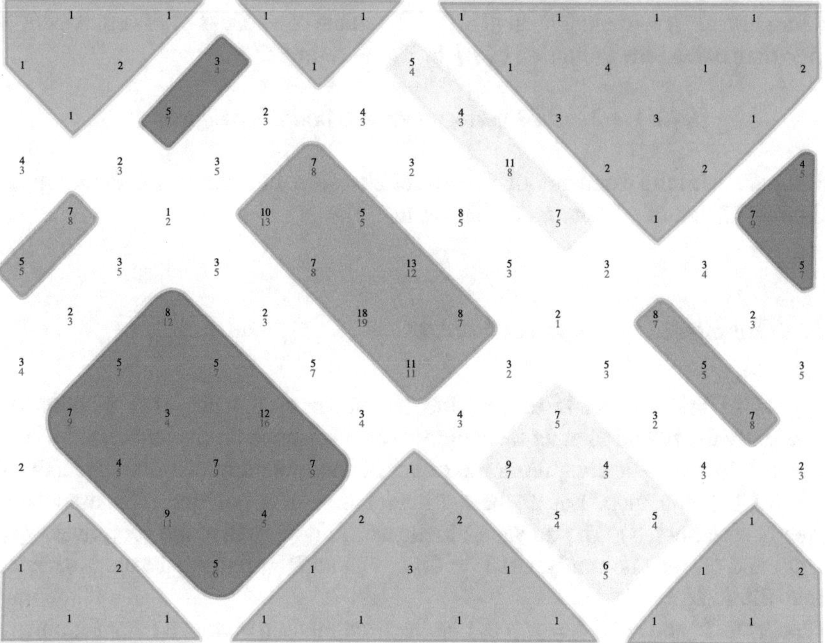

Fig. 7 Frieze pattern of Example 3.2. Red entries: after flip of diagonal 1

mutation as a mutation at an entry of value 1, namely the one sitting in the position of the indecomposable object $T_i[1]$.

We describe how, using graphic calculus, we can obtain each entry of the frieze $F(T')$ independently and directly from the frieze $F(T)$, thus effectively introducing the concept of mutations of friezes at entries of value 1 that do not lie in one of the two constant rows of 1s bounding the frieze pattern.

We are able to give an explicit formula of how each entry in the frieze $F(T)$ changes under mutation at the entry corresponding to T_i, see Theorem 4.7 below. We observe that each frieze can be divided into four separate regions, relative to the entry of value 1 at which we want to mutate. Each of these regions gets affected differently by mutation. The formula of the theorem relies solely on the shape of the frieze and the entry at which we mutate. It determines how each entry of the frieze individually changes under mutation.

In Sect. 4.2, we will describe the four separate regions and introduce the necessary notation before stating the theorem.

4.1 Frieze Category

We extend $\operatorname{ind}\mathscr{C}$ by adding an indecomposable for each boundary segment of the polygon and denote the resulting category by $\mathscr{C}_f$. Then, $\mathscr{C}_f$ is the Frobenius category of maximal CM-modules categorifying the cluster algebra structure of the coordinate ring of the (affine cone of the) Grassmannian Gr(2,n) as studied in [11] and for general Grassmannians in [2, 16]. The stable category of $\mathscr{C}_f$ is equivalent to $\mathscr{C}$. We then extend the definition of ρ_T to $\mathscr{C}_f$ by setting

$$\rho_T(M) = 1 \quad \text{if } M \text{ corresponds to a boundary segment.}$$

This agrees with the extension of the cluster character to Frobenius category given by Fu and Keller, cf. Theorem [15, Theorem 3.3].

4.2 The Effect of Flips on Friezes

The goal of this section is to describe the effect of the flip of a diagonal or equivalently the mutation at an indecomposable projective on the associated frieze. We give a formula for computing the effect of the mutation using the specialized Caldero Chapoton map. Let $\mathscr{T}$ be a triangulation of a polygon with associated quiver Q (see Sect. 3). The quiver Q looks as in Fig. 3, where the subquivers Q_b, Q_c, Q_d and Q_e may be empty. Let $T = \oplus_{x\in T} P_x$ and $B = \operatorname{End}_{\mathscr{C}} T$ be the associated cluster-tilted algebra.

Take $a \in \mathscr{T}$ and let $\mathscr{T}' = \mu_a(\mathscr{T})$ be the triangulation obtained from flipping a, with quiver $Q' = \mu_a(Q)$ (Fig. 8).

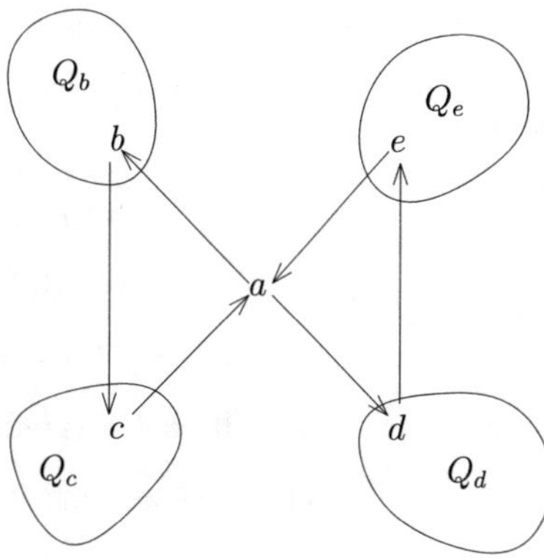

Fig. 8 Quiver after flipping diagonal a

Let B' be the algebra obtained through this, it is the cluster-tilted algebra for $T' = \oplus_{x \in \mathscr{T}'} P_x$. If M is an indecomposable B-module, we write M' for $\mu_a(M)$ in the sense of [12]. If M is an indecomposable B-module, the *entry of* M in the frieze $F(T)$ is the entry at the position of M in the frieze.

Definition 4.1 Let $\mathscr{T}$ be a triangulation of a polygon, $a \in \mathscr{T}$ and M an indecomposable object of $\mathscr{C}_f$. Then, we define the *frieze difference (w.r.t. mutation at a)* $\delta_a : \operatorname{ind} \mathscr{C}_f \to \mathbb{Z}$ by:

$$\delta_a(M) = \rho_{\mathscr{T}}(M) - \rho_{\mathscr{T}'}(M') \in \mathbb{Z}$$

In Sect. 4.3, we first describe the effect mutation has on the regions in the frieze. This gives us all the necessary tools to compute the frieze difference δ_a (Sect. 4.4).

4.3 Mutation of Regions

Here, we describe how mutation affects the regions (Sect. 3.3) of the frieze $F(T)$. Let $\mathscr{T}$, a, B and $\mathscr{T}'$, B' be as above. When mutating at a, the change in support of the indecomposable modules can be described explicitly in terms of the local quiver around a. This is what we will do here. We first describe the regions in the AR quiver of $\mathscr{C}_f$ for B'.

If x is a diagonal or a boundary segment, we write

$$\mathscr{X}' = \{M \in \operatorname{ind} B' \mid \operatorname{Hom}(P_x, M) \neq 0\}$$

for the indecomposable modules supported on x.

After mutating a, the regions in the AR quiver are still determined by the projective indecomposables corresponding to the framing diagonals (or edges) b, c, d, e. The relative positions of a, b, c, d and e have changed; however, it follows from [12] that except for vertex a the support of an indecomposable module at all other vertices remains the same. Therefore, the regions are now described as follows:

$$\mathcal{B}' \cap \mathcal{E}' = \{M \in \operatorname{ind} B' \mid M \text{ is supported on } e \to a \to b\}$$
$$\mathcal{C}' \cap \mathcal{D}' = \{M \in \operatorname{ind} B' \mid M \text{ is supported on } c \to a \to d\}$$
$$\mathcal{B}' \cap \mathcal{C}' = \{M \in \operatorname{ind} B' \mid M \text{ is supported on } b \to c\}$$
$$\mathcal{D}' \cap \mathcal{E}' = \{M \in \operatorname{ind} B' \mid M \text{ is supported on } d \to e\}$$
$$\mathcal{B}' \cap \mathcal{D}' = \{M \in \operatorname{ind} B' \mid M \text{ is supported on } b \leftarrow a \to d\}$$
$$\mathcal{C}' \cap \mathcal{E}' = \{M \in \operatorname{ind} B' \mid M \text{ is supported on } c \to a \leftarrow e\}$$

Under the mutation at a, if a module M lies on one of the rays $\mathfrak{b}_a$, $\mathfrak{d}_a$ $\mathfrak{c}^a$ and $\mathfrak{e}^a$, then M' is obtained from M by removing support at vertex a. The modules lying on the remaining four rays gain support at vertex a after the mutation.

4.4 Mutation of Frieze

We next present the main result of this section, the effect of flip on the generalized Caldero Chapoton map, i.e. the description of the frieze difference δ_a. We begin by introducing the necessary notation.

Depending on the position of an indecomposable object M, we define several projection maps sending M to objects on the eight rays from Sect. 3.2.

Let $M \in \operatorname{ind} B$, and let $\mathfrak{i}$ be one of the sectional paths defined in Sect. 3.2. Suppose $M \notin \mathfrak{i}$, then we denote by $M_{\mathfrak{i}}$ a module on $\mathfrak{i}$ if there exists a sectional path $M_{\mathfrak{i}} \to \cdots \to M$ or $M \to \cdots \to M_{\mathfrak{i}}$ in $\mathcal{C}_f$; otherwise, we let $M_{\mathfrak{i}} = 0$. If $M \in \mathfrak{i}$, then we let $M_{\mathfrak{i}} = M$. In the case when it is well defined, we call $M_{\mathfrak{i}}$ the *projection* of M onto the path $\mathfrak{i}$.

It will be convenient to write these projections in a uniform way.

Definition 4.2 (Projections) If (x, y) is one of the pairs $\{(b, c), (d, e), (b, e), (c, d)\}$, the region $\mathcal{X} \cap \mathcal{Y}$ has two paths along its boundary and two paths further backwards or forwards met along the two sectional paths through any vertex M of $\mathcal{X} \cap \mathcal{Y}$. We call the backwards projection onto the first path $\pi_1^-(M)$ and the projection onto the second path $\pi_2^-(M)$. The forwards projection onto the first path is denoted by $\pi_1^+(M)$ and the one onto the second path $\pi_2^+(M)$.

Figure 9 illustrates these projections in the case $(x, y) \in \{(b, c), (d, e)\}$.

The remaining two regions will be treated together with the surrounding paths.

Definition 4.3 The *closure of* $\mathcal{C} \cap \mathcal{E}$ is the Hom-hammock

$$\overline{\mathcal{C} \cap \mathcal{E}} = \operatorname{ind}(\operatorname{Hom}_{\mathcal{C}_f}(P_a[1], -) \cap \operatorname{Hom}_{\mathcal{C}_f}(-, S_a))$$

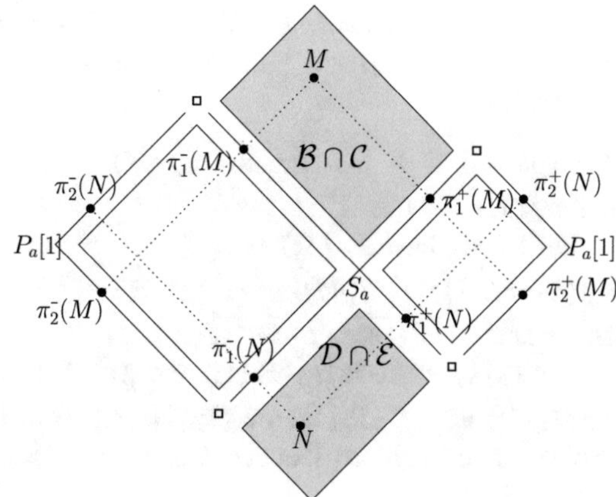

Fig. 9 Projections for $\mathscr{B}\cap\mathscr{C}$, $\mathscr{D}\cap\mathscr{E}$

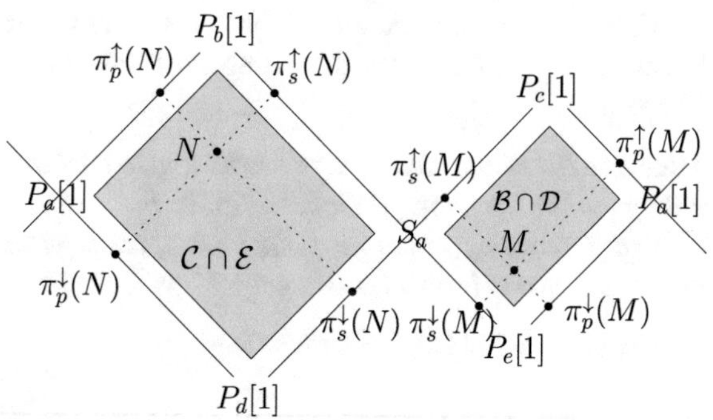

Fig. 10 Projections for $\mathscr{B}\cap\mathscr{D}$, $\mathscr{C}\cap\mathscr{E}$

in $\mathscr{C}_f$ starting at $P_a[1]$ and ending at S_a. Similarly, the *closure of* $\mathscr{B}\cap\mathscr{D}$ is the Hom-hammock

$$\overline{\mathscr{B}\cap\mathscr{D}} = \mathrm{ind}(\mathrm{Hom}_{\mathscr{C}_f}(S_a,-)\cap\mathrm{Hom}_{\mathscr{C}_f}(-,P_a[1]))$$

in $\mathscr{C}_f$ starting at S_a and ending at $P_a[1]$. For $(x,y)\in\{(c,e),(b,d)\}$, the *boundary of* $\overline{\mathscr{X}\cap\mathscr{Y}}$ (or of $\mathscr{X}\cap\mathscr{Y}$) is $\overline{\mathscr{X}\cap\mathscr{Y}}\setminus(\mathscr{X}\cap\mathscr{Y})$.

Note that $\overline{\mathscr{C}\cap\mathscr{E}}$ is the union of $\mathscr{C}\cap\mathscr{E}$ with the surrounding rays and the shifted projectives $\{P_b[1], P_d[1]\}$. Analogously, $\overline{\mathscr{B}\cap\mathscr{D}}$ contains $\{P_c[1], P_e[1]\}$.

Definition 4.4 (Projections, Continued) If M is a vertex of one of the two closures, we define four projections for M onto the four different "edges" of the boundary of its region: We denote the projections onto the paths starting or ending next to $P_a[1]$ by $\pi_p^{\uparrow}$, $\pi_p^{\downarrow}$ and the projections onto the paths starting or ending next to S_a by $\pi_s^{\uparrow}$ and $\pi_s^{\downarrow}$, respectively. We choose the upwards arrow to refer to the paths ending/starting near $P_b[1]$ or $P_c[1]$ and the downwards arrow to refer to paths ending/starting near $P_d[1]$ or $P_e[1]$. See Fig. 10.

Remark 4.5 The statement of Theorem 4.7 is independent of the choice of $\uparrow$ (paths near $P_b[1]$ or $P_c[1]$) and $\downarrow$ in Definition 4.4 as the formula is symmetric in these expressions.

Example 4.6 If $M \in \mathfrak{e}$, we have $\pi_p^{\uparrow}(M) = M$, $\pi_s^{\uparrow}(M) = P_b[1]$, $\pi_p^{\downarrow}(M) = P_a[1]$ and $\pi_s^{\downarrow}(M) = M_{\mathfrak{e}^a}$.

For S_a, we have $\pi_s^{\uparrow}(S_a) = \pi_s^{\downarrow}(S_a) = S_a$ whereas the two modules $\pi_p^{\uparrow}(S_a)$ and $\pi_p^{\downarrow}(S_a)$ are $\{P_b[1], P_d[1]\}$ or $\{P_c[1], P_e[1]\}$ depending on whether S_a is viewed as an element of $\overline{\mathscr{C} \cap \mathscr{E}}$ or of $\overline{\mathscr{B} \cap \mathscr{D}}$.

For $P_a[1]$, we have $\pi_p^{\uparrow}(P_a[1]) = \pi_p^{\downarrow}(P_a[1]) = P_a[1]$ whereas the two modules $\pi_s^{\uparrow}(P_a[1])$ and $\pi_s^{\downarrow}(P_a[1])$ are $\{P_b[1], P_d[1]\}$ or $\{P_c[1], P_e[1]\}$. These four shifted projectives evaluate to 1 under s, and so in Theorem 4.7, this ambiguity does not play a role.

With this notation, we are ready to state the theorem, proved in [3, Section 6]. Recall that $s(M)$ denotes the number of submodules of a module M, cf. also Theorem 2.1.

Theorem 4.7 *Consider a frieze associated to a triangulation of a polygon. Let a be a diagonal in the triangulation. Consider the mutation of the frieze at a. Then, the frieze difference $\delta_a(M)$ at the point corresponding to the indecomposable object M in the associated Frobenius category $\mathscr{C}_f$ is given by:*

If $M \in (\mathscr{B} \cap \mathscr{C}) \cup (\mathscr{D} \cap \mathscr{E})$, then

$$\delta_a(M) = (s(\pi_1^+(M)) - s(\pi_2^+(M)))\,(s(\pi_1^-(M)) - s(\pi_2^-(M));$$

If $M \in (\mathscr{B} \cap \mathscr{E}) \cup (\mathscr{C} \cap \mathscr{D})$, then

$$\delta_a(M) = -(s(\pi_2^+(M)) - 2s(\pi_1^+(M)))\,(s(\pi_2^-(M)) - 2s(\pi_1^-(M));$$

If $M \in \overline{\mathscr{C} \cap \mathscr{E}} \cup \overline{\mathscr{B} \cap \mathscr{D}}$, then

$$\delta_a(M) = s(\pi_s^{\downarrow}(M))s(\pi_p^{\downarrow}(M)) + s(\pi_s^{\uparrow}(M))s(\pi_p^{\uparrow}(M)) - 3\,s(\pi_p^{\downarrow}(M))s(\pi_p^{\uparrow}(M));$$

If $M \in \mathscr{F}$, then

$$\delta_a(M) = 0.$$

Note that given a frieze and an indecomposable M in one of the six regions $\mathscr{X} \cap \mathscr{Y}$, it is easy to locate the entries required to compute the frieze difference $\delta_a(M)$. We simply need to find projections onto the appropriate rays in the frieze. In this way, we do not need to know the precise shape of the modules appearing in the formulas of Theorem 4.7.

Example 4.8 Let $\mathscr{C}_f$ be the category given in Example 3.2. We consider three possibilities for M below.

If $M = \begin{smallmatrix} & 4 \\ 10 & 1 \\ 11 & 2 \end{smallmatrix}$, then we know by Fig. 7 that $s(M) = 11$ and $s(M') = 8$. On the other hand, we see from Fig. 6 that $M \in \mathscr{B} \cap \mathscr{C}$. Theorem 4.7 implies that

$$\begin{aligned}\delta_a(M) = s(M) - s(M') &= (s(M_{b_a}) - s(M_b))(s(M_{c^a}) - s(M_c)) \\ &= (s(\begin{smallmatrix} 4 \\ 1 \end{smallmatrix}) - s(4))(s(\begin{smallmatrix} 10 & 1 \\ 11 & 2 \end{smallmatrix}) - s(\begin{smallmatrix} 10 & \\ 11 & 2 \end{smallmatrix})) \\ &= (3-2)(8-5) = 3.\end{aligned}$$

Similarly, if $M = \begin{smallmatrix} 8 & & 2 \\ & 3 & \end{smallmatrix}$, then $M \in \mathscr{C} \cap \mathscr{D}$ with $s(M) = 5$ and $s(M') = 7$. The same theorem implies that

$$\begin{aligned}\delta_a(M) = s(M) - s(M') &= -(s(M_{c^a}) - 2s(M_c))\,(s(M_{d^a}) - 2s(M_d)) \\ &= -(s(\begin{smallmatrix} 1 \\ 2 \end{smallmatrix}) - 2s(2))(s(\begin{smallmatrix} 8 \\ 3 \\ 1 \end{smallmatrix}) - 2s(\begin{smallmatrix} 8 \\ 3 \end{smallmatrix})) \\ &= -(3-4)(4-6) = -2.\end{aligned}$$

Finally, if $M = \begin{smallmatrix} 10 & 1 & \\ & 2 & 5 \\ & & 6 \end{smallmatrix}$, then $M \in \mathscr{C} \cap \mathscr{E}$. We also know that $s(M) = s(M') = 11$. By the third formula in Theorem 4.7, we have

$$\begin{aligned}\delta_a(M) = s(M) - s(M') &= s(M_{e^a})s(M_c) + s(M_{c^a})s(M_e) - 3s(M_e)s(M_c) \\ &= s(\begin{smallmatrix} 1 \\ 5 \\ 6 \end{smallmatrix})s(\begin{smallmatrix} 10 \\ 2 \end{smallmatrix}) + s(\begin{smallmatrix} 10 & 1 \\ 2 & \end{smallmatrix})s(\begin{smallmatrix} 5 \\ 6 \end{smallmatrix}) - 3s(\begin{smallmatrix} 5 \\ 6 \end{smallmatrix})s(\begin{smallmatrix} 10 \\ 2 \end{smallmatrix}) \\ &= 4 \cdot 3 + 5 \cdot 3 - 3 \cdot 3 \cdot 3 = 0.\end{aligned}$$

Acknowledgments We thank AWM for encouraging us to write this summary and giving us opportunity to continue this work. We also thank the referees for useful comments on the paper.

EF, KS and GT received support from the AWM Advance grant to attend the symposium.

KB was supported by the FWF grant W1230.

KS was supported by NSF Postdoctoral Fellowship MSPRF-1502881.

References

1. C. Amiot, Cluster categories for algebras of global dimension 2 and quivers with potential. Ann. Inst. Fourier (Grenoble) **59**(6), 2525–2590 (2009). http://aif.cedram.org/item?id=AIF_2009__59_6_2525_0
2. K. Baur, A.D. King, R.J. Marsh, Dimer models and cluster categories of Grassmannians. Proc. Lond. Math. Soc. (3) **113**(2), 213–260 (2016). http://dx.doi.org/10.1112/plms/pdw029
3. K. Baur, E. Faber, S. Gratz, K. Serhiyenko, G. Todorov, Mutation of friezes. Bull. Sci. Math. **142**, 1–48 (2018). https://doi.org/10.1016/j.bulsci.2017.09.004

4. A.B. Buan, R. Marsh, M. Reineke, I. Reiten, G. Todorov, Tilting theory and cluster combinatorics. Adv. Math. **204**(2), 572–618 (2006). http://dx.doi.org/10.1016/j.aim.2005.06.003
5. A.B. Buan, R.J. Marsh, I. Reiten, Cluster-tilted algebras of finite representation type. J. Algebra **306**(2), 412–431 (2006). http://dx.doi.org/10.1016/j.jalgebra.2006.08.005
6. P. Caldero, F. Chapoton, Cluster algebras as Hall algebras of quiver representations. Comment. Math. Helv. **81**(3), 595–616 (2006)
7. P. Caldero, F. Chapoton, R. Schiffler, Quivers with relations arising from clusters (A_n case). Trans. Am. Math. Soc. **358**(3), 1347–1364 (2006). http://dx.doi.org/10.1090/S0002-9947-05-03753-0
8. J.H. Conway, H.S. Coxeter, Triangulated polygons and frieze patterns. Math. Gaz. **57**(401), 175–183 (1973)
9. J.H. Conway, H.S.M. Coxeter, Triangulated polygons and frieze patterns. Math. Gaz. **57**(400), 87–94 (1973)
10. H.S.M. Coxeter, Frieze patterns. Acta Arith. **18**, 297–310 (1971)
11. L. Demonet, X. Luo, Ice quivers with potential associated with triangulations and Cohen-Macaulay modules over orders. Trans. Am. Math. Soc. **368**(6), 4257–4293 (2016). http://dx.doi.org/10.1090/tran/6463
12. H. Derksen, J. Weyman, A. Zelevinsky, Quivers with potentials and their representations. I. Mutations. Sel. Math. (N.S.) **14**(1), 59–119 (2008)
13. S. Fomin, A. Zelevinsky, Cluster algebras. I. Foundations. J. Am. Math. Soc. **15**(2), 497–529 (electronic) (2002). http://dx.doi.org/10.1090/S0894-0347-01-00385-X
14. S. Fomin, A. Zelevinsky, Cluster algebras. II. Finite type classification. Invent. Math. **154**(1), 63–121 (2003). http://dx.doi.org/10.1007/s00222-003-0302-y
15. C. Fu, B. Keller, On cluster algebras with coefficients and 2-Calabi-Yau categories. Trans. Am. Math. Soc. **362**(2), 859–895 (2010). http://dx.doi.org/10.1090/S0002-9947-09-04979-4
16. B.T. Jensen, A.D. King, X. Su, A categorification of Grassmannian cluster algebras. Proc. Lond. Math. Soc. (3) **113**(2), 185–212 (2016). http://dx.doi.org/10.1112/plms/pdw028
17. S. Morier-Genoud, Coxeter's frieze patterns at the crossroads of algebra, geometry and combinatorics. Bull. Lond. Math. Soc. **47**(6), 895–938 (2015). http://dx.doi.org/10.1112/blms/bdv070

Orbit Decompositions of Unipotent Elements in the Generalized Symmetric Spaces of $\mathrm{SL}_2(\mathbb{F}_q)$

Catherine Buell, Vicky Klima, Jennifer Schaefer, Carmen Wright, and Ellen Ziliak

Abstract In this chapter, we determine the orbits of the fixed-point group on the unipotent elements in the generalized symmetric space for each involution of $\mathrm{SL}_2(\mathbb{F}_q)$ with $\mathrm{char}\left(\mathbb{F}_q\right) \neq 2$. We discuss how the generalized symmetric spaces can be decomposed into semisimple elements and unipotent elements, and why this decomposition allows the orbits of the fixed-point group on the entire generalized symmetric space to be more easily classified. We conclude by providing a description of and a count for the orbits of the fixed-point group on the unipotent elements in the generalized symmetric space for each involution of $\mathrm{SL}_2(\mathbb{F}_q)$.

1 Introduction

Élie Cartan [6] first introduced symmetric spaces as a special class of homogeneous Riemannian manifolds. Symmetric k-varieties [9] are defined as the space G_k/H_k, where G_k (resp., H_k) are the k-rational points of an algebraic group G (resp., fixed-point group). These k-varieties generalize both real reductive symmetric spaces and symmetric varieties to homogeneous spaces defined over general fields of characteristic not 2. Real reductive symmetric spaces are homogeneous spaces $Q_{\mathbb{R}} \cong G_{\mathbb{R}}/H_{\mathbb{R}}$ with $G_{\mathbb{R}}$ a reductive Lie group and $H_{\mathbb{R}}$ the fixed-point group

C. Buell
Fitchburg State University, Fitchburg, MA, USA

V. Klima
Appalachian State University, Boone, NC, USA

J. Schaefer
Dickinson College, Carlisle, PA, USA

C. Wright (✉)
Jackson State University, Jackson, MS, USA
e-mail: carmen.m.wright@jsums.edu

E. Ziliak
Benedictine University, Lisle, IL, USA

A. Deines et al. (eds.), *Advances in the Mathematical Sciences*, Association for Women in Mathematics Series 15, https://doi.org/10.1007/978-3-319-98684-5_5

associated with an involution of $G_{\mathbb{R}}$. Recently, the study of non-Riemannian symmetric spaces and generalizations of these real reductive spaces to other base fields has led to exciting applications in many areas including representation theory, geometry, and singularity theory [1, 7, 8, 13, 14, 18, 19].

Understanding the orbits of the fixed-point group H acting on the generalized symmetric space Q is essential to the study of symmetric k-varieties and their representations. For reductive algebraic groups defined over algebraically closed fields and the real numbers, these orbits of various subgroups of G_k over G_k/H_k have been well studied. For k algebraically closed, the orbits of the Borel subgroup were characterized by Springer [17] and the orbits of a general parabolic group were characterized by Brion and Helminck [3, 10]. For $k = \mathbb{R}$, the orbits of a minimal parabolic k-subgroup were characterized by Matsuki [15] and Rossmann [16]. For general fields, the orbits of a minimal parabolic k-subgroup on the symmetric variety G_k/H_k were characterized by Helminck and Wang [11].

In this chapter, we extend previous work by characterizing the orbits of the fixed-point group on the unipotent elements in the generalized symmetric space for each involution of $\mathrm{SL}_2(k)$ where k is a finite field of order q with $\mathrm{char}(k) \neq 2$. In Sect. 2, we provide the fixed-point group corresponding to each involution, the generalized symmetric space corresponding to each involution, and a few basic results regarding their structure. In Sect. 3, we discuss how the generalized symmetric space Q can be decomposed into semisimple elements and unipotent elements and why this decomposition allows us to more easily classify the orbits of H on Q. Finally in Sect. 4, we provide a description and a count for the orbits of H on the unipotent elements in Q.

2 Preliminaries

We begin with a few definitions. Generalized symmetric spaces are defined as the homogeneous spaces G/H with G an arbitrary group and $H = G^\theta = \{g \in G | \theta(g) = g\}$ the fixed-point group of an order n-automorphism θ. Of special interest are automorphisms of order 2, also called *involutions*. The following theorem classifies the involutions for $\mathrm{SL}_2(k)$ for k a finite field with $\mathrm{char}(k) \neq 2$.

Theorem 1 (Helminck and Wu [12, Theorem 1]) *Let $G = \mathrm{SL}_2(k)$ where k is a finite field with $\mathrm{char}(k) \neq 2$ and let $\theta \in \mathrm{Aut}\,(G)$ be an involution. Then, θ acts as conjugation by $X_m = \begin{pmatrix} 0 & 1 \\ m & 0 \end{pmatrix}$ where $m \in k^*$. Furthermore, if $m, n \in k^*$ are in the same square class, then conjugation by X_m is isomorphic to conjugation by X_n.*

Using this theorem and the definition of the fixed-point group determined by $(\mathrm{SL}_2(k), \theta)$, one can explicitly state the elements in H as follows:

$$H = \left\{ \begin{pmatrix} \alpha & \beta \\ m\beta & \alpha \end{pmatrix} \;\middle|\; \alpha^2 - m\beta^2 = 1, \alpha, \beta \in k \right\}.$$

As we plan to characterize the H-orbits on a subset of Q, it will be important to know the size of H.

Lemma 2 (Beun [2, Lemma 4.2.2]) *Let k be a finite field with* $\mathrm{char}(k) \neq 2$ *and θ_m be the involution of* $\mathrm{SL}_2(k)$ *that acts as conjugation by X_m. Then:*

$$|H| = \begin{cases} q-1 & \text{if } m \text{ is square} \\ q+1 & \text{if } m \text{ is not square}. \end{cases}$$

For involutions, there is a natural embedding of the homogeneous spaces G/H in the group G as follows. Let $\tau : G \to G$ be a morphism of G given by $\tau(g) = g\theta(g)^{-1}$ for $g \in G$ where θ is an involution of G. The image is $\tau(G) = Q = \{g\theta(g)^{-1} \mid g \in G\}$. The map τ induces an isomorphism of the coset space G/H onto $\tau(G) = Q$. Thus, $G/H \cong Q$. If G is an algebraic group defined over an algebraically closed field, then Q is a closed subvariety of G. We take Q as our definition of the generalized symmetric space determined by (G, θ).

Definition 3 The **generalized symmetric space** determined by (G, θ) is $G/H \cong Q = \{g\theta(g)^{-1} \mid g \in G\}$.

Note, if one considers H acting on G from the left, we define $\tau(g) = g^{-1}\theta(g)$. We can also define the extended symmetric space determined by (G, θ).

Definition 4 The **extended symmetric space** determined by (G, θ) is $R = \{g \in G \mid \theta(g) = g^{-1}\}$.

The group G acts on its extended symmetric space R via θ-twisted conjugation, defined as $g.r = gr\theta(g)^{-1}$ for $g \in G$ and $r \in R$. Let $G.x$ represent the orbit of G containing x and observe that under this action $G.e_G = \left\{g\theta(g)^{-1} | g \in G\right\} = Q$ where e_G is the identity in G. Typically, the action of θ-twisted conjugation on R is not transitive and thus $Q \subseteq R$ but $Q \neq R$. For example, the involution $\theta : \mathrm{SL}_2(\mathbb{R}) \to \mathrm{SL}_2(\mathbb{R})$ defined by $\theta(g) = \left(g^{-1}\right)^T$ finds Q as the set of matrices of the form $\{gg^T \mid g \in \mathrm{SL}_2(\mathbb{R})\}$ and $R = \{g \in \mathrm{SL}_2(\mathbb{R}) \mid g = g^T\}$. Here, R is the familiar set of symmetric matrices in $\mathrm{SL}_2(\mathbb{R})$, but Q is the set of symmetric positive definite matrices in $\mathrm{SL}_2(\mathbb{R})$. Clearly, $Q \subseteq R$ but $Q \neq R$. However, we have shown in [4] that equality of these spaces can be obtained for $\mathrm{SL}_2(k)$ when k is a finite field of characteristic not equal to two.

Theorem 5 (Theorem 2.2 [4]) *Let k be a finite field with* $\mathrm{char}(k) \neq 2$. *Then, $R = Q$ for any involution of* $\mathrm{SL}_2(k)$.

By this result and the definition of the extended symmetric space, every element in Q can be represented as follows:

$$\begin{aligned} Q = R &= \left\{ g \in \mathrm{SL}_2(k) | \theta(g) = g^{-1} \right\} \\ &= \left\{ \begin{pmatrix} a & b \\ c & d \end{pmatrix} \in \mathrm{SL}_2(k) \;\middle|\; \begin{pmatrix} 0 & 1 \\ m & 0 \end{pmatrix} \begin{pmatrix} a & b \\ c & d \end{pmatrix} \begin{pmatrix} 0 & 1 \\ m & 0 \end{pmatrix}^{-1} = \begin{pmatrix} a & b \\ c & d \end{pmatrix}^{-1} \right\} \\ &= \left\{ \begin{pmatrix} a & b \\ -mb & d \end{pmatrix} \;\middle|\; ad + mb^2 = 1 \right\}. \end{aligned}$$

As it will be useful when determining the decomposition of Q in Sect. 3, we also provide the size of Q.

Theorem 6 *For $G = \mathrm{SL}_2(k)$ with k a finite field with* $\mathrm{char}(k) \neq 2$ *and θ_m the involution that acts as conjugation by X_m,*

$$|Q| = |R| = \begin{cases} q^2 + q & \text{if } m \text{is square} \\ q^2 - q & \text{if } m \text{ is not square} \end{cases}.$$

Proof Using the structure of Q from Theorem 5, we split the counting argument into two cases.

Case 1: Suppose $ad = 0$. This implies $mb^2 = 1$ and, for only the case when m is a square, $b = \pm\sqrt{m^{-1}}$. When $a = 0$, we have q choices for d. When $a \neq 0$ and $d = 0$, we have $q - 1$ additional choices for a. Therefore, when m is a square, there are $2(q + (q - 1))$ elements of this form.

Case 2: Suppose $ad \neq 0$. Consider $a, d \neq 0$ and $b = 0$. Then, $ad = 1 \mod q$ gives $q - 1$ choices for a, each of which fixes d. Suppose $b \neq 0$. When m is a square, $b \neq \pm\sqrt{m^{-1}}$. Thus, we have $q - 3$ choices for b. For each b, there are $q - 1$ choices for a which determine d. Hence, there are $(q - 1)(q - 3)$ elements when m is a square. When m is not a square, we know $mb^2 \neq 1$ for any b, so there are $q - 1$ choices for b. For each b, we have $q - 1$ choices for a, each of which fix d. Hence, there are $(q - 1)(q - 1)$ elements when m is not a square.

Therefore,

$$|Q| = \begin{cases} 2(q + (q-1)) + (q-1) + (q-1)(q-3) = q^2 + q & \text{if } m \text{ is a square} \\ (q-1) + (q-1)(q-1) = q^2 - q & \text{if } m \text{ is not a square} \end{cases}.$$

□

3 Decomposition of Q

In this section, we prove that every element of Q can be factored into a product of a semisimple element and a unipotent element. Let Q^{ss} represent the set of semisimple elements in Q and Q^u represent the set of unipotent elements in Q. Consider $H.x$ for $x \in Q$. Then, assuming $x = x_{ss}x_u$ for some $x_{ss} \in Q^{ss}$ and $x_u \in Q^u$ and $h \in H$, $h.x = hxh^{-1} = hx_{ss}x_uh^{-1} = hx_{ss}h^{-1}hx_uh^{-1}$. Thus, if the representatives of the H-orbits are known in both Q^{ss} and Q^u, then the resulting orbit for any elements of Q can be characterized.

We begin with a classification and count for Q^{ss}. Let $x \in \mathrm{SL}_2(k)$. In our discussion, $c_x(\lambda)$ represents the characteristic polynomial of x, $m_x(\lambda)$ represents the minimal polynomial of x, $\mathrm{Discr}(p(\lambda))$ represents the discriminant of the quadratic polynomial $p(\lambda)$, and $\mathrm{tr}(x)$ represents the trace of the matrix x.

We know $x \in Q^{ss}$ if and only if $x \in Q$ and $m_x(\lambda)$ splits into distinct linear factors over $\overline{k}$. For $x \neq \pm I$ where I is the identity element in $\mathrm{SL}_2(k)$, $m_x(\lambda) = c_x(\lambda)$ is quadratic and splits into distinct linear factors if and only if $\mathrm{Discr}(c_x(\lambda)) = \mathrm{tr}(x)^2 - 4 \neq 0$. Thus, $Q^{ss} = \{\pm I\} \cup \{x \in Q | \, \mathrm{tr}(x) \neq \pm 2\}$.

With a clear description of Q^{ss}, we can now investigate the cardinality of this set. In lieu of considering the set directly, we look at its complement in Q, namely,

$$Q - Q^{ss} = \{x \in Q | \, \mathrm{tr}(x) = 2, x \neq I\} \cup \{x \in Q | \, \mathrm{tr}(x) = -2, x \neq -I\}.$$

Let $x \in Q - Q^{ss}$. Since $x \in Q = R$, there exist $a, b, d \in k$ such that

$$x = \begin{pmatrix} a & b \\ -mb & d \end{pmatrix} \text{ with } ad + mb^2 = 1. \tag{1}$$

Recall that m in this expression is determined by our involution and that we have two isomorphism classes of involutions—one when m is a square and one when m is not a square.

When $\mathrm{tr}(x) = 2$, $d = 2 - a$ and thus $mb^2 = (a-1)^2$. Similarly, if $\mathrm{tr}(x) = -2$, then $d = -2 - a$ and $mb^2 = (a+1)^2$. Thus, mb^2 is square for every element of $Q - Q^{ss}$. Hence, $Q - Q^{ss}$ is empty when m is not square.

Suppose m is a square. Combining our expressions for d with the description of $x \in Q$ given in Eq. (1), we have

$$\{x \in Q | \, \mathrm{tr}(x) = 2\} = \left\{ \begin{pmatrix} a & \frac{a-1}{\pm\sqrt{m}} \\ \mp\sqrt{m}(a-1) & 2-a \end{pmatrix} \middle| \, a \in k \right\}$$

and

$$\{x \in Q | \, \mathrm{tr}(x) = -2\} = \left\{ \begin{pmatrix} a & \frac{a+1}{\pm\sqrt{m}} \\ \mp\sqrt{m}(a+1) & -2-a \end{pmatrix} \middle| \, a \in k \right\}.$$

Both sets have cardinality $2q - 1$ and thus $|Q - Q^{ss}| = 2(2q - 1) - 2 = 4q - 4$ when m is a square. Theorem 7 summarizes our findings regarding Q^{ss}.

Theorem 7 *Let k be a finite field with* $\text{char}(k) \neq 2$ *and let Q^{ss} be the set of all elements in the generalized symmetric space determined by* $(\text{SL}_2(k), \theta_m)$ *that are semisimple. Then:*

$$Q^{ss} = \begin{cases} Q & \textit{if } m \textit{ is not a square} \\ \{x \in Q \mid tr(x) \neq \pm 2\} \cup \{\pm I\} & \textit{if } m \textit{ is a square} \end{cases}$$

and

$$|Q^{ss}| = \begin{cases} q^2 - q & \textit{if } m \textit{ is not a square} \\ q^2 - 3q + 4 & \textit{if } m \textit{ is a square.} \end{cases}$$

Proof This theorem follows directly from the cardinality of Q given in Theorem 6 and our previous discussion. □

Now that we understand the semisimple elements of Q, we turn our attention to the unipotent elements in Q, denoted Q^u. We have seen that $Q = Q^{ss}$ when m is not square. Only the identity matrix is both unipotent and semisimple and thus $Q^u = \{I\}$ for m a non-square. Suppose m is a square and let $x \in Q^u$. Since $\det(x) = 1$, we know $c_x(\lambda) = \lambda^2 - \text{tr}(x)\lambda + 1$. However, x is also unipotent and so $c_x(\lambda) = (\lambda - 1)^2 = \lambda^2 - 2\lambda + 1$. Thus, $Q^u = \{x \in Q \mid \text{tr}(x) = 2\}$, the set of cardinality $2q - 1$ described above. We summarize our results in Theorem 8.

Theorem 8 *Let k be a finite field with* $\text{char}(k) \neq 2$ *and let Q^u be the set of all elements in the generalized symmetric space determined by* $(\text{SL}_2(k), \theta_m)$ *that are unipotent. Then:*

$$Q^u = \begin{cases} \{I\} & \textit{if } m \textit{ is not a square} \\ \{x \in Q \mid tr(x) = 2\} & \textit{if } m \textit{ is square} \end{cases}$$

and

$$|Q^u| = \begin{cases} 1 & \textit{if } m \textit{ is not a square} \\ 2q - 1 & \textit{if } m \textit{ is square} \end{cases}.$$

Our descriptions of Q^{ss} and Q^u easily lead to the desired factorization of Q.

Theorem 9 *Let k be a finite field with $char(k) \neq 2$ and let Q be the generalized symmetric space determined by* $(\text{SL}_2(k), \theta_m)$. *Then, $Q \subseteq Q^{ss} Q^u$ where Q^{ss} represents the semisimple elements of Q and Q^u represents the unipotent elements of Q.*

Proof We need only focus on elements of Q with nontrivial factorizations, that is elements of Q that lie outside of $Q^{ss} \cup Q^u$. When m is not a square, $Q = Q^{ss}$ by Theorem 7 and no such elements exist. Suppose m is square. Combining our description of Q^{ss} from Theorem 7 with our description of Q^u given in Theorem 8, we see that $Q - (Q^{ss} \cup Q^u) = \{x \in Q \mid \mathrm{tr}(x) = -2, x \neq -I\}$. Let $x \in Q - (Q^{ss} \cup Q^u)$. Then, $\mathrm{tr}(-x) = 2$ and thus $-x \in Q^u$. Certainly, $-I \in Q^{ss}$ and $x = (-I)(-x)$. Therefore, we have found our factorization. □

4 H-orbits of Q^u

Given that each $x \in Q$ appears as an element of either Q^{ss}, Q^u, or $-Q^u$, understanding the H-orbits on Q^u and Q^{ss} independently will allow us to classify the H-orbits of Q. In [5], we consider the orbits of H on Q^{ss}. Combining these results with the work of this chapter will give the classification of the H-orbits of Q.

As noted in Sect. 3, $Q^u = \{I\}$ when m is a non-square. Thus, we will focus on the case when m is a square, since the H-orbits of Q^u will be nontrivial. Recall that the elements of Q^u can be described as follows:

$$Q^u = \left\{ \begin{pmatrix} a & \frac{a-1}{\pm\sqrt{m}} \\ \mp\sqrt{m}(a-1) & 2-a \end{pmatrix} \middle| \, a \in k \right\}.$$

Observe that this expression contains a $\pm$ sign. It will be useful for our analysis to decompose Q^u a bit further by considering these terms separately. Consider the two elements $\begin{pmatrix} a & \frac{a-1}{\pm\sqrt{m}} \\ \mp\sqrt{m}(a-1) & 2-a \end{pmatrix}$ in Q^u. We define a^+ as:

$$a^+ = \begin{pmatrix} a & \frac{a-1}{\sqrt{m}} \\ -\sqrt{m}(a-1) & 2-a \end{pmatrix}, \tag{2}$$

choosing the (1,2) entry to be positive and the (2,1) entry to be negative and define a^- as:

$$a^- = \begin{pmatrix} a & \frac{a-1}{-\sqrt{m}} \\ \sqrt{m}(a-1) & 2-a \end{pmatrix}. \tag{3}$$

Let Q^{u^+} be the set of all such elements a^+ and Q^{u^-} be the set of all such elements a^-. One can readily observe from these definitions that $Q^u = Q^{u^+} \cup Q^{u^-}$ and $Q^{u^+} \cap Q^{u^-} = \{I\}$. Lemma 10 then shows that the orbits of Q^u appear as subsets of Q^{u^+} and Q^{u-}.

Lemma 10 *Let Q be the generalized symmetric space for the involution θ_m, Q^{u^+} be the set of all elements a^+ as defined in Eq. (2), and Q^{u^-} be the set of all elements a^- as defined in Eq. (3). Then, $H.a^+ \subseteq Q^{u^+}$ and $H.a^- \subseteq Q^{u^-}$ for all $a^+ \in Q^{u^+}$ and $a^- \in Q^{u^-}$.*

Proof Consider an element $a^+ \in Q^{u^+}$ and an element $h \in H$ with $h = \begin{pmatrix} \alpha & \beta \\ m\beta & \alpha \end{pmatrix}$ and $\alpha^2 - m\beta^2 = 1$. Then:

$$\begin{aligned} ha^+h^{-1} &= \begin{pmatrix} a(\alpha-\beta\sqrt{m})^2 + 2\beta\sqrt{m}(\alpha-\beta\sqrt{m}) & \sqrt{m}^{-1}(a-1)(\alpha-\beta\sqrt{m})^2 \\ -\sqrt{m}(a-1)(\alpha-\beta\sqrt{m})^2 & -a(\alpha-\beta\sqrt{m})^2 + 2\alpha(\alpha-\beta\sqrt{m}) \end{pmatrix} \\ &= \begin{pmatrix} v & \frac{v-1}{\sqrt{m}} \\ -\sqrt{m}(v-1) & 2-v \end{pmatrix} \\ &\in Q^{u^+} \text{ for } v = a(\alpha-\beta\sqrt{m})^2 + 2\beta\sqrt{m}(\alpha-\beta\sqrt{m}). \end{aligned}$$

Similarly, $ha^-h^{-1} \in Q^{u^-}$ for all $a^- \in Q^{u^-}$ and $h \in H$. In this case,

$$\begin{aligned} ha^-h^{-1} &= \begin{pmatrix} a(\alpha+\beta\sqrt{m})^2 - 2\beta\sqrt{m}(\alpha+\beta\sqrt{m}) & -\sqrt{m}^{-1}(a-1)(\alpha+\beta\sqrt{m})^2 \\ \sqrt{m}(a-1)(\alpha+\beta\sqrt{m})^2 & -a(\alpha+\beta\sqrt{m})^2 + 2\alpha(\alpha+\beta\sqrt{m}) \end{pmatrix} \\ &= \begin{pmatrix} w & -\frac{w-1}{\sqrt{m}} \\ \sqrt{m}(w-1) & 2-w \end{pmatrix} \text{ for } w = a(\alpha+\beta\sqrt{m})^2 - 2\beta\sqrt{m}(\alpha+\beta\sqrt{m}). \end{aligned}$$

□

To characterize the orbits of H on Q^u, we will use the descriptions of Q^{u^+}, Q^{u^-}, and H to find representatives of the orbits.

Theorem 11 *Let Q be the generalized symmetric space for the involution θ_m. When m is not a square, H acts trivially on Q^u. When m is a square, there are 4 nontrivial orbits of H on Q^u of size $\frac{q-1}{2}$.*

Proof By Theorem 8, the case when m is not a square is clear. Suppose m is a square. Then, Q^u is nontrivial with $Q^u = Q^{u^+} \cup Q^{u^-}$ and $Q^{u^+} \cap Q^{u^-} = \{I\}$. By Lemma 10, it follows that $H.a^+ \subseteq Q^{u^+}$ for $a^+ \in Q^{u^+}$ and $H.a^- \subseteq Q^{u^-}$ for $a^- \in Q^{u^+}$. Thus, we will consider the H-orbits of the nonidentity elements in Q^{u^+} and Q^{u^-}, separately.

First, we consider the orbits of H on Q^{u^+}. Let $a^+ \in Q^{u^+} - \{I\}$. Then, $H.a^+ = \{ha^+h^{-1} \mid h \in H\}$. Based on our proof of Lemma 10, the off-diagonal entries of ha^+h^{-1} differ from the off-diagonal entries of a^+ by a factor of $(\alpha - \beta\sqrt{m})^2$ and so these entries remain in the same square class under conjugation.

Because m is a square, we know that $|H| = q - 1$ by Lemma 2. In addition, it can be verified that $ha^+h^{-1} = (-h)a^+(-h)^{-1}$ and $ha^+h^{-1} \neq ka^+k^{-1}$ for all $k \neq \pm h$. Thus, $|H.a^+| = (q-1)/2$. Therefore by $|Q^{u^+} - \{I\}| = q - 1$, we have 2 orbits in $Q^{u^+} - \{I\}$: one which contains all of the nonidentity elements of Q^{u^+} with a square in the (1,2) entry and one which contains all of the nonidentity elements of Q^{u^+} with a non-square in the (1,2) entry.

Given that the off-diagonal entries of ha^-h^{-1} differ from the off-diagonal entries of a^- by a factor of $(\alpha+\beta\sqrt{m})^2$ for all $a^- \in Q^{u^-} - \{I\}$, a similar argument follows for the orbits of H on Q^{u^-}.

□

References

1. A. Beilinson, J. Bernstein, Localisation de g-modules. C.R. Acad. Sci. Paris **292**(I), 1518 (1981)
2. S. Beun, On the classification of orbits of minimal parabolic k-subgroups acting on symmetric k-varieties. Ph.D. thesis, North Carolina State University, 2008
3. M. Brion, A. Helminck, On orbit closures of symmetric subgroups in flag varieties. Can. J. Math. **52**(2), 265–292 (2000)
4. C. Buell, A.G. Helminck, V. Kilma, J. Schaefer, C. Wright, E. Ziliak, On the structure of generalized symmetric spaces of SL(2, $\mathbb{F}_q$). Note Mat. **37**(2), 1–10 (2017)
5. C. Buell, A.G. Helminck, V. Kilma, J. Schaefer, C. Wright, E. Ziliak, Orbit decomposition of generalized symmetric spaces of $\mathrm{SL}_2(\mathbb{F}_q)$. Submitted for publication (2018)
6. E. Cartan, Groupes simples clos et ouvert et geometrie Riemannienne. J. Math. Pure Appl. **8**, 1–33 (1929)
7. C. De Concini, C. Procesi, *Complete Symmetric Varieties, Invariant Theory*. Lecture Notes in Mathematics, vol. 996 (Springer, Berlin, 1983), pp. 1–44
8. C. De Concini, C. Procesi, Complete symmetric varieties II, intersection theory, in *Algebraic Groups and Related Topics* (North-Holland, Amsterdam, 1984), pp. 481–513
9. A.G. Helminck, Symmetric k-varieties. Proc. Sympos. Pure Math **56**, 233–279 (1993)
10. A. Helminck, Combinatorics related to orbit closures of symmetric subgroups in flag varieties, in *Invariant Theory in All Characteristics* (2004), pp. 71–90
11. A.G. Helminck, S.P. Wang, On rationality properties of involutions of reductive groups. Adv. Math **99**(1), 26–96 (1993)
12. A.G. Helminck, L. Wu, Classification of involutions of SL(2, k). Commun. Algebra **30**(1), 193–203 (2002)
13. F. Hirzebruch, P.J. Slodowy, Elliptic genera, involutions, and homogeneous spin manifolds. Geom. Dedicata **35**(1–3), 309–343 (1990)
14. G. Lusztig, D.A. Vogan, Singularities of closures of K-orbits on flag manifolds. Invent. Math. **71**, 65–379 (1983)
15. T. Matsuki, The orbits of affine symmetric spaces under the action of minimal parabolic subgroups. J. Math. Soc. Japan **31**(2), 331–357 (1979)
16. W. Rossmann, The structure of semisimple symmetric spaces. Can. J. Math. **31**(1), 157–180 (1979)
17. T.A. Springer, Some results on algebraic groups with involutions, in *Algebraic Groups and Related Topics (Kyoto/Nagoya, 1983)* (1985), pp. 525–543
18. D.A. Vogan, Irreducible characters of semi-simple Lie groups IV, character-multiplicity duality. Duke Math. J. **49**, 943–1073 (1982)
19. D.A. Vogan, Irreducible characters of semi-simple Lie groups III. Proof of the Kazhdan-Lusztig conjectures in the integral case. Invent. Math. **71**, 381–417 (1983)

A Characterization of the $U(\Omega, m)$ Sets of a Hyperelliptic Curve as Ω and m Vary

Christelle Vincent

Abstract In this chapter, we consider a certain distinguished set $U(\Omega, m) \subseteq \{1, 2, \ldots, 2g+1, \infty\}$ that can be attached to a marked hyperelliptic curve of genus g equipped with a small period matrix Ω for its polarized Jacobian. We show that as Ω and the marking m vary, this set ranges over all possibilities prescribed by an argument of Poor.

1 Introduction and Statement of Results

Let X be a hyperelliptic curve of genus g defined over $\mathbb{C}$, and let $J(X)$ be its polarized Jacobian. In Definition 2.2, we associate to $J(X)$ a *small period matrix* Ω, which is an element of the Siegel upper half-space $\mathbb{H}_g$ with the property that there is an isomorphism

$$J(X)(\mathbb{C}) \cong \mathbb{C}^g / L_\Omega, \tag{1.1}$$

where L_Ω is the rank $2g$ lattice generated by the columns of Ω and the standard basis $\{e_i\}$ of $\mathbb{C}^g$.

After this choice, we may define an analytic theta function:

$$\vartheta(z, \Omega) \colon \mathbb{C}^g \to \mathbb{C}, \tag{1.2}$$

whose exact definition is given in Definition 2.13. While this function is not well defined on $J(X)(\mathbb{C})$, it is quasiperiodic with respect to the lattice L_Ω, and so its zero set on the Jacobian is well defined. In this chapter, we study how a certain combinatorial characterization of this zero set depends on the choice of small period matrix (since the theta function itself depends on the small period matrix) and on a further choice we now make.

C. Vincent (✉)
Department of Mathematics and Statistics, University of Vermont, Burlington, VT, USA
e-mail: christelle.vincent@uvm.edu

A. Deines et al. (eds.), *Advances in the Mathematical Sciences*, Association for Women in Mathematics Series 15, https://doi.org/10.1007/978-3-319-98684-5_6

Since X is hyperelliptic, there is a morphism $\pi : X \to \mathbb{P}^1$ of degree two, branched at $2g + 2$ points. Suppose further that X is given a *marking of its branch points*, denoted m, by which we mean that the branch points of π are numbered $1, 2, \ldots, 2g + 1, \infty$. As we explain in Proposition 2.8, this choice gives a bijection between sets

$$S \subseteq \{1, 2, \ldots, 2g + 1, \infty\}, \quad \#S \equiv 0 \pmod 2, \tag{1.3}$$

up to the equivalence $S \sim S^c$, where ^{c} denotes taking the complement within $\{1, 2, \ldots, 2g + 1, \infty\}$, and the two-torsion in $J(X)(\mathbb{C})$.

Then, we have the following theorem, which we will repeat and make more precise in Theorem 2.14:

Theorem 1.1 (Riemann Vanishing Theorem) *Let X be a hyperelliptic curve, m be a marking of its branch points, and let Ω be a small period matrix associated to its polarized Jacobian. Then, there is a distinguished set Θ on $J(X)(\mathbb{C})$ (defined in Definition 2.12) and the zero set of the theta function $\vartheta(z, \Omega)$, considered as a subset of $J(X)(\mathbb{C})$, is exactly the set Θ translated by an element of the two-torsion of $J(X)$.*

Under the correspondence given above, this two-torsion point corresponds to a set which we denote $T(\Omega, m)$. Note that the set $T(\Omega, m)$ is only well defined up to the equivalence $S \sim S^c$, where as before ^{c} denotes the complement.

This theorem gives rise to the following distinguished set:

Definition 1.2 Let X be a hyperelliptic curve of genus g, Ω a choice of small period matrix associated to its Jacobian via the process described in Definition 2.2, and m a marking of the branch points of X. Let $U(\Omega, m) \subset \{1, 2, \ldots, 2g+1, \infty\}$ be defined up to the equivalence $S \sim S^c$ by the following formula:

$$U(\Omega, m) = \begin{cases} T(\Omega, m) & \text{if } g \text{ is odd, and} \\ T(\Omega, m) \circ \{\infty\} & \text{if } g \text{ is even,} \end{cases} \tag{1.4}$$

where $\circ$ here denotes the symmetric difference of sets (see Definition 2.6). To fix one set in this equivalence class, we take $U(\Omega, m)$ to be the set containing ∞.

Remark We note that Mumford [6] adopts the opposite convention and chooses $U(\Omega, m)$ to be the member of the equivalence class that does not contain ∞. In this respect, we follow the convention adopted by Poor [7].

The significance of this set $U(\Omega, m)$ is especially salient in computational applications; we invite the reader to consult Sect. 2.3 for a further account of its role. This set first appeared in work of Mumford [6], where given a marked hyperelliptic curve X, the author constructs a certain small period matrix Ω and computes the set $U(\Omega, m)$ explicitly. In this example, it is the case that

$$\#U(\Omega, m) = g + 1, \tag{1.5}$$

where as before g is the genus of the curve. In the theorems following this computation (in particular Mumford's version of Theorem 2.15, Theorem 9.1 of [6], which is the most important of those from our point of view), the set $U(\Omega, m)$ is always assumed to have this cardinality.

However, in later work of Poor [7], the same set $U(\Omega, m)$ is shown to have the property that

$$\#U(\Omega, m) \equiv g + 1 \pmod 4 \tag{1.6}$$

(see [7, Proposition 1.4.9]). This raises the following interesting question: Does the set $U(\Omega, m)$ always have cardinality $g + 1$, or do other cardinalities occur? We answer this question completely:

Theorem 1.3 *Let $g \geq 1$ and X be a hyperelliptic curve of genus g defined over $\mathbb{C}$. Then, for any set $U \subseteq \{1, 2, \ldots, 2g + 1, \infty\}$ containing ∞ such that*

$$\#U \equiv g + 1 \pmod 4, \tag{1.7}$$

there exists a small period matrix Ω associated to the Jacobian of X via the process described in Definition 2.2, and a marking m of the branch points of X such that

$$U = U(\Omega, m). \tag{1.8}$$

In other words, every possible set U occurs as the set $U(\Omega, m)$ for a given hyperelliptic curve X, and Poor's characterization of $U(\Omega, m)$ is sharp.

2 Preliminaries

Let X be a hyperelliptic curve, by which we mean a smooth complete curve of genus g defined over $\mathbb{C}$ admitting a map $\pi : X \to \mathbb{P}^1$ of degree 2. Throughout, we denote its Jacobian variety by $J(X)$.

2.1 *The Small Period Matrix of the Jacobian of a Curve*

We give here standard facts about abelian varieties and Jacobians. We refer the reader to [2] for further background and proofs.

We begin by giving an analytic space associated to polarized abelian varieties of dimension g:

Definition 2.1 Let $g \geq 1$. The *Siegel upper half-space* $\mathbb{H}_g$ is the set of symmetric $g \times g$ complex matrices M such that the imaginary part of M (obtained by taking the imaginary part of each entry in M) is positive definite.

Although much of the discussion below would apply to general polarized abelian varieties, in this chapter we focus our attention to Jacobians of curves equipped with their principal polarization. To simplify matters, at this time we restrict our attention to these objects. In this setting, the connection between this space and Jacobians is through the following object:

Definition 2.2 Let X be a curve of genus g defined over $\mathbb{C}$, and let $J(X)$ be its principally polarized Jacobian. To $J(X)$, we can associate matrices $\Omega \in \mathbb{H}_g$ in the following manner: Let A_i, B_i, $i = 1, \ldots, g$, be a basis for the homology group $H_1(J(X), \mathbb{Z}) \cong H_1(X, \mathbb{Z})$, which is a $2g$-dimensional vector space over $\mathbb{C}$. Assume further that this basis is symplectic with respect to the cup product. There exists a unique basis $\omega_1, \omega_2, \ldots, \omega_g$ of $\Omega^1(J(X)) \cong \Omega^1(X)$, the space of holomorphic 1-forms on $J(X)$ or X, such that

$$\int_{B_i} \omega_j = \delta_{ij}, \tag{2.1}$$

where δ_{ij} is the Kronecker delta function. Then, the matrix given by $\int_{A_i} \omega_j$ belongs to $\mathbb{H}_g$ and is called a *small period matrix* for $J(X)$.

Let $\mathrm{Sp}_{2g}(\mathbb{Z})$ be the group of $2g \times 2g$ matrices with coefficients in $\mathbb{Z}$ and symplectic with respect to the bilinear form given by the matrix:

$$\begin{pmatrix} 0 & \mathbb{1}_g \\ -\mathbb{1}_g & 0 \end{pmatrix}, \tag{2.2}$$

where $\mathbb{1}_g$ is the $g \times g$ identity matrix. We note that two elements of $\mathbb{H}_g$ can be associated to isomorphic polarized abelian varieties if and only if they differ by a matrix in $\mathrm{Sp}_{2g}(\mathbb{Z})$, where the action of $\mathrm{Sp}_{2g}(\mathbb{Z})$ on $\mathbb{H}_g$ is given in the following manner: Let

$$\gamma = \begin{pmatrix} A & B \\ C & D \end{pmatrix} \in \mathrm{Sp}_{2g}(\mathbb{Z}), \tag{2.3}$$

where A, B, C, and D are four $g \times g$ matrices. Then,

$$\gamma \cdot \Omega = (A\Omega + B)(C\Omega + D)^{-1}, \tag{2.4}$$

where on the right multiplication and addition are the usual operations on $g \times g$ matrices.

We can further define an Abel–Jacobi map for a principally polarized Jacobian variety $J(X)$:

Definition 2.3 Let X be a curve of genus g defined over $\mathbb{C}$, let $J(X)$ be its principally polarized Jacobian, and fix A_i, B_i, $i = 1, \ldots, g$, a symplectic basis for the homology group $H_1(X, \mathbb{Z})$. Let $\omega_1, \omega_2, \ldots, \omega_g$ be the basis of $\Omega^1(X)$ described in Definition 2.2, Ω be the small period matrix attached to $J(X)$ via this choice of symplectic basis for homology, and let L_Ω be the rank $2g$ lattice generated by the columns of Ω and the standard basis $\{e_i\}$ of $\mathbb{C}^g$. Then, there is an isomorphism called the *Abel–Jacobi map*:

$$AJ\colon J(X) \to \mathbb{C}^g/L_\Omega, \tag{2.5}$$

given by the map:

$$D = \sum_{k=1}^{s} P_k - \sum_{k=1}^{s} Q_k \mapsto \left(\sum_{k=1}^{s} \int_{Q_k}^{P_k} \omega_i\right)_i, \tag{2.6}$$

where the P_ks and Q_ks are points on X. This map is well defined since the value of each integral on X is well defined up to the value of integrating the differentials ω_i along the basis elements A_i, B_i, and thus up to elements of L_Ω.

We will in fact need a slightly modified version of this Abel–Jacobi map for our purposes:

Definition 2.4 Let X be a curve of genus g defined over $\mathbb{C}$, let $J(X)$ be its principally polarized Jacobian, and fix A_i, B_i, $i = 1, \ldots, g$, a symplectic basis for the homology group $H_1(X, \mathbb{Z})$. Let Ω be the small period matrix attached to $J(X)$ via this choice of symplectic basis for homology and let L_Ω be the rank $2g$ lattice generated by the columns of Ω and the standard basis $\{e_i\}$ of $\mathbb{C}^g$. This gives rise to an isomorphism:

$$\mathbb{C}^g/L_\Omega \to \mathbb{R}^{2g}/\mathbb{Z}^{2g}, \tag{2.7}$$

given by writing an element of $\mathbb{C}^g/L_\Omega$ as a linear combination of the columns of Ω and the standard basis $\{e_i\}$ of $\mathbb{C}^g$ and sending the element to the coefficients of the linear combination. Composing this isomorphism with the Abel–Jacobi map defined in Definition 2.3, we obtain the *modified Abel–Jacobi map*:

$$AJ_c\colon J(X) \to \mathbb{R}^{2g}/\mathbb{Z}^{2g}, \tag{2.8}$$

which gives the coordinates of a point of $J(X)$ under the Abel–Jacobi map.

In this paper, we will need to know how a change of symplectic basis for $H_1(X, \mathbb{Z})$ affects the image of the Abel–Jacobi map and the coordinates of a point of $J(X)$ under the Abel–Jacobi map. We have

Proposition 2.5 (Adapted from Section 1.4 of [7]) *Let X be a curve of genus g defined over $\mathbb{C}$, let $J(X)$ be its principally polarized Jacobian, and let A_i, B_i be a symplectic basis for $H_1(X, \mathbb{Z})$ from which arises the small period matrix Ω, the Abel–Jacobi map AJ, and the modified Abel–Jacobi map AJ_c. Let $\gamma \in \mathrm{Sp}_{2g}(\mathbb{Z})$ act on the elements A_i, B_i. Since $\mathrm{Sp}_{2g}(\mathbb{Z})$ preserves the cup pairing, the images $\tilde{A}_i$, $\tilde{B}_i$*

give rise to a second Abel–Jacobi map $\widetilde{AJ}$. *If*

$$\gamma = \begin{pmatrix} A & B \\ C & D \end{pmatrix}, \tag{2.9}$$

where A, B, C, and D are $g \times g$ *matrices, then*

$$\widetilde{AJ} = (C\Omega + D)^{-T} AJ, \tag{2.10}$$

where M^{-T} *is the inverse of the transpose of the matrix M. Furthermore, we have*

$$\widetilde{AJ}_c = \gamma^{-T} AJ_c. \tag{2.11}$$

2.2 The Two-Torsion on the Jacobian of a Hyperelliptic Curve

We now turn our attention to the two-torsion of the Jacobian of a hyperelliptic curve of genus g defined over $\mathbb{C}$. As a group, it is isomorphic to C_2^{2g}, where C_2 is the cyclic group with two elements.

Throughout, let $B = \{1, 2, \ldots, 2g+1, \infty\}$. When $S \subseteq B$, we let S^c be the complement of S in B.

Definition 2.6 Let S_1 and S_2 be any two subsets of B. We define

$$S_1 \circ S_2 = (S_1 \cup S_2) - (S_1 \cap S_2), \tag{2.12}$$

the *symmetric difference* of S_1 and S_2.

This binary operation on subsets in turn gives rise to the following group:

Proposition 2.7 *The set*

$$\{S \subseteq B : \#S \equiv 0 \pmod 2\}/\{S \sim S^c\} \tag{2.13}$$

is a commutative group under the operation $\circ$, *of order* 2^{2g}, *with identity* $\emptyset \sim B$. *Since* $S \circ S = \emptyset$ *for all* $S \subseteq B$, *this is a group of exponent 2. Therefore, this group, which we denote* G_B, *is isomorphic to* C_2^{2g}.

If the hyperelliptic curve X is equipped with a marking of its branch points (recall that this means that we label the $2g+2$ branch points of the degree two map $\pi : X \to \mathbb{P}^1$, $P_1, P_2, \ldots, P_{2g+1}, P_\infty$), there is in fact an explicit isomorphism between G_B and $J(X)[2]$, the two-torsion on the Jacobian of X:

Proposition 2.8 (Corollary 2.11 of [6]) *To each set* $S \subseteq B$ *such that* $\#S \equiv 0 \pmod 2$, *associate the divisor class of the divisor*

$$e_S = \sum_{i \in S} P_i - (\#S)P_\infty. \tag{2.14}$$

This association is a group isomorphism between $J(X)[2]$ and G_B.

We may now compose the isomorphism of Proposition 2.8 with the modified Abel–Jacobi map given in Definition 2.4.

Definition 2.9 We denote by $\eta_{\Omega,m}$ the isomorphism:

$$\eta_{\Omega,m} \colon \{S \subseteq B : \#S \equiv 0 \pmod 2\}/\{S \sim S^c\} \to (\frac{1}{2}\mathbb{Z})^{2g}/\mathbb{Z}^{2g} \tag{2.15}$$

given by composing the isomorphism $G_B \to J(X)[2]$ given in Proposition 2.8 and the map AJ_c given in Definition 2.4.

Remark We note that in Poor's work [7], this is the *class* of the map η, which is an equivalence class of maps to $(\frac{1}{2}\mathbb{Z})^{2g}$. In this work, we will not need the distinction between the "true" η-map and its class, and therefore by a slight abuse of notation we consider the map above to be the η-map.

This map $\eta_{\Omega,m}$ will allow us to give a more concrete definition of the set $U(\Omega, m)$, which we will use in our proof in Sect. 3. We first need one more notion.

Definition 2.10 If $x \in \mathbb{C}^{2g}$, let $x = (x_1, x_2)$, with $x_i \in \mathbb{C}^g$; in other words, let x_1 denote the vector of the first g entries of x, and x_2 denote the vector of the last g entries of x. Furthermore, for $x_i \in \mathbb{C}^g$, let x_i^T denote the transpose of x_i. Then for $\xi \in (\frac{1}{2}\mathbb{Z})^{2g}$, we define

$$e_*(\xi) = \exp(4\pi i \xi_1^T \xi_2) \tag{2.16}$$

to be the *parity* of ξ. Note that e_* is also well defined on $(\frac{1}{2}\mathbb{Z})^{2g}/\mathbb{Z}^{2g}$.

Proposition 2.11 (Lemma 1.4.13 of [7]) *Let X be a hyperelliptic curve of genus g equipped with a marking m of its branch points, and let $J(X)$ be equipped with a choice of small period matrix Ω via the process described in Definition 2.2. Then, the set $U(\Omega, m)$ of Definition 1.2 is given by:*

$$\{i \in \{1, 2, \ldots, 2g+1\} : e_*(\eta_{\Omega,m}(\{i, \infty\})) = -1\} \cup \{\infty\}. \tag{2.17}$$

In other words, if we consider the distinguished elements $D_i = P_i - P_\infty \in J(X)[2]$ for $i = 1, 2, \ldots, 2g+1, \infty$, the set $U(\Omega, m)$ can be made to contain ∞ as well as i such that the coordinates of D_i under the Abel–Jacobi map are odd, for $i = 1, 2, \ldots, 2g+1$.

2.3 Mumford and Poor's Vanishing Theorem

We now turn our attention to explaining the significance of the set $U(\Omega, m)$. As we explained briefly in the introduction, the set connects the vanishing set of an analytic theta function to a distinguished divisor Θ on the Jacobian $J(X)$ of a marked hyperelliptic curve X.

We begin by defining this divisor:

Definition 2.12 Let X be a curve of genus g defined over $\mathbb{C}$ and P_∞ be a basepoint on X. Then, we define the *theta divisor* Θ on $J(X)$ to be the subset of divisor classes of the form:

$$\sum_{i=1}^{g-1} Q_i - (g-1)P_\infty. \tag{2.18}$$

Note that if X is a marked hyperelliptic curve and we choose P_∞ to be the branch point of X labeled ∞, this gives a unique choice of theta divisor on $J(X)$. We therefore call it "the" theta divisor on the marked curve X.

We now define the theta function whose zeroes we will study:

Definition 2.13 For $z \in \mathbb{C}^g$ and $\Omega \in \mathbb{H}_g$, we define the *theta function*

$$\vartheta(z, \Omega) = \sum_{n \in \mathbb{Z}^g} \exp(\pi i n^T \Omega n + 2\pi i n^T z). \tag{2.19}$$

Remark As noted in the introduction, this function is quasiperiodic for the lattice L_Ω in the coordinate z. Indeed, if $k \in \mathbb{Z}^g$, by Mumford [5, p. 120], we have

$$\vartheta(z + k, \Omega) = \vartheta(z, \Omega) \tag{2.20}$$

and

$$\vartheta(z + \Omega k, \Omega) = \exp(-i\pi k^T \Omega k - 2\pi i k^T z)\vartheta(z, \Omega). \tag{2.21}$$

However, since the automorphy factor is nonzero, the zero set of ϑ is well defined as a subset of $\mathbb{C}^g / L_\Omega$.

For the convenience of the reader, we repeat the Riemann Vanishing Theorem now that all terms have been defined:

Theorem 2.14 (Riemann Vanishing Theorem, or Theorem 5.3 of [6]) *Let X be a hyperelliptic curve, m be a marking of its branch points, and let Ω be a small period matrix associated to its Jacobian via the process described in Definition 2.2. If $\Theta \in J(X)$ is as in Definition 2.12, then the zero set of the theta function $\vartheta(z, \Omega)$, considered as a subset of $J(X)(\mathbb{C})$ is a translate of Θ by a two-torsion point of $J(X)$.*

From the introduction, we recall that this gives rise to the set $U(\Omega, m)$ in the following manner: Given a marking m and a small period matrix Ω, the Riemann Vanishing Theorem singles out a divisor on the Jacobian of X (the zero locus of the function ϑ). As this is a translate of Θ by a two-torsion point, this gives in turn a distinguished two-torsion point on $J(X)$. Recall that Proposition 2.8 gives an isomorphism between the group G_B defined in Proposition 2.7 and the two-torsion of $J(X)$. Therefore, via this isomorphism, we obtain an element of the group G_B. Finally, since the elements of G_B are equivalence classes of certain subsets of B (where the equivalence consists in taking the complement in $B = \{1, 2, \ldots, 2g + 1, \infty\}$), we obtain a certain (equivalence class of) subset of B, which we denote by $T(\Omega, m)$ here.

We then define the set $U(\Omega, m)$ to be the element of the equivalence class of

$$\begin{cases} T(\Omega, m) & \text{if } g \text{ is odd, and} \\ T(\Omega, m) \circ \{\infty\} & \text{if } g \text{ is even} \end{cases} \tag{2.22}$$

that contains ∞, as noted in Definition 1.2.

This definition is motivated by the proof of Proposition 6.2 of [6]: Under the correspondence given in part a) of this Proposition, the set $T(\Omega, m)$ when g is odd, or $T(\Omega, m) \circ \{\infty\}$ when g is even, corresponds to the translate $\Theta + e_{T(\Omega,m)}$ and to the characteristic $\delta + \eta_{T(\Omega,m)}$ (in our notation $\eta_{T(\Omega,m)}$ is $\eta_{\Omega,m}(T(\Omega, m))$). Since $\eta_{T(\Omega,m)} = \delta$ and $\delta \in \frac{1}{2}L_\Omega$, $T(\Omega, m)$ when g is odd, or $T(\Omega, m) \circ \{\infty\}$ when g is even, corresponds to 0 and is therefore the set $U(\Omega, m)$ defined here.

We end by giving part of the Vanishing Criterion for hyperelliptic small period matrices, which highlights how truly central the set $U(\Omega, m)$ is to the computational theory of hyperelliptic curves.

Theorem 2.15 (Main Theorem 2.6.1 of [7]) *Let X be a hyperelliptic curve of genus g, with a marking of its branch points m and let Ω be a small period matrix associated to its Jacobian $J(X)$ via the process described in Definition 2.2. Then for $S \subseteq B$ with $\#S \equiv 0 \pmod 2$, we have*

$$\vartheta(AJ(e_S), \Omega) = 0 \tag{2.23}$$

if and only if

$$\#(S \circ U(\Omega, m)) \neq g + 1. \tag{2.24}$$

We stress that here we have only stated part of the Vanishing Criterion for hyperelliptic matrices, and that the important part of this Vanishing Criterion for computational purposes is a strengthening of the statement which allows one to give a converse for general curves. This converse then allows the detection of hyperelliptic small period matrices among all small period matrices. We refer the reader to Poor's work [7], notably Definition 1.4.11 for a complete account of this converse with proofs, or to [1] for a shorter exposition.

3 The Proof

The proof of Theorem 1.3 has two main parts. In the first part, for a fixed $g \geq 1$ we count the number of different sets U satisfying $U \subseteq \{1, 2, \ldots, 2g+1, \infty\}$, $\infty \in U$ and $\#U \equiv g+1 \pmod 4$ (this is Proposition 3.4). In the second part, we count how many different sets $U(\Omega, m)$ arise as we vary among all possible small period matrices Ω that can be associated to the Jacobian of a hyperelliptic curve X via the process described in Definition 2.2 and all possible markings m of its branch points (this is Proposition 3.11). Since these two numbers are equal, we conclude that every allowable set U must arise $U(\Omega, m)$ for some choice of Ω and m.

3.1 *Counting the Allowable Sets U*

Counting the sets such that $U \subseteq \{1, 2, \ldots, 2g+1, \infty\}$, $\#U \equiv g+1 \pmod 4$, and $\infty \in U$ is equivalent to counting the sets satisfying the following two conditions:

- $\tilde{U} \subseteq \{1, 2, \ldots, 2g+1\}$, and
- $\#\tilde{U} \equiv g \pmod 4$.

We turn to this task.

Definition 3.1 Let $n \geq 1$, $d \geq 0$ and $m \geq 2$ be integers. We define the sum:

$$S(n, d, m) = \sum_{\substack{0 \leq k \leq n \\ k \equiv d \pmod m}} \binom{n}{k}. \tag{3.1}$$

This is the number of subsets of $\{1, \ldots, n\}$ of any cardinality $k \equiv d \pmod m$.

We are interested in computing the quantity $S(2g+1, g, 4)$. We first note the following well-known result:

Proposition 3.2 *Let n be any positive integer, then*

$$S(n, 0, 2) = S(n, 1, 2) = 2^{n-1}. \tag{3.2}$$

In other words, for any n, of the 2^n subsets of $\{1, \ldots, n\}$, half of them have even cardinality, and half have odd cardinality.

Lemma 3.3 *We have*

$$S(n, d, 4) = S(n-1, d, 4) + S(n-1, d-1, 4). \tag{3.3}$$

Proof This follows from Pascal's identity, which says that for $n \geq 1$ and $k \geq 0$, we have

$$\binom{n}{k} = \binom{n-1}{k} + \binom{n-1}{k-1}. \tag{3.4}$$

Here, we use the usual convention that $\binom{n}{k} = 0$ if $k < 0$.

□

This is enough to show

Proposition 3.4 *Let $g \geq 1$, then*

$$S(2g+1, g, 4) = 2^{g-1}(2^g + 1). \tag{3.5}$$

Proof The proof is done by induction on g. The case of $g = 1$ is the claim that $S(3, 1, 4) = 3$. Indeed, of the subsets of $\{1, 2, 3\}$, three of them have cardinality congruent to 1 modulo 4 (and therefore actually equal to 1, since there are no subsets of $\{1, 2, 3\}$ of cardinality greater than or equal to 5).

We now assume that $S(2g-1, g-1, 4) = 2^{g-2}(2^{g-1}+1)$ and $g \geq 2$. We have

$$\begin{aligned} S(2g+1, g, 4) &= S(2g, g, 4) + S(2g, g-1, 4) && (3.6)\\ &= (S(2g-1, g, 4) + S(2g-1, g-1, 4)) && (3.7)\\ &\quad + (S(2g-1, g-1, 4) + S(2g-1, g-2, 4)) \\ &= S(2g-1, g, 4) + S(2g-1, g-2, 4) && (3.8)\\ &\quad + 2S(2g-1, g-1, 4). \end{aligned}$$

We now note that if g is even, then

$$S(2g-1, g, 4) + S(2g-1, g-2, 4) = S(2g-1, 0, 2), \tag{3.9}$$

and if g is odd, then

$$S(2g-1, g, 4) + S(2g-1, g-2, 4) = S(2g-1, 1, 2). \tag{3.10}$$

In either case, by Proposition 3.2,

$$S(2g-1, g, 4) + S(2g-1, g-2, 4) = 2^{2g-2}. \tag{3.11}$$

Furthermore, by induction $S(2g-1, g-1, 4) = 2^{g-2}(2^{g-1}+1)$.

Therefore, we have

$$\begin{aligned} S(2g+1, g, 4) &= 2^{2g-2} + 2 \cdot 2^{g-2}(2^{g-1}+1) && (3.12)\\ &= 2^{g-1}(2^{g-1} + 2^{g-1} + 1) && (3.13)\\ &= 2^{g-1}(2^g + 1). && (3.14) \end{aligned}$$

This completes the proof. □

3.2 Counting the Different Sets $U(\Omega, m)$ for a Hyperelliptic Curve

Here, we show that in fact, given a hyperelliptic curve X with a marking m of its branch points, every allowable U-set is realized as $U(\Omega, m)$ as we vary the small period matrix Ω associated to its Jacobian $J(X)$ by Definition 2.2. This certainly implies our main theorem. Thus, we begin by fixing a marking m on the branch points of X.

The proof is carried out by considering the action of $\mathrm{Sp}_{2g}(\mathbb{Z})$ on Ω and considering which matrices fix the set $U(\Omega, m)$. We will see in Proposition 3.9 that they are exactly a subgroup of $\mathrm{Sp}_{2g}(\mathbb{Z})$ denoted $\Gamma_{1,2}$:

Definition 3.5 Let $\Gamma_{1,2}$ be the subgroup of $\mathrm{Sp}_{2g}(\mathbb{Z})$ containing the matrices that fix the parity of every element of $(\frac{1}{2}\mathbb{Z})^{2g}$. In other words, $\gamma \in \Gamma_{1,2}$ if and only if

$$e_*(\gamma\xi) = e_*(\xi) \tag{3.15}$$

for all $\xi \in (\frac{1}{2}\mathbb{Z})^{2g}$, where e_* is as in Definition 2.10 and $\gamma\xi$ is the usual matrix-vector multiplication.

We will need two further characterizations of these matrices below. First, we have:

Proposition 3.6 *Let* $\gamma \in \mathrm{Sp}_{2g}(\mathbb{Z})$ *with*

$$\gamma = \begin{pmatrix} A & B \\ C & D \end{pmatrix} \tag{3.16}$$

where A, B, C, *and* D *are four* $g \times g$ *matrices. Then,* $\gamma \in \Gamma_{1,2}$ *if and only if the diagonals of the matrices* $A^T C$ *and* $B^T D$ *have all even entries.*

Proof This can be verified directly, or found in [5, page 189]. □

The second characterization of these matrices relies on an important property of the vectors $\eta_{\Omega,m}(\{i, \infty\})$ for $i = 1, 2, \ldots, 2g + 1$:

Proposition 3.7 *Let* X *be a marked hyperelliptic curve,* $J(X)$ *its Jacobian, and* Ω *a small period matrix associated to* $J(X)$ *via the process outlined in Definition 2.2. Furthermore, given this data, let* $\eta_{\Omega,m}$ *be the map given in Definition 2.9. Then, the set*

$$\{\eta_{\Omega,m}(\{i, \infty\}) : i = 1, \ldots, 2g + 1\} \tag{3.17}$$

contains a basis of the $\mathbb{F}_2$*-vector space* $(\frac{1}{2}\mathbb{Z})^{2g}/\mathbb{Z}^{2g}$.

Proof By the proof Lemma 1.4.13 of [7], the set

$$\{\eta_{\Omega,m}(\{i, \infty\}) : i = 1, \ldots, 2g + 1\} \tag{3.18}$$

is an azygetic basis of $(\frac{1}{2}\mathbb{Z})^{2g}/\mathbb{Z}^{2g}$, and by Definition 1.4.12 of *ibid*, therefore spans the vector space $(\frac{1}{2}\mathbb{Z})^{2g}/\mathbb{Z}^{2g}$. Therefore, it contains a basis of the space. □

We can now prove the following:

Lemma 3.8 *A matrix $\gamma \in \mathrm{Sp}_{2g}(\mathbb{Z})$ belongs to $\Gamma_{1,2}$ if and only if it fixes the parity of $\eta_{\Omega,m}(\{i, \infty\})$ for $i = 1, 2, \ldots, 2g + 1$.*

Proof It is clear that if $\gamma \in \Gamma_{1,2}$, then it will fix the parity of $\eta_{\Omega,m}(\{i, \infty\})$ for $i = 1, 2, \ldots, 2g + 1$. Therefore, we assume that $\gamma \in \mathrm{Sp}_{2g}(\mathbb{Z})$ fixes the parity of $\eta_{\Omega,m}(\{i, \infty\})$ for $i = 1, 2, \ldots, 2g + 1$ and show that $\gamma \in \Gamma_{1,2}$.

We first establish some notation: For $\xi \in (\frac{1}{2}\mathbb{Z})^{2g}$, let

$$q(\xi) = \xi_1^T \xi_2 \tag{3.19}$$

be the quadratic form associated to the parity function e_* defined in Definition 2.10. We note that

$$q(\xi) \equiv q(\zeta) \pmod{(\frac{1}{2}\mathbb{Z})^{2g}}, \tag{3.20}$$

if and only if

$$e_*(\xi) = e_*(\zeta). \tag{3.21}$$

Let also

$$b(\xi, \zeta) = \xi^T J \zeta, \tag{3.22}$$

be the bilinear form associated to the matrix J, where as before

$$J = \begin{pmatrix} 0 & \mathbb{1}_g \\ -\mathbb{1}_g & 0 \end{pmatrix}, \tag{3.23}$$

and $\mathbb{1}_g$ is the $g \times g$ identity matrix.

A quick computation shows that for any $\xi, \zeta \in (\frac{1}{2}\mathbb{Z})^{2g}$

$$q(\xi + \zeta) \equiv q(\xi) + q(\zeta) + b(\xi, \zeta) \pmod{(\frac{1}{2}\mathbb{Z})^{2g}}. \tag{3.24}$$

Now, let $\gamma \in \mathrm{Sp}_{2g}(\mathbb{Z})$. We have then that

$$b(\gamma\xi, \gamma\zeta) = b(\xi, \zeta), \tag{3.25}$$

by definition of b and $\mathrm{Sp}_{2g}(\mathbb{Z})$. Therefore, for any $\xi, \zeta \in (\frac{1}{2}\mathbb{Z})^{2g}$

$$q(\gamma(\xi+\zeta)) = q(\gamma\xi+\gamma\zeta) \equiv q(\gamma\xi)+q(\gamma\zeta)+b(\gamma\xi,\gamma\zeta) \pmod{(\frac{1}{2}\mathbb{Z})^{2g}} \tag{3.26}$$

$$\equiv q(\gamma\xi)+q(\gamma\zeta)+b(\xi,\zeta) \pmod{(\frac{1}{2}\mathbb{Z})^{2g}}. \tag{3.27}$$

As a result, if

$$q(\gamma\xi) \equiv q(\xi) \pmod{(\frac{1}{2}\mathbb{Z})^{2g}} \tag{3.28}$$

and

$$q(\gamma\zeta) \equiv q(\zeta) \pmod{(\frac{1}{2}\mathbb{Z})^{2g}}, \tag{3.29}$$

then

$$q(\gamma(\xi+\zeta)) \equiv q(\xi+\zeta) \pmod{(\frac{1}{2}\mathbb{Z})^{2g}}. \tag{3.30}$$

From this discussion, we conclude that if $e_*(\gamma\xi) = e_*(\xi)$ and $e_*(\gamma\zeta) = e_*(\zeta)$, it follows that

$$e_*(\gamma(\xi+\zeta)) = e_*(\xi+\zeta). \tag{3.31}$$

The result now follows from the fact that the set

$$\{\eta_{\Omega,m}(\{i,\infty\}) : i = 1,\ldots,2g+1\} \tag{3.32}$$

contains a basis of the $\mathbb{F}_2$-vector space $(\frac{1}{2}\mathbb{Z})^{2g}/\mathbb{Z}^{2g}$ by Proposition 3.7. Therefore, if a matrix γ fixes the parity of each element of this basis, it must fix the parity of each element of the vector space. □

We are now in a position to show:

Proposition 3.9 *Let X be a marked hyperelliptic curve, $J(X)$ its Jacobian, and Ω a small period matrix associated to $J(X)$ via the process outlined in Definition 2.2. Furthermore, given this data, let $\eta_{\Omega,m}$ be the map given in Definition 2.9 and $U(\Omega,m)$ be the set defined in Definition 1.2.*

Let $\gamma \in \mathrm{Sp}_{2g}(\mathbb{Z})$. Then, the matrix $\gamma\cdot\Omega$ is another small period matrix for $J(X)$, to which we may similarly attach a map $\eta_{\gamma\cdot\Omega,m}$ and a set $U(\gamma\cdot\Omega,m)$.

In that case, we have

$$U(\gamma\cdot\Omega,m) = U(\Omega,m) \tag{3.33}$$

if and only if

$$\gamma \in \Gamma_{1,2}. \tag{3.34}$$

Proof Recall from Proposition 2.11 that $U(\Omega, m)$ can be described as the set

$$\{i \in \{1, 2, \ldots, 2g+1\} : e_*(\eta_{\Omega,m}(\{i, \infty\})) = -1\} \cup \{\infty\}. \tag{3.35}$$

Since $\eta_{\Omega,m}(\{i, \infty\}) \in (\frac{1}{2}\mathbb{Z})^{2g}/\mathbb{Z}^{2g}$ is none other than $AJ_c(e_{\{i,\infty\}})$, by Proposition 2.5, we have

$$\eta_{\gamma\cdot\Omega,m}(\{i, \infty\}) = \gamma^{-T}\eta_{\Omega,m}(\{i, \infty\}), \tag{3.36}$$

Therefore, we have that

$$U(\gamma \cdot \Omega, m) = U(\Omega, m) \tag{3.37}$$

if and only if multiplication by γ^{-T} does not change the parity of any $\eta_{\Omega,m}(\{i, \infty\})$ for $i = 1, 2, \ldots, 2g+1$. By Lemma 3.8, this is the case if and only if $\gamma^{-T} \in \Gamma_{1,2}$.

To finish the proof, we must show that $\gamma^{-T} \in \Gamma_{1,2}$ if and only if $\gamma \in \Gamma_{1,2}$. Note that since $\gamma \in \mathrm{Sp}_{2g}(\mathbb{Z})$, we have

$$\gamma^{-T} = \begin{pmatrix} D & -C \\ -B & A \end{pmatrix}. \tag{3.38}$$

By Proposition 3.6, it suffices thus to show that the diagonals of the matrices

$$D^T(-B) = (-B^T D)^T \tag{3.39}$$

and

$$(-C)^T A = (-A^T C)^T \tag{3.40}$$

have all even entries if and only if the diagonals of the matrices $A^T C$ and $B^T D$ have all even entries, which is true. □

As a direct consequence, we now have:

Theorem 3.10 *The number of different sets $U(\Omega, m)$ that arise, as Ω varies over all small period matrices that can be attached to the polarized Jacobian of a marked hyperelliptic curve X with the process outlined in Definition 2.2, is equal to the cardinality of the quotient group*

$$\mathrm{Sp}_{2g}(\mathbb{Z})/\Gamma_{1,2}. \tag{3.41}$$

Proof As described in Sect. 2, the group $\mathrm{Sp}_{2g}(\mathbb{Z})$ acts transitively on the set of small period matrices that can be associated to $J(X)$ via the process described in Definition 2.2. This action changes $U(\Omega, m)$ if and only if $\gamma \in \Gamma_{1,2}$ by Proposition 3.9, which completes the proof. □

We now compute the cardinality of this quotient group, which will give us the number of different sets $U(\Omega, m)$ attached to a fixed hyperelliptic curve X with a marking of its branch points m as Ω is allowed to vary over all possible small period matrices that can be associated to its Jacobian $J(X)$ via the process described in Definition 2.2.

Proposition 3.11 *We have that*

$$\#\,\mathrm{Sp}_{2g}(\mathbb{Z})/\,\Gamma_{1,2} = 2^{g-1}(2^g + 1). \tag{3.42}$$

Proof To compute the cardinality of this quotient group, we use the following facts: First, by the Third Group Isomorphism Theorem,

$$\mathrm{Sp}_{2g}(\mathbb{Z})/\,\Gamma_{1,2} \cong \frac{\mathrm{Sp}_{2g}(\mathbb{Z})/\,\Gamma(2)}{\Gamma_{1,2}/\,\Gamma(2)}, \tag{3.43}$$

where

$$\Gamma(2) = \left\{\gamma \in \mathrm{Sp}_{2g}(\mathbb{Z}) : \gamma \equiv \mathbb{1}_{2g} \pmod 2\right\}. \tag{3.44}$$

Furthermore, we have

$$\mathrm{Sp}_{2g}(\mathbb{Z})/\,\Gamma(2) \cong \mathrm{Sp}_{2g}(\mathbb{F}_2), \tag{3.45}$$

where $\mathrm{Sp}_{2g}(\mathbb{F}_2)$ is the group of matrices with coefficients in $\mathbb{F}_2$ and symplectic with respect to the bilinear form given by the matrix:

$$\begin{pmatrix} 0 & \mathbb{1}_g \\ \mathbb{1}_g & 0 \end{pmatrix}, \tag{3.46}$$

and

$$\Gamma_{1,2}/\,\Gamma(2) \cong \mathrm{SO}_{2g}(\mathbb{F}_2, +1), \tag{3.47}$$

where $\mathrm{SO}_{2g}(\mathbb{F}_2, +1)$ is the special orthogonal group of matrices with entries in $\mathbb{F}_2$ and preserving the quadratic form:

$$Q(x_1, x_2, \ldots, x_{2g-1}, x_{2g}) = \sum_{i=1}^{g} x_i x_{g+i}. \tag{3.48}$$

(These last two facts are implicit in the discussion in [5], Appendix to Chapter 5.)

There are therefore

$$\frac{\#\,\mathrm{Sp}_{2g}(\mathbb{F}_2)}{\#\,\mathrm{SO}_{2g}(\mathbb{F}_2, +1)} \tag{3.49}$$

different sets $U(\Omega, m)$ as Ω varies over all small period matrices that can be associated to $J(X)$ via the process described in Definition 2.2.

We have

$$\#\,\mathrm{Sp}_{2g}(\mathbb{F}_2) = 2^{g^2} \prod_{i=1}^{g} (2^{2i} - 1), \tag{3.50}$$

(see, for example, [3, Theorem 3.12]) and

$$\#\,\mathrm{SO}_{2g}(\mathbb{F}_2, +1) = 2 \cdot 2^{g(g-1)}(2^g - 1) \prod_{i=1}^{g-1} (2^{2i} - 1), \tag{3.51}$$

(see, for example, [4, Table 2.1C]). Computing the quotient gives the result we sought. □

Acknowledgements The author would like to thank Sorina Ionica for many discussions that led her to understand the set $U(\Omega, m)$ more deeply, and the referees for their careful reading of the manuscript and thoughtful comments.

References

1. J.S. Balakrishnan, S. Ionica, K. Lauter, C. Vincent, Constructing genus-3 hyperelliptic Jacobians with CM. Lond. Math. Soc. J. Comput. Math. **19**(suppl. A), 283–300 (2016)
2. C. Birkenhake, H. Lange, *Complex Abelian Varieties*. Grundlehren der Mathematischen Wissenschaften, vol. 302, 2nd edn. (Springer, Berlin, 2004)
3. L.C. Grove, *Classical Groups and Geometric Algebra*. Graduate Studies in Mathematics, vol. 39 (American Mathematical Society, Providence, 2002)
4. P. Kleidman, M. Liebeck, *The Subgroup Structure of the Finite Classical Groups*. London Mathematical Society Lecture Note Series, vol. 129 (Cambridge University Press, Cambridge, 1990)
5. D. Mumford, *Tata Lectures on Theta. I*. Modern Birkhäuser Classics (Birkhäuser, Boston, 2007)
6. D. Mumford, *Tata Lectures on Theta. II*. Modern Birkhäuser Classics (Birkhäuser, Boston, 2007)
7. C. Poor, The hyperelliptic locus. Duke Math. J. **76**(3), 809–884 (1994)

A First Step Toward Higher Order Chain Rules in Abelian Functor Calculus

Christina Osborne and Amelia Tebbe

Abstract One of the fundamental tools of undergraduate calculus is the chain rule. The notion of higher order directional derivatives was developed by Huang, Marcantognini, Young in Huang et al. (Math. Intell. 28(2):61–69, 2006), along with a corresponding higher order chain rule. When Johnson and McCarthy established abelian functor calculus, they proved a chain rule for functors that is analogous to the directional derivative chain rule when $n = 1$. In joint work with Bauer, Johnson, and Riehl, we defined an analogue of the iterated directional derivative and provided an inductive proof of the analogue to the chain rule of Huang et al.

This paper consists of the initial investigation of the chain rule found in Bauer et al., which involves a concrete computation of the case when $n = 2$. We describe how to obtain the second higher order directional derivative chain rule for functors of abelian categories. This proof is fundamentally different in spirit from the proof given in Bauer et al. as it relies only on properties of cross effects and the linearization of functors.

1 Introduction

In this paper, we consider *abelian functor calculus*, the calculus of functors of abelian categories established by Brenda Johnson and Randy McCarthy (see [4]). Functor calculus enjoys certain properties that are analogous to the results in undergraduate calculus. This paper is a companion to [1], in which many of these analogies are made explicit. In [1], one of the main results [1, Theorem 8.1] provides

C. Osborne (✉)
University of Virginia, Charlottesville, VA, USA
e-mail: cdo5bv@virginia.edu

A. Tebbe
Indiana University Kokomo, Kokomo, IN, USA
e-mail: antebbe@iu.edu

A. Deines et al. (eds.), *Advances in the Mathematical Sciences*, Association for Women in Mathematics Series 15, https://doi.org/10.1007/978-3-319-98684-5_7

a chain rule for the nth higher order directional derivative, denoted as Δ_n [1, Definition 7.3], associated to a functor between abelian categories. A summary of previously developed chain rules for general functor calculus is given in [1].

In order to arrive at this chain rule, we started with an explicit calculation in the case when $n = 2$. It became clear that this method of calculation would not lend itself well to an inductive proof for a general n, which is why the proof of [1, Theorem 8.1] is different in spirit from the approach in this paper. Because this result is quite technical and lengthy, it warranted independent documentation. However, the groundwork—including the definitions and properties of most of the functors we will use—is already documented in [1]. For this reason, we will heavily cite [1] throughout this paper.

The goal of this paper is a chain rule for the second-order directional derivative of a functor F, which is stated in the following theorem.

Theorem 2.11 *Given two composable functors of abelian categories $G : \mathrm{A} \to Ch\mathrm{B}$ and $F : \mathrm{B} \to Ch\mathrm{C}$ with objects x, v, and w in* A, *there is a chain homotopy equivalence*

$$\varDelta_2(F \circ G)(w, v; x) \simeq \varDelta_2 F(\varDelta_2 G(w, v; x), \varDelta_1 G(v; x); G(x)).$$

The directional derivatives are defined in Sect. 2.3. The left and right sides of this equivalence are written in terms of the smallest component parts of the functors in Sects. 3 and 4, respectively. These smallest components are the cross effects of F and G; cross effects of functors are defined in Sect. 2.2. The proof of Theorem 2.11 is concluded in Sect. 5 by matching terms (2) through (32) from Sect. 3 with homotopy equivalent terms (35) through (66) from Sect. 4. Note that a single term from Sect. 3 will be matched to two of the terms from Sect. 4 via a further decomposition, while the rest of the terms have a one-to-one correspondence.

2 Categorical Context, Cross Effects, Linearization, and Directional Derivatives

In this section, we provide the foundational tools and motivation for the main result, which is the second higher order directional derivative chain rule (Theorem 2.11). The construction of higher directional derivatives is possible once cross effects and linearizations of functors and some of their key properties are obtained. We begin by defining the categorical setting in which our results take place.

2.1 Categorical Context

In studying abelian functor calculus, we frequently consider functors that are valued in chain complexes of an abelian category. In particular, the linearization functor we will construct is traditionally defined as sending a functor $F : \mathrm{A} \to \mathrm{B}$ to a functor $D_1 F : \mathrm{A} \to Ch\mathrm{B}$. If we iterate such a construction, we would arrive at functors valued in multicomplexes of increasing dimension. Additionally, we run into issues of composability. Intuitively, we would expect a relationship between $D_1(F \circ G)$ and $D_1 F \circ D_1 G$ (this will be made explicit in Lemmas 5.1 and 5.2). However, a priori, if $G : \mathrm{A} \to \mathrm{B}$ and $F : \mathrm{B} \to \mathrm{C}$, then $D_1 F : \mathrm{B} \to Ch\mathrm{C}$ and $D_1 G : \mathrm{A} \to Ch\mathrm{B}$ are not composable.

To resolve these issues, we work in the following Kleisli category. A more detailed development can be found in [1, Section 3].

Definition 2.1 ([1, Definition 3.2]) There is a large category AbCat_{Ch} whose:

- objects are abelian categories;
- morphisms $\mathrm{A} \rightsquigarrow \mathrm{B}$ are natural isomorphism classes of functors $\mathrm{A} \to Ch\mathrm{B}$;
- identity morphisms $\mathrm{A} \rightsquigarrow \mathrm{A}$ are the functors $\deg_0 : \mathrm{A} \to Ch\mathrm{A}$; and in which
- composition of morphisms $\mathrm{B} \rightsquigarrow \mathrm{C}$ and $\mathrm{A} \rightsquigarrow \mathrm{B}$, corresponding to the pair of functors $F : \mathrm{B} \to Ch\mathrm{C}$ and $G : \mathrm{A} \to Ch\mathrm{B}$, is defined by:

$$F \circ G : \mathrm{A} \xrightarrow{G} Ch\mathrm{B} \xrightarrow{ChF} ChCh\mathrm{C} \xrightarrow{\mathrm{Tot}} Ch\mathrm{C},$$

where ChF is defined by:

$$Ch\mathcal{B} \xrightarrow[\simeq]{K} \mathcal{B}^{\Delta^{op}} \xrightarrow{F^{\Delta^{op}}} (ChC)^{\Delta^{op}} \xrightarrow[\simeq]{N} ChChC \xrightarrow{Tot} ChC,$$

with K and N denoting the inverse functors of the Dold–Kan equivalence, and Tot the totalization functor.

It is necessary to use this procedure to define composition, rather than simply applying F degreewise, to ensure that chain homotopy equivalences are preserved. In particular, as shown in [1, Lemma 3.4], natural chain homotopy equivalences are preserved by this definition of composition.

In effect, working in the Kleisli category AbCat_{Ch} allows us to treat all functors of abelian categories as landing in chain complexes. In the case where the functor of interest, $F : \mathrm{A} \to \mathrm{B}$, does not land in chain complexes, the outputs are treated as chain complexes concentrated in degree zero. We will continue to use the notation $F : \mathrm{A} \rightsquigarrow \mathrm{B}$ for what is really a functor $F : \mathrm{A} \to Ch\mathrm{B}$. Given this setting, when we define the linearization functor we will incorporate the totalization functor so that we again land in chain complexes rather than bicomplexes.

2.2 Cross Effects and Linearization

The following definition appears as [1, Definition 2.1].

Definition 2.2 ([2]) The *nth cross effect* of a functor $F\colon \mathrm{A} \to \mathrm{B}$ between two abelian categories, where the zero object of A is denoted by 0, is the n variable functor

$$\mathrm{cr}_n F\colon \mathrm{A}^n \to \mathrm{B}$$

defined recursively by:

$$F(x) \cong F(0) \oplus \mathrm{cr}_1 F(x)$$

$$\mathrm{cr}_1 F(x_1 \oplus x_2) \cong \mathrm{cr}_1 F(x_1) \oplus \mathrm{cr}_1 F(x_2) \oplus \mathrm{cr}_2 F(x_1, x_2)$$

and in general,

$$\mathrm{cr}_{n-1} F(x_1 \oplus x_2, x_3, \ldots, x_n) \cong \mathrm{cr}_{n-1} F(x_1, x_3, \ldots, x_n) \oplus \mathrm{cr}_{n-1} F(x_2, x_3, \ldots, x_n) \oplus \mathrm{cr}_n F(x_1, x_2, \ldots, x_n),$$

where $\oplus$ denotes the biproduct in both categories A and B (a common abuse of notation).

Definition 2.2 is functorial in the sense that a natural transformation $F \to G$ induces a naturally defined map $\mathrm{cr}_n F \to \mathrm{cr}_n G$.

The cross effect functor is *strictly multi-reduced* in the following sense.

Proposition 2.3 ([4, Proposition 1.2]) *For a functor of abelian categories* F : $\mathrm{A} \to \mathrm{B}$ *and objects* $x_1, \ldots, x_n$ *in* A, *if any* $x_i = 0$, *then*

$$\mathrm{cr}_n F(x_1, \ldots, x_n) \cong 0.$$

Let $C_n F$ denote the nth cross effect of F, cr_n, composed with the diagonal functor. That is, $C_n F(x) := \mathrm{cr}_n F(x, \ldots, x)$. Corollary 2.7 of [1] shows that cr_n is right adjoint to pre-composition with the diagonal functor, so that C_n is a comonad. Let ϵ denote the counit of this comonad.

Functor calculus studies approximations of functors that behave like degree n polynomials, up to chain homotopy equivalence. Let $\simeq$ denote chain homotopy equivalence. The following definition makes the idea of polynomial degree n functors precise.

Definition 2.4 A functor of abelian categories $F : \mathrm{A} \rightsquigarrow \mathrm{B}$ is *degree n* if

$$\mathrm{cr}_{n+1} F \simeq 0.$$

In particular, F is degree 1 if $\mathrm{cr}_2 F \simeq 0$. If F is also reduced, meaning that $F(0) \simeq 0$, then we say that F is *linear*. We call F *strictly reduced* if $F(0) \cong 0$.

Definition 2.5 [1, Definition 5.1] The *linearization* of a functor of abelian categories $F\colon \mathrm{A} \rightsquigarrow \mathrm{B}$ is the functor of abelian categories $D_1F\colon \mathrm{A} \rightsquigarrow \mathrm{B}$ given as totalization of the explicit chain complex of chain complexes (D_1F_*, ∂_*), where:

$$(D_1F)_k := \begin{cases} C_2^{\times k}F & k \geq 1 \\ \mathrm{cr}_1 F & k = 0 \\ 0 & \text{otherwise} \end{cases}$$

and $C_2^{\times k}$ is the functor $C_2 = \mathrm{cr}_2 \circ diag$ composed with itself k times.

The chain differential $\partial_1\colon (D_1F)_1 \to (D_1F)_0$ is given by the map ϵ of [1, Remark 2.8], and the chain differential $\partial_k\colon (D_1F)_k \to (D_1F)_{k-1}$ is given by $\sum_{i=1}^{k}(-1)^i C_2^{\times i}\epsilon$ when $k \geq 1$, where ϵ is the counit of the adjunction of [1, Corollary 2.7].

The linearization functor satisfies the following properties:

Lemma 2.6 ([1, Lemma 5.6])

i *For any functor of abelian categories $F\colon \mathrm{A} \rightsquigarrow \mathrm{B}$, the functor $D_1F\colon \mathrm{A} \rightsquigarrow \mathrm{B}$ is strictly reduced, and for any $x, y \in \mathrm{A}$, the natural map*

$$D_1F(x) \oplus D_1F(y) \to D_1F(x \oplus y)$$

is a chain homotopy equivalence. In particular, D_1F is linear.

ii *In* AbCat_{Ch}, *pointwise chain homotopy equivalence classes in $Fun(\mathrm{A}, \mathrm{B})$ are denoted by $[\mathrm{A}, \mathrm{B}]$. The functor $D_1\colon [\mathrm{A}, \mathrm{B}] \to [\mathrm{A}, \mathrm{B}]$ is linear in the sense that $D_1 0 \cong 0$ and for any pair of functors $F, G \in [\mathrm{A}, \mathrm{B}]$,*

$$D_1F \oplus D_1G \cong D_1(F \oplus G).$$

We will follow Convention 5.11 in [1]. In particular, given $F : \mathrm{A}^n \rightsquigarrow \mathrm{B}$, consider $F_i : \mathrm{A} \rightsquigarrow \mathrm{B}$ defined by:

$$F_i(y) := F(x_1, \ldots, x_{i-1}, y, x_{i+1}, \ldots, x_n),$$

where $x_1, \ldots, x_{i-1}, x_{i+1}, \ldots, x_n$ are fixed objects of A. We will write $D_1^i F(x_1, \ldots, x_n)$ for $D_1F_i(x_i)$. In cases where a single variable x_i occurs in multiple inputs of a multivariable functor F, and we wish to indicate simultaneous multi-linearization of all occurrences of x_i, we will use the notation $D_1^{x_i}F$. Let us look at a specific example to see how $D_1^{x_i}F$ works.

Example 2.7 Let $F : \mathrm{A}^4 \to \mathrm{B}$ and consider $D_1^x F(x, y, x, z)$. Define $G : \mathrm{A} \to \mathrm{B}$ as:

$$\begin{aligned} G(x) &:= F(x, y, x, z) \\ &= F(-, y, -, z) \circ diag(x) \end{aligned}$$

where $diag$ is the diagonal functor. Then, $D_1^x F(x, y, x, z) := D_1G(x)$.

We will frequently use these two notations when dealing with cross effects. For example, in the proof of Lemma 2.13 we consider the sequential linearizations $D_1^1 D_1^2 \mathrm{cr}_2 F(a, b)$ of the two-variable functor $\mathrm{cr}_2 F$. In this case, we linearize this functor in each variable separately. On the other hand, in Lemma 3.3, we consider functors such as $D_1^x \mathrm{cr}_2 F(x, x)$, which is the linearization of the functor $\mathrm{cr}_2 F$ in its two variables simultaneously. These two linearization processes produce quite different results.

2.3 Higher Order Directional Derivatives

Recall that the directional derivative of a differentiable function $f : \mathbb{R}^n \to \mathbb{R}^m$ at the point $x \in \mathbb{R}^n$ in the direction $v \in \mathbb{R}^n$ measures how the value of f at x changes while translating along the infinitesimal vector from x in the direction v. One way to make this idea precise is to define $\nabla f(v; x)$ to be the derivative of the composite function, substituting the affine linear function $t \mapsto x + tv$ into the argument of f, evaluated at $t = 0$:

$$\Delta_1 f(v; x) := \nabla f(v; x) = \frac{\partial}{\partial t} f(x + tv)\Big|_{t=0}.$$

In [3], it was shown that the first directional derivative has a chain rule:

$$\Delta_1(f \circ g)(v; x) = \Delta_1 f(\Delta_1 g(v; x); g(x)).$$

Using the first directional derivative, we can define the second directional derivative:

$$\Delta_2 f(w, v; x) := \frac{\partial}{\partial t} \Delta_1 f(v + tw; x + tv)\Big|_{t=0},$$

which also has a chain rule [3]:

$$\Delta_2(f \circ g)(w, v; x) = \Delta_2 f(\Delta_2 g(w, v; x), \Delta_1 g(v; x); g(x)).$$

Generally speaking, for any n there is a higher order directional derivative along with a corresponding chain rule (see [3, Theorem 3]).

When Johnson and McCarthy established abelian functor calculus, they constructed an analog to the first directional derivative along with a chain rule [4, Proposition 5.6]. The formula for this chain rule from [4] mirrors the case when $n = 1$ in the directional derivative chain rule for functions found in [3]. These similarities provide the motivation to pursue higher order directional derivatives of functors of abelian categories in the hopes of acquiring an analogous higher order chain rule.

When defining higher order directional derivatives for functors in [1], the goal was to imitate the iterative process used to define the higher directional derivatives of functions in [3]. This iterative process can be seen in the case of the second directional derivative as:

$$\begin{aligned}\Delta_2 f(w, v; x) &= \frac{\partial}{\partial t}\Delta_1 f\,(v + tw; x + tv)\bigg|_{t=0} \\ &= \Delta_1(\Delta_1 f)((w; v); (v; x)),\end{aligned}$$

which motivates the following definition for functors of abelian categories.

Definition 2.8 ([1, Definition 7.3]) Consider a functor of abelian categories $F : \mathrm{A} \rightsquigarrow \mathrm{B}$ and objects x, v, and w in A. The *higher order directional derivatives* of F are defined recursively by:

$$\begin{aligned}\Delta_0 F(x) &:= F(x), \\ \Delta_1 F(v; x) &:= D_1 F(x \oplus -)(v), \\ \Delta_2 F(w, v; x) &:= \Delta_1(\Delta_1 F)\left((w; v); (v; x)\right).\end{aligned}$$

We say that $\Delta_1 F(v; x)$ is the *first directional derivative* of F at x in the direction v. Similarly, we say that $\Delta_2 F(w, v; x)$ is the *second higher order directional derivative* of F at x in the directions v and w.

The notation ∇ could be used in place of Δ_1 in order to further highlight the analogy with the directional derivative of undergraduate calculus. We have elected to use Δ_1 here for simplicity.

Remark 2.9 As in [3], we can continue and define the nth directional derivative, but in this paper we will stop at $n = 2$. The full definition can be found in [1, Definition 7.3].

It was shown in [4, Proposition 5.6] that the first directional derivative has a chain rule up to quasi-isomorphism. The chain rule is now strengthened to a chain homotopy equivalence.

Theorem 2.10 ([1, Theorem 6.5(v)]) *Given two composable functors of abelian categories $G : \mathrm{A} \rightsquigarrow \mathrm{B}$ and $F : \mathrm{B} \rightsquigarrow \mathrm{C}$, there is a chain homotopy equivalence*

$$\Delta_1(F \circ G)(v; x) \simeq \Delta_1 F(\Delta_1 G(v; x); G(x)).$$

This brings us to the formulation of our main theorem.

Theorem 2.11 *Given two composable functors of abelian categories $G : \mathrm{A} \rightsquigarrow \mathrm{B}$ and $F : \mathrm{B} \rightsquigarrow \mathrm{C}$ with objects x, v, and w in A, there is a chain homotopy equivalence*

$$\Delta_2(F \circ G)(w, v; x) \simeq \Delta_2 F(\Delta_2 G(w, v; x), \Delta_1 G(v; x); G(x)).$$

We prove Theorem 2.11 by expanding both sides and showing that they are equivalent. The idea of the proof is to break each side down into direct sums of the smallest component parts. These smallest component parts are linearizations of compositions of cross effects for F and G. The left-hand side is expanded in Sect. 3, the right-hand side is expanded in Sect. 4, and the terms from both sides are aligned to finish the proof in Sect. 5.

Before moving on to the expansions of each side, equivalent formulations of the first and second directional derivatives are stated.

We will use the following chain homotopy equivalent formulation of Δ_1.

Lemma 2.12 ([1, Lemma 6.3]) *For a functor $F : \mathrm{A} \rightsquigarrow \mathrm{B}$ between abelian categories and any pair of objects $x, v \in \mathrm{A}$, there is a chain homotopy equivalence*

$$\Delta_1 F(w; x) \simeq D_1 F(w) \oplus D_1^1 \mathrm{cr}_2 F(w, x).$$

Lemma 2.12 is proved using the chain rule for D_1, namely that $D_1(G \circ H) \simeq D_1 G \circ D_1 H \oplus D_1 \mathrm{cr}_2 G(H(0), \mathrm{cr}_1 H)$, for functors F and $w \oplus - : \mathrm{A} \to \mathrm{A}$ and simplifying the resulting cross effects.

Similarly, there is an equivalent formulation for Δ_2.

Lemma 2.13 *For a functor $F : \mathrm{A} \rightsquigarrow \mathrm{B}$ between abelian categories, there is a chain homotopy equivalence,*

$$\Delta_2 F(w, v; x) \simeq D_1 F(w) \oplus D_1^1 \mathrm{cr}_2 F(w, x) \oplus D_1^1 D_1^2 \mathrm{cr}_2 F(v, v) \oplus D_1^1 D_1^2 \mathrm{cr}_3 F(v, v, x).$$

Proof By [1, Lemma 6.9], which is proved using the definition of cross effect, the linearity of D_1 and the fact that cross effects are multi-reduced, for objects a, b, c, and d in A, we have

$$\Delta_1(\Delta_1 F)((d; c); (b; a))$$
$$\simeq D_1 F(d) \oplus D_1^1 \mathrm{cr}_2 F(d, a) \oplus D_1^1 D_1^2 \mathrm{cr}_2 F(b, c) \oplus D_1^1 D_1^2 \mathrm{cr}_3 F(b, c, a).$$

Letting $a := x$, $b := v$, $c := v$, and $d := w$, we get our desired result:

$$\begin{aligned}\Delta_2(F \circ G)(w, v; x) &= \Delta_1(\Delta_1 F)((w, v); (v; x)) \\ &\simeq D_1 F(w) \oplus D_1^1 \mathrm{cr}_2 F(w, x) \oplus D_1^1 D_1^2 \mathrm{cr}_2 F(v, v) \\ &\oplus D_1^1 D_1^2 \mathrm{cr}_3 F(v, v, x),\end{aligned}$$

where, for example, $D_1^1 \mathrm{cr}_2 F(w, x) := D_1(\mathrm{cr}_2 F(-, x))(w)$.

3 The Second-Order Directional Derivative of a Composition

In this section, we expand the left-hand side of Theorem 2.11. We begin by applying Lemma 2.13. We then use Lemma 3.1 to rewrite the cross effects of the composition of functors in terms that are more manageable. Finally, we use Lemma 3.3, which shows that a majority of the terms are contractible.

Before explicitly computing $\Delta_2(F \circ G)$ in terms of cross effects, it will be useful to understand how we can rewrite terms such as $D_1^{x_2}\mathrm{cr}_2(F \circ G)(x_1, x_2)$. Recall that [4] provides a formula for the pth cross effects of a composition of functors.

Lemma 3.1 ([4, Proof of Proposition 1.6]) *Let $G : \mathrm{A} \to \mathrm{B}$ and $F : \mathrm{B} \to \mathrm{C}$ be functors between abelian categories. Let $x_1, \ldots, x_p$ be objects in* A *and $\langle p \rangle = \{1, 2, \ldots, p\}$. For $U = \{s_1, \ldots, s_t\} \subseteq \langle p \rangle$, let $\mathrm{cr}_U G$ denote $\mathrm{cr}_t G(x_{s_1}, \ldots, x_{s_t})$ and for $U = \emptyset$, let $\mathrm{cr}_U G = G(0)$. Then,*

$$\mathrm{cr}_p(F \circ G)(x_1, \ldots, x_p) \cong \bigoplus_{\{U_1,\ldots,U_k | U_i \neq U_j\} \subseteq \mathrm{P}(\langle p \rangle), \cup_{i=1}^k U_i = \langle p \rangle} \mathrm{cr}_k F(\mathrm{cr}_{U_1} G, \ldots, \mathrm{cr}_{U_k} G).$$

The isomorphism of Lemma 3.1 comes from applying the natural isomorphism in [4, Proposition 1.2],

$$H(X_1 \oplus \cdots \oplus X_n) \cong H(0) \oplus \left(\bigoplus_{p=1}^{n} \left(\bigoplus_{j_1 < \cdots < j_p} \mathrm{cr}_p H(X_{j_1}, \ldots X_{j_p} \right) \right)$$

and applying the definition of cross effect.

Let us see explicitly what this formula gives in a simple case.

Example 3.2 Consider the case when $p = 2$. Note that we can cover the set $\langle p \rangle$ with up to four distinct subsets, since the cardinality of $\mathrm{P}(\langle p \rangle)$ is 4. To cover the set $\langle p \rangle$ with one set, it must be itself. To cover $\langle p \rangle$ with two subsets, there are four possibilities: $\{\{1, 2\} \emptyset\}$, $\{\{1\}, \{2\}\}$, $\{\{1\}, \{1, 2\}\}$, and $\{\{2\}, \{1, 2\}\}$. Similarly, there are four different ways to cover $\langle p \rangle$ with three subsets, and one way to cover $\langle p \rangle$ with four subsets. Applying the formula from Lemma 3.1 gives

$$\begin{aligned}
&\mathrm{cr}_2(F \circ G)(x_1, x_2) \\
&\quad \cong \mathrm{cr}_1 F(\mathrm{cr}_2 G(x_1, x_2)) \oplus \mathrm{cr}_2 F(\mathrm{cr}_2 G(x_1, x_2), G(0)) \\
&\quad \oplus \mathrm{cr}_2 F(\mathrm{cr}_1 G(x_1), \mathrm{cr}_1 G(x_2)) \\
&\quad \oplus \mathrm{cr}_2 F(\mathrm{cr}_1 G(x_1), \mathrm{cr}_2 G(x_1, x_2)) \\
&\quad \oplus \mathrm{cr}_2 F(\mathrm{cr}_1 G(x_2), \mathrm{cr}_2 G(x_1, x_2)) \\
&\quad \oplus \mathrm{cr}_3 F(\mathrm{cr}_1 G(x_1), \mathrm{cr}_1 G(x_2), G(0)) \\
&\quad \oplus \mathrm{cr}_3 F(\mathrm{cr}_1 G(x_1), \mathrm{cr}_2 G(x_1, x_2), G(0))
\end{aligned}$$

$$\oplus \operatorname{cr}_3 F(\operatorname{cr}_1 G(x_2), \operatorname{cr}_2 G(x_1, x_2), G(0))$$
$$\oplus \operatorname{cr}_3 F(\operatorname{cr}_1 G(x_1), \operatorname{cr}_1 G(x_2), \operatorname{cr}_2 G(x_1, x_2))$$
$$\oplus \operatorname{cr}_4 F(\operatorname{cr}_1 G(x_1), \operatorname{cr}_1 G(x_2), \operatorname{cr}_2 G(x_1, x_2), G(0)).$$

In order to simplify the expansion, we will use Lemma 3.3 to conclude that some of the summands are in fact contractible.

Lemma 3.3 ([1, Corollary 5.13]) *Suppose the functor of abelian categories* $F : \mathrm{A} \rightsquigarrow \mathrm{B}$ *factors as*

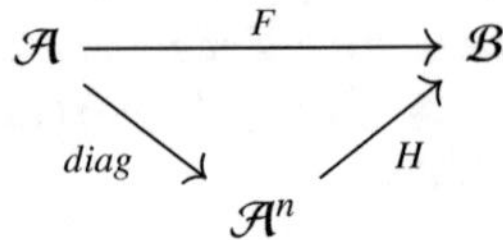

where $diag : \mathrm{A} \to \mathrm{A}^n$ *is the diagonal functor (with* $n \geq 2$*) and* H *is strictly multi-reduced. Then,* $D_1 F$ *is contractible.*

Recall that cross effects are strictly multi-reduced functors. Note that the composition of strictly multi-reduced functors is still strictly multi-reduced.

Example 3.4 Using the expansion from Example 3.2, let us compute $D_1^{x_2} \operatorname{cr}_2(F \circ G)(x_1, x_2)$. As noted previously, we can distribute $D_1^{x_2}$ to each summand. Since we are linearizing with respect to x_2 and since the functor $\operatorname{cr}_2 F(\operatorname{cr}_1 G(-), \operatorname{cr}_2 G(x_1, -))$ is strictly multi-reduced, Lemma 3.3 tells us that, for example,

$$D_1^{x_2} \operatorname{cr}_2 F(\operatorname{cr}_1 G(x_2), \operatorname{cr}_2 G(x_1, x_2)) = D_1^{x_2} \operatorname{cr}_2 F(\operatorname{cr}_1 G(-), \operatorname{cr}_2 G(x_1, -)) \circ diag(x_2)$$
$$\simeq 0,$$

where $diag : \mathrm{A} \to \mathrm{A} \times \mathrm{A}$ is the diagonal functor $x_2 \mapsto (x_2, x_2)$. Hence,

$$D_1^{x_2} \operatorname{cr}_2(F \circ G)(x_1, x_2)$$
$$\simeq D_1^{x_2} \operatorname{cr}_1 F(\operatorname{cr}_2 G(x_1, x_2))$$
$$\oplus D_1^{x_2} \operatorname{cr}_2 F(\operatorname{cr}_2 G(x_1, x_2), G(0))$$
$$\oplus D_1^{x_2} \operatorname{cr}_2 F(\operatorname{cr}_1 G(x_1), \operatorname{cr}_1 G(x_2))$$
$$\oplus D_1^{x_2} \operatorname{cr}_2 F(\operatorname{cr}_1 G(x_1), \operatorname{cr}_2 G(x_1, x_2))$$
$$\oplus D_1^{x_2} \operatorname{cr}_3 F(\operatorname{cr}_1 G(x_1), \operatorname{cr}_1 G(x_2), G(0))$$
$$\oplus D_1^{x_2} \operatorname{cr}_3 F(\operatorname{cr}_1 G(x_1), \operatorname{cr}_2 G(x_1, x_2), G(0)).$$

We proceed with the expansion of the left-hand side of 2.11. To start, applying Lemma 2.13 gives

$$\begin{aligned}\Delta_2(F \circ G)(w, v; x) \simeq & D_1(F \circ G)(w) \oplus D_1^2 \mathrm{cr}_2(F \circ G)(x, w) \\ & \oplus D_1^1 D_1^2 \mathrm{cr}_2(F \circ G)(v, \bar{v}) \oplus D_1^2 D_1^3 \mathrm{cr}_3(F \circ G)(x, v, \bar{v}),\end{aligned} \tag{1}$$

where the new variable $\bar{v} := v$ is introduced to better illustrate the computations. This expression provides the foundation for expanding the left-hand side. We expand the second, third, and fourth terms of (1) using the methods illustrated in Examples 3.2 and 3.4. Specifically, Lemma 3.1 is used to rewrite the cross effect, then Lemma 2.6 is used to distribute the linearization functor(s) to each summand, and finally Lemma 3.3 is applied to find the terms that are contractible. The first term of (1) will be addressed later in Sect. 5.

For the second summand of (1):

$$\begin{aligned}& D_1^2 \mathrm{cr}_2(F \circ G)(w, x) \\ & \quad = D_1^w \mathrm{cr}_2(F \circ G)(w, x) \\ & \quad \simeq D_1^w \mathrm{cr}_1 F(\mathrm{cr}_2 G(w, x)) \\ & \qquad \oplus D_1^w \mathrm{cr}_2 F(\mathrm{cr}_2 G(w, x), G(0)) \\ & \qquad \oplus D_1^w \mathrm{cr}_2 F(\mathrm{cr}_1 G(w), \mathrm{cr}_1 G(x)) \\ & \qquad \oplus D_1^w \mathrm{cr}_2 F(\mathrm{cr}_2 G(w, x), \mathrm{cr}_1 G(x)) \\ & \qquad \oplus D_1^w \mathrm{cr}_3 F(\mathrm{cr}_1 G(w), \mathrm{cr}_1 G(x), G(0)) \\ & \qquad \oplus D_1^w \mathrm{cr}_3 F(\mathrm{cr}_2 G(w, x), \mathrm{cr}_1 G(x), G(0)).\end{aligned}$$

For the third summand of (1):

$$\begin{aligned}& D_1^1 D_1^2 \mathrm{cr}_2(F \circ G)(v, \bar{v}) \\ & \quad = D_1^v D_1^{\bar{v}} \mathrm{cr}_2(F \circ G)(v, \bar{v}) \\ & \quad \simeq D_1^v D_1^{\bar{v}} \mathrm{cr}_1 F(\mathrm{cr}_2 G(v, \bar{v})) \\ & \qquad \oplus D_1^v D_1^{\bar{v}} \mathrm{cr}_2 F(\mathrm{cr}_2 G(v, \bar{v}), G(0)) \\ & \qquad \oplus D_1^v D_1^{\bar{v}} \mathrm{cr}_2 F(\mathrm{cr}_1 G(v), \mathrm{cr}_1 G(\bar{v})) \\ & \qquad \oplus D_1^v D_1^{\bar{v}} \mathrm{cr}_3 F(\mathrm{cr}_1 G(v), \mathrm{cr}_1 G(\bar{v}), G(0)).\end{aligned}$$

For the fourth summand of (1):

$$\begin{aligned}& D_1^2 D_1^3 \mathrm{cr}_3(F \circ G)(v, \bar{v}, x) \\ & \quad = D_1^v D_1^{\bar{v}} \mathrm{cr}_3(F \circ G)(v, \bar{v}, x) \\ & \quad \simeq D_1^v D_1^{\bar{v}} \mathrm{cr}_1 F(\mathrm{cr}_3 G(v, \bar{v}, x))\end{aligned}$$

$$
\begin{aligned}
&\oplus D_1^v D_1^{\bar{v}} \mathrm{cr}_2 F(\mathrm{cr}_3 G(v, \bar{v}, x), G(0)) \oplus D_1^v D_1^{\bar{v}} \mathrm{cr}_2 F(\mathrm{cr}_2 G(v, \bar{v}), \mathrm{cr}_1 G(x)) \\
&\oplus D_1^v D_1^{\bar{v}} \mathrm{cr}_2 F(\mathrm{cr}_3 G(v, \bar{v}, x), \mathrm{cr}_1 G(x)) \oplus D_1^v D_1^{\bar{v}} \mathrm{cr}_2 F(\mathrm{cr}_1 G(v), \mathrm{cr}_2 G(\bar{v}, x)) \\
&\oplus D_1^v D_1^{\bar{v}} \mathrm{cr}_2 F(\mathrm{cr}_1 G(\bar{v}), \mathrm{cr}_2 G(v, x)) \\
&\oplus D_1^v D_1^{\bar{v}} \mathrm{cr}_2 F(\mathrm{cr}_2 G(v, x), \mathrm{cr}_2 G(\bar{v}, x)) \\
&\oplus D_1^v D_1^{\bar{v}} \mathrm{cr}_3 F(\mathrm{cr}_2 G(v, \bar{v}), \mathrm{cr}_1 G(x), G(0)) \\
&\oplus D_1^v D_1^{\bar{v}} \mathrm{cr}_3 F(\mathrm{cr}_3 G(v, \bar{v}, x), \mathrm{cr}_1 G(x), G(0)) \\
&\oplus D_1^v D_1^{\bar{v}} \mathrm{cr}_3 F(\mathrm{cr}_1 G(v), \mathrm{cr}_2 G(\bar{v}, x), G(0)) \\
&\oplus D_1^v D_1^{\bar{v}} \mathrm{cr}_3 F(\mathrm{cr}_1 G(\bar{v}), \mathrm{cr}_2 G(v, x), G(0)) \\
&\oplus D_1^v D_1^{\bar{v}} \mathrm{cr}_3 F(\mathrm{cr}_2 G(v, x), \mathrm{cr}_2 G(\bar{v}, x), G(0)) \\
&\oplus D_1^v D_1^{\bar{v}} \mathrm{cr}_3 F(\mathrm{cr}_1 G(v), \mathrm{cr}_1 G(\bar{v}), \mathrm{cr}_1 G(x)) \\
&\oplus D_1^v D_1^{\bar{v}} \mathrm{cr}_3 F(\mathrm{cr}_1 G(v), \mathrm{cr}_2 G(\bar{v}, x), \mathrm{cr}_1 G(x)) \\
&\oplus D_1^v D_1^{\bar{v}} \mathrm{cr}_3 F(\mathrm{cr}_1 G(\bar{v}), \mathrm{cr}_2 G(v, x), \mathrm{cr}_1 G(x)) \\
&\oplus D_1^v D_1^{\bar{v}} \mathrm{cr}_3 F(\mathrm{cr}_2 G(v, x), \mathrm{cr}_2 G(\bar{v}, x), \mathrm{cr}_1 G(x)) \\
&\oplus D_1^v D_1^{\bar{v}} \mathrm{cr}_4 F(\mathrm{cr}_1 G(v), \mathrm{cr}_1 G(\bar{v}), \mathrm{cr}_1 G(x), G(0)) \\
&\oplus D_1^v D_1^{\bar{v}} \mathrm{cr}_4 F(\mathrm{cr}_1 G(v), \mathrm{cr}_2 G(\bar{v}, x), \mathrm{cr}_1 G(x), G(0)) \\
&\oplus D_1^v D_1^{\bar{v}} \mathrm{cr}_4 F(\mathrm{cr}_1 G(\bar{v}), \mathrm{cr}_2 G(v, x), \mathrm{cr}_1 G(x), G(0)) \\
&\oplus D_1^v D_1^{\bar{v}} \mathrm{cr}_4 F(\mathrm{cr}_2 G(v, x), \mathrm{cr}_2 G(\bar{v}, x), \mathrm{cr}_1 G(x), G(0)).
\end{aligned}
$$

All together, the expansion of the left-hand side of Theorem 2.11 is

$$
\begin{aligned}
&\Delta_2(F \circ G)(w, v; x) \simeq \\
&D_1(F \circ G)(w) && (2)\\
&\oplus D_1^w \mathrm{cr}_1 F(\mathrm{cr}_2 G(w, x)) && (3)\\
&\oplus D_1^w \mathrm{cr}_2 F(\mathrm{cr}_2 G(w, x), G(0)) && (4)\\
&\oplus D_1^w \mathrm{cr}_2 F(\mathrm{cr}_1 G(w), \mathrm{cr}_1 G(x)) && (5)\\
&\oplus D_1^w \mathrm{cr}_2 F(\mathrm{cr}_2 G(w, x), \mathrm{cr}_1 G(x)) && (6)\\
&\oplus D_1^w \mathrm{cr}_3 F(\mathrm{cr}_1 G(w), \mathrm{cr}_1 G(x), G(0)) && (7)\\
&\oplus D_1^w \mathrm{cr}_3 F(\mathrm{cr}_2 G(w, x), \mathrm{cr}_1 G(x), G(0)) && (8)\\
&\oplus D_1^v D_1^{\bar{v}} \mathrm{cr}_1 F(\mathrm{cr}_2 G(v, \bar{v})) && (9)
\end{aligned}
$$

$$\oplus D_1^v D_1^{\bar{v}} \mathrm{cr}_2 F(\mathrm{cr}_2 G(v, \bar{v}), G(0)) \tag{10}$$

$$\oplus D_1^v D_1^{\bar{v}} \mathrm{cr}_2 F(\mathrm{cr}_1 G(v), \mathrm{cr}_1 G(\bar{v})) \tag{11}$$

$$\oplus D_1^v D_1^{\bar{v}} \mathrm{cr}_3 F(\mathrm{cr}_1 G(v), \mathrm{cr}_1 G(\bar{v}), G(0)) \tag{12}$$

$$\oplus D_1^v D_1^{\bar{v}} \mathrm{cr}_1 F(\mathrm{cr}_3 G(v, \bar{v}, x)) \tag{13}$$

$$\oplus D_1^v D_1^{\bar{v}} \mathrm{cr}_2 F(\mathrm{cr}_3 G(v, \bar{v}, x), G(0)) \tag{14}$$

$$\oplus D_1^v D_1^{\bar{v}} \mathrm{cr}_2 F(\mathrm{cr}_2 G(v, \bar{v}), \mathrm{cr}_1 G(x)) \tag{15}$$

$$\oplus D_1^v D_1^{\bar{v}} \mathrm{cr}_2 F(\mathrm{cr}_3 G(v, \bar{v}, x), \mathrm{cr}_1 G(x)) \tag{16}$$

$$\oplus D_1^v D_1^{\bar{v}} \mathrm{cr}_2 F(\mathrm{cr}_1 G(v), \mathrm{cr}_2 G(\bar{v}, x)) \tag{17}$$

$$\oplus D_1^v D_1^{\bar{v}} \mathrm{cr}_2 F(\mathrm{cr}_1 G(\bar{v}), \mathrm{cr}_2 G(v, x)) \tag{18}$$

$$\oplus D_1^v D_1^{\bar{v}} \mathrm{cr}_2 F(\mathrm{cr}_2 G(v, x), \mathrm{cr}_2 G(\bar{v}, x)) \tag{19}$$

$$\oplus D_1^v D_1^{\bar{v}} \mathrm{cr}_3 F(\mathrm{cr}_2 G(v, \bar{v}), \mathrm{cr}_1 G(x), G(0)) \tag{20}$$

$$\oplus D_1^v D_1^{\bar{v}} \mathrm{cr}_3 F(\mathrm{cr}_3 G(v, \bar{v}, x), \mathrm{cr}_1 G(x), G(0)) \tag{21}$$

$$\oplus D_1^v D_1^{\bar{v}} \mathrm{cr}_3 F(\mathrm{cr}_1 G(v), \mathrm{cr}_2 G(\bar{v}, x), G(0)) \tag{22}$$

$$\oplus D_1^v D_1^{\bar{v}} \mathrm{cr}_3 F(\mathrm{cr}_1 G(\bar{v}), \mathrm{cr}_2 G(v, x), G(0)) \tag{23}$$

$$\oplus D_1^v D_1^{\bar{v}} \mathrm{cr}_3 F(\mathrm{cr}_2 G(v, x), \mathrm{cr}_2 G(\bar{v}, x), G(0)) \tag{24}$$

$$\oplus D_1^v D_1^{\bar{v}} \mathrm{cr}_3 F(\mathrm{cr}_1 G(v), \mathrm{cr}_1 G(\bar{v}), \mathrm{cr}_1 G(x)) \tag{25}$$

$$\oplus D_1^v D_1^{\bar{v}} \mathrm{cr}_3 F(\mathrm{cr}_1 G(v), \mathrm{cr}_2 G(\bar{v}, x), \mathrm{cr}_1 G(x)) \tag{26}$$

$$\oplus D_1^v D_1^{\bar{v}} \mathrm{cr}_3 F(\mathrm{cr}_1 G(\bar{v}), \mathrm{cr}_2 G(v, x), \mathrm{cr}_1 G(x)) \tag{27}$$

$$\oplus D_1^v D_1^{\bar{v}} \mathrm{cr}_3 F(\mathrm{cr}_2 G(v, x), \mathrm{cr}_2 G(\bar{v}, x), \mathrm{cr}_1 G(x)) \tag{28}$$

$$\oplus D_1^v D_1^{\bar{v}} \mathrm{cr}_4 F(\mathrm{cr}_1 G(v), \mathrm{cr}_1 G(\bar{v}), \mathrm{cr}_1 G(x), G(0)) \tag{29}$$

$$\oplus D_1^v D_1^{\bar{v}} \mathrm{cr}_4 F(\mathrm{cr}_1 G(v), \mathrm{cr}_2 G(\bar{v}, x), \mathrm{cr}_1 G(x), G(0)) \tag{30}$$

$$\oplus D_1^v D_1^{\bar{v}} \mathrm{cr}_4 F(\mathrm{cr}_1 G(\bar{v}), \mathrm{cr}_2 G(v, x), \mathrm{cr}_1 G(x), G(0)) \tag{31}$$

$$\oplus D_1^v D_1^{\bar{v}} \mathrm{cr}_4 F(\mathrm{cr}_2 G(v, x), \mathrm{cr}_2 G(\bar{v}, x), \mathrm{cr}_1 G(x), G(0)). \tag{32}$$

Notice that each term is labeled individually. In Sect. 5, these terms will be aligned with terms (35) through (66), which come from the right-hand side of Theorem 2.11. Note that (2) will be matched to (35) $\oplus$ (55) via a further decomposition, while the rest of the terms have a one-to-one correspondence. We turn our attention to the expansion of the right-hand side in the next section.

4 A Composition of Directional Derivatives

The right-hand side of Theorem 2.11, which is a composition of directional derivatives, can also be expanded. Most of the results needed for this expansion were discussed previously in Sect. 2.3. Specifically, Lemmas 2.12 and 2.13 reformulate the expression as a direct sum of linearizations of cross effects, rather than directional derivatives. In addition, Lemma 4.1 is necessary to expand the right-hand side into its smallest component parts in order to align the terms with the left-hand side expansion.

The first step in expanding the right-hand side of Theorem 2.11,

$$\Delta_2 F(\Delta_2 G(w, v; x), \Delta_1 G(v; x); G(x)),$$

is to rewrite the second directional derivative of F, $\Delta_2 F$, in terms of linearizations of F using Lemma 2.13:

$$\begin{aligned}
&\Delta_2 F(\Delta_2 G(w, v; x), \Delta_1 G(v; x); G(x)) \\
&\quad \simeq D_1^1 F(\Delta_2 G(w, v; x)) \oplus D_1^1 D_1^2 \mathrm{cr}_2 F(\Delta_1 G(v; x), \Delta_1 G(v; x)) \qquad (33) \\
&\quad \oplus D_1^1 D_1^2 \mathrm{cr}_3 F(\Delta_1 G(v; x), \Delta_1 G(v; x), G(x)) \oplus D_1^1 \mathrm{cr}_2 F(\Delta_2 G(w, v; x), G(x))
\end{aligned}$$

Notice that $G(x)$ appears as a variable in this expansion. But, in the complete expansion of the left-hand side at the end of Sect. 3, the term $G(x)$ never appears as a variable of one of the cross effects of F. Instead, it is observed that $\mathrm{cr}_1 G(x)$ appears. In order to get a clear correspondence between the expansions of the two sides, a description of the relationship between the occurrence of $G(x)$ versus $\mathrm{cr}_1 G(x)$ as a variable of $\mathrm{cr}_k F$ is required.

Lemma 4.1 *Let $G : \mathrm{A} \to \mathrm{B}$ and $F : \mathrm{B} \to \mathrm{C}$ be two composable functors of abelian categories. For $k > 0$,*

$$\begin{aligned}
\mathrm{cr}_k F(x_1, \dots, x_{k-1}, G(x)) \cong\; & \mathrm{cr}_k F(x_1, \dots, x_{k-1}, G(0)) \oplus \mathrm{cr}_k F(x_1, \dots, x_{k-1}, \mathrm{cr}_1 G(x)) \\
& \oplus \mathrm{cr}_{k+1} F(x_1, \dots, x_{k-1}, G(0), \mathrm{cr}_1 G(x))
\end{aligned}$$

Proof Using the definition of the first cross effect,

$$G(x) \cong \mathrm{cr}_1 G(x) \oplus G(0)$$

and the definition of the $(k+1)$st cross effect,

$$\begin{aligned}
\mathrm{cr}_k F(x_1, \dots, x_{k-1}, x_k \oplus x_{k+1}) \cong\; & \mathrm{cr}_k F(x_1, \dots, x_{k-1}, x_k) \oplus \mathrm{cr}_k F(x_1, \dots, x_{k-1}, x_{k+1}) \\
& \oplus \mathrm{cr}_{k+1} F(x_1, \dots, x_{k-1}, x_k, x_{k+1}),
\end{aligned}$$

it follows that

$$\begin{aligned}
\mathrm{cr}_k F(x_1, \dots, x_{k-1}, G(x)) &\cong \mathrm{cr}_k F(x_1, \dots, x_{k-1}, \mathrm{cr}_1 G(x) \oplus G(0)) \\
&\cong \mathrm{cr}_k F(x_1, \dots, x_{k-1}, \mathrm{cr}_1 G(x)) \\
&\oplus \mathrm{cr}_k F(x_1, \dots, x_{k-1}, G(0)) \\
&\quad \oplus \mathrm{cr}_{k+1} F(x_1, \dots, x_{k-1}, \mathrm{cr}_1 G(x), G(0)).
\end{aligned}$$

Applying Lemma 4.1 to the third and fourth summands of (33), we obtain

$$\begin{aligned}
&\Delta_2 F(\Delta_2 G(w, v; x), \Delta_1 G(v; x); G(x)) \\
&\simeq D_1^1 F(\Delta_2 G(w, v; x)) \oplus D_1^1 D_1^2 \mathrm{cr}_2 F(\Delta_1 G(v; x), \Delta_1 G(v; x)) \qquad (34) \\
&\quad \oplus D_1^1 D_1^2 \mathrm{cr}_3 F(\Delta_1 G(v; x), \Delta_1 G(v; x), G(0)) \\
&\quad \oplus D_1^1 D_1^2 \mathrm{cr}_3 F(\Delta_1 G(v; x), \Delta_1 G(v; x), \mathrm{cr}_1 G(x)) \\
&\quad \oplus D_1^1 D_1^2 \mathrm{cr}_4 F(\Delta_1 G(v; x), \Delta_1 G(v; x), \mathrm{cr}_1 G(x), G(0)) \oplus D_1^1 \mathrm{cr}_2 F(\Delta_2 G(w, v; x), G(0)) \\
&\quad \oplus D_1^1 \mathrm{cr}_2 F(\Delta_2 G(w, v; x), \mathrm{cr}_1 G(x)) \oplus D_1^1 \mathrm{cr}_3 F(\Delta_2 G(w, v; x), \mathrm{cr}_1 G(x), G(0)).
\end{aligned}$$

We will further expand the right-hand side by working with each of the eight summands of (34) individually. Note that by using Lemmas 2.12 and 2.13, we can rewrite $\Delta_1 G(v; x)$ and $\Delta_2 G(w, v; x)$, respectively, as:

$$\Delta_1 G(v; x) \simeq D_1 G(v) \oplus D_1^1 \mathrm{cr}_2 G(v, x),$$

and

$$\Delta_2 G(w, v; x) \simeq D_1 G(w) \oplus D_1^1 D_1^2 \mathrm{cr}_2 G(v, v) \oplus D_1^1 D_1^2 \mathrm{cr}_3 G(v, v, x) \oplus D_1^1 \mathrm{cr}_2 G(w, x).$$

Applying these reformulations of $\Delta_1 G(v; x)$ and $\Delta_2 G(w, v; x)$ as well as Lemma 2.6, the first summand of (34) is

$$\begin{aligned}
&D_1^1 F(\Delta_2 G(w, v; x)) \\
&\quad \simeq D_1^1 F\left(D_1 G(w) \oplus D_1^1 D_1^2 \mathrm{cr}_2 G(v, v) \oplus D_1^1 D_1^2 \mathrm{cr}_3 G(v, v, x) \oplus D_1^1 \mathrm{cr}_2 G(w, x)\right) \\
&\quad \simeq D_1^1 F\left(D_1 G(w)\right) \oplus D_1^1 F\left(D_1^1 D_1^2 \mathrm{cr}_2 G(v, v)\right) \\
&\qquad \oplus D_1^1 F\left(D_1^1 D_1^2 \mathrm{cr}_3 G(v, v, x)\right) \oplus D_1^1 F\left(D_1^1 \mathrm{cr}_2 G(w, x)\right),
\end{aligned}$$

the second summand of (34) is

$$
\begin{aligned}
&D_1^1 D_1^2 \mathrm{cr}_2 F(\Delta_1 G(v;x), \Delta_1 G(v;x)) \\
&\quad \simeq D_1^1 D_1^2 \mathrm{cr}_2 F\left(D_1 G(v) \oplus D_1^1 \mathrm{cr}_2 G(v,x), D_1 G(v) \oplus D_1^1 \mathrm{cr}_2 G(v,x)\right) \\
&\quad \simeq D_1^1 D_1^2 \mathrm{cr}_2 F\left(D_1 G(v), D_1 G(v) \oplus D_1^1 \mathrm{cr}_2 G(v,x)\right) \\
&\qquad \oplus D_1^1 D_1^2 \mathrm{cr}_2 F\left(D_1^1 \mathrm{cr}_2 G(v,x), D_1 G(v) \oplus D_1^1 \mathrm{cr}_2 G(v,x)\right) \\
&\quad \simeq D_1^1 D_1^2 \mathrm{cr}_2 F\left(D_1 G(v), D_1 G(v)\right) \\
&\qquad \oplus D_1^1 D_1^2 \mathrm{cr}_2 F\left(D_1 G(v), D_1^1 \mathrm{cr}_2 G(v,x)\right) \\
&\qquad \oplus D_1^1 D_1^2 \mathrm{cr}_2 F\left(D_1^1 \mathrm{cr}_2 G(v,x), D_1 G(v)\right) \\
&\qquad \oplus D_1^1 D_1^2 \mathrm{cr}_2 F\left(D_1^1 \mathrm{cr}_2 G(v,x), D_1^1 \mathrm{cr}_2 G(v,x)\right),
\end{aligned}
$$

the third summand of (34) is

$$
\begin{aligned}
&D_1^1 D_1^2 \mathrm{cr}_3 F(\Delta_1 G(v;x), \Delta_1 G(v;x), G(0)) \\
&\quad \simeq D_1^1 D_1^2 \mathrm{cr}_3 F\left(D_1 G(v) \oplus D_1^1 \mathrm{cr}_2 G(v,x), D_1 G(v) \oplus D_1^1 \mathrm{cr}_2 G(v,x), G(0)\right) \\
&\quad \simeq D_1^1 D_1^2 \mathrm{cr}_3 F\left(D_1 G(v), D_1 G(v) \oplus D_1^1 \mathrm{cr}_2 G(v,x), G(0)\right) \\
&\qquad \oplus D_1^1 D_1^2 \mathrm{cr}_3 F\left(D_1^1 \mathrm{cr}_2 G(v,x), D_1 G(v) \oplus D_1^1 \mathrm{cr}_2 G(v,x), G(0)\right) \\
&\quad \simeq D_1^1 D_1^2 \mathrm{cr}_3 F\left(D_1 G(v), D_1 G(v), G(0)\right) \\
&\qquad \oplus D_1^1 D_1^2 \mathrm{cr}_3 F\left(D_1 G(v), D_1^1 \mathrm{cr}_2 G(v,x), G(0)\right) \\
&\qquad \oplus D_1^1 D_1^2 \mathrm{cr}_3 F\left(D_1^1 \mathrm{cr}_2 G(v,x), D_1 G(v), G(0)\right) \\
&\qquad \oplus D_1^1 D_1^2 \mathrm{cr}_3 F\left(D_1^1 \mathrm{cr}_2 G(v,x), D_1^1 \mathrm{cr}_2 G(v,x), G(0)\right),
\end{aligned}
$$

the fourth summand of (34) is

$$
\begin{aligned}
&D_1^1 D_1^2 \mathrm{cr}_3 F(\Delta_1 G(v;x), \Delta_1 G(v;x), \mathrm{cr}_1 G(x)) \\
&\quad \simeq D_1^1 D_1^2 \mathrm{cr}_3 F\left(D_1 G(v) \oplus D_1^1 \mathrm{cr}_2 G(v,x), D_1 G(v) \oplus D_1^1 \mathrm{cr}_2 G(v,x), \mathrm{cr}_1 G(x)\right) \\
&\quad \simeq D_1^1 D_1^2 \mathrm{cr}_3 F\left(D_1 G(v), D_1 G(v) \oplus D_1^1 \mathrm{cr}_2 G(v,x), \mathrm{cr}_1 G(x)\right)
\end{aligned}
$$

$$\oplus\, D_1^1 D_1^2 \mathrm{cr}_3 F\left(D_1^1\mathrm{cr}_2 G(v,x),\, D_1 G(v)\oplus D_1^1\mathrm{cr}_2 G(v,x),\, \mathrm{cr}_1 G(x)\right)$$
$$\simeq D_1^1 D_1^2 \mathrm{cr}_3 F(D_1 G(v),\, D_1 G(v),\, \mathrm{cr}_1 G(x))$$
$$\oplus\, D_1^1 D_1^2 \mathrm{cr}_3 F(D_1 G(v),\, D_1^1\mathrm{cr}_2 G(v,x),\, \mathrm{cr}_1 G(x))$$
$$\oplus\, D_1^1 D_1^2 \mathrm{cr}_3 F(D_1^1\mathrm{cr}_2 G(v,x),\, D_1 G(v),\, \mathrm{cr}_1 G(x))$$
$$\oplus\, D_1^1 D_1^2 \mathrm{cr}_3 F(D_1^1\mathrm{cr}_2 G(v,x),\, D_1^1\mathrm{cr}_2 G(v,x),\, \mathrm{cr}_1 G(x)),$$

the fifth summand of (34) is

$$D_1^1 D_1^2 \mathrm{cr}_4 F(\Delta_1 G(v;x),\, \Delta_1 G(v;x),\, \mathrm{cr}_1 G(x),\, G(0))$$
$$\simeq D_1^1 D_1^2 \mathrm{cr}_4 F\left(D_1 G(v)\oplus D_1^1\mathrm{cr}_2 G(v,x),\, D_1 G(v)\oplus D_1^1\mathrm{cr}_2 G(v,x),\, \mathrm{cr}_1 G(x),\, G(0)\right)$$
$$\simeq D_1^1 D_1^2 \mathrm{cr}_4 F\left(D_1 G(v),\, D_1 G(v)\oplus D_1^1\mathrm{cr}_2 G(v,x),\, \mathrm{cr}_1 G(x),\, G(0)\right)$$
$$\oplus\, D_1^1 D_1^2 \mathrm{cr}_4 F\left(D_1^1\mathrm{cr}_2 G(v,x),\, D_1 G(v)\oplus D_1^1\mathrm{cr}_2 G(v,x),\, \mathrm{cr}_1 G(x),\, G(0)\right)$$
$$\simeq D_1^1 D_1^2 \mathrm{cr}_4 F(D_1 G(v),\, D_1 G(v),\, \mathrm{cr}_1 G(x),\, G(0))$$
$$\oplus\, D_1^1 D_1^2 \mathrm{cr}_4 F(D_1 G(v),\, D_1^1\mathrm{cr}_2 G(v,x),\, \mathrm{cr}_1 G(x),\, G(0))$$
$$\oplus\, D_1^1 D_1^2 \mathrm{cr}_4 F(D_1^1\mathrm{cr}_2 G(v,x),\, D_1 G(v),\, \mathrm{cr}_1 G(x),\, G(0))$$
$$\oplus\, D_1^1 D_1^2 \mathrm{cr}_4 F(D_1^1\mathrm{cr}_2 G(v,x),\, D_1^1\mathrm{cr}_2 G(v,x),\, \mathrm{cr}_1 G(x),\, G(0)),$$

the sixth summand of (34) is

$$D_1^1\mathrm{cr}_2 F(\Delta_2 G(w,v;x),\, G(0))$$
$$\simeq D_1^1\mathrm{cr}_2 F(D_1 G(w)\oplus D_1^1\mathrm{cr}_2 G(w,x)\oplus D_1^1 D_1^2\mathrm{cr}_2 G(v,v)\oplus D_1^1 D_1^2\mathrm{cr}_3 G(v,v,x),\, G(0))$$
$$\simeq D_1^1\mathrm{cr}_2 F(D_1 G(w),\, G(0))\oplus D_1^1\mathrm{cr}_2 F(D_1^1\mathrm{cr}_2 G(w,x),\, G(0))$$
$$\oplus\, D_1^1\mathrm{cr}_2 F(D_1^1 D_1^2\mathrm{cr}_2 G(v,v),\, G(0))\oplus D_1^1\mathrm{cr}_2 F(D_1^1 D_1^2\mathrm{cr}_3 G(v,v,x),\, G(0)),$$

the seventh summand of (34) is

$$D_1^1\mathrm{cr}_2 F(\Delta_2 G(w,v;x),\, \mathrm{cr}_1 G(x))$$
$$\simeq D_1^1\mathrm{cr}_2 F(D_1 G(w)\oplus D_1^1\mathrm{cr}_2 G(w,x)\oplus D_1^1 D_1^2\mathrm{cr}_2 G(v,v)\oplus D_1^1 D_1^2\mathrm{cr}_3 G(v,v,x),\, \mathrm{cr}_1 G(x))$$
$$\simeq D_1^1\mathrm{cr}_2 F(D_1 G(w),\, \mathrm{cr}_1 G(x))\oplus D_1^1\mathrm{cr}_2 F(D_1^1\mathrm{cr}_2 G(w,x),\, \mathrm{cr}_1 G(x))$$
$$\oplus\, D_1^1\mathrm{cr}_2 F(D_1^1 D_1^2\mathrm{cr}_2 G(v,v),\, \mathrm{cr}_1 G(x))\oplus D_1^1\mathrm{cr}_2 F(D_1^1 D_1^2\mathrm{cr}_3 G(v,v,x),\, \mathrm{cr}_1 G(x)),$$

and the eighth summand of (34) is

$$
\begin{aligned}
&D_1^1\mathrm{cr}_3F(\Delta_2G(w,v;x),\mathrm{cr}_1G(x),G(0))\\
&\quad\simeq D_1^1\mathrm{cr}_3F(D_1G(w)\oplus D_1^1\mathrm{cr}_2G(w,x)\oplus D_1^1D_1^2\mathrm{cr}_2G(v,v)\\
&\qquad\oplus D_1^1D_1^2\mathrm{cr}_3G(v,v,x),\mathrm{cr}_1G(x),G(0))\\
&\quad\simeq D_1^1\mathrm{cr}_3F(D_1G(w),\mathrm{cr}_1G(x),G(0))\oplus D_1^1\mathrm{cr}_3F(D_1^1\mathrm{cr}_2G(w,x),\mathrm{cr}_1G(x),G(0))\\
&\qquad\oplus D_1^1\mathrm{cr}_3F(D_1^1D_1^2\mathrm{cr}_2G(v,v),\mathrm{cr}_1G(x),G(0))\\
&\qquad\oplus D_1^1\mathrm{cr}_3F(D_1^1D_1^2\mathrm{cr}_3G(v,v,x),\mathrm{cr}_1G(x),G(0)).
\end{aligned}
$$

Putting all of these expansions together, the expansion for the right-hand side is

$$\Delta_2F(\Delta_2G(w,v;x),\Delta_1G(v;x);G(x))\simeq$$

$$D_1F(D_1G(w)) \tag{35}$$

$$\oplus\, D_1F(D_1^1D_1^2\mathrm{cr}_2G(v,v)) \tag{36}$$

$$\oplus\, D_1F(D_1^1D_1^2\mathrm{cr}_3G(v,v,x)) \tag{37}$$

$$\oplus\, D_1F(D_1^1\mathrm{cr}_2G(w,x)) \tag{38}$$

$$\oplus\, D_1^1D_1^2\mathrm{cr}_2F\,(D_1G(v),D_1G(v)) \tag{39}$$

$$\oplus\, D_1^1D_1^2\mathrm{cr}_2F\left(D_1G(v),D_1^1\mathrm{cr}_2G(v,x)\right) \tag{40}$$

$$\oplus\, D_1^1D_1^2\mathrm{cr}_2F\left(D_1^1\mathrm{cr}_2G(v,x),D_1G(v)\right) \tag{41}$$

$$\oplus\, D_1^1D_1^2\mathrm{cr}_2F\left(D_1^1\mathrm{cr}_2G(v,x),D_1^1\mathrm{cr}_2G(v,x)\right) \tag{42}$$

$$\oplus\, D_1^1D_1^2\mathrm{cr}_3F(D_1G(v),D_1G(v),G(0)) \tag{43}$$

$$\oplus\, D_1^1D_1^2\mathrm{cr}_3F(D_1G(v),D_1^1\mathrm{cr}_2G(v,x),G(0)) \tag{44}$$

$$\oplus\, D_1^1D_1^2\mathrm{cr}_3F(D_1^1\mathrm{cr}_2G(v,x),D_1G(v),G(0)) \tag{45}$$

$$\oplus\, D_1^1D_1^2\mathrm{cr}_3F(D_1^1\mathrm{cr}_2G(v,x),D_1^1\mathrm{cr}_2G(v,x),G(0)) \tag{46}$$

$$\oplus\, D_1^1D_1^2\mathrm{cr}_3F(D_1G(v),D_1G(v),\mathrm{cr}_1G(x)) \tag{47}$$

$$\oplus\, D_1^1D_1^2\mathrm{cr}_3F(D_1G(v),D_1^1\mathrm{cr}_2G(v,x),\mathrm{cr}_1G(x)) \tag{48}$$

$$\oplus\, D_1^1D_1^2\mathrm{cr}_3F(D_1^1\mathrm{cr}_2G(v,x),D_1G(v),\mathrm{cr}_1G(x)) \tag{49}$$

$$\oplus\, D_1^1D_1^2\mathrm{cr}_3F(D_1^1\mathrm{cr}_2G(v,x),D_1^1\mathrm{cr}_2G(v,x),\mathrm{cr}_1G(x)) \tag{50}$$

$$\oplus\, D_1^1D_1^2\mathrm{cr}_4F(D_1G(v),D_1G(v),\mathrm{cr}_1G(x),G(0)) \tag{51}$$

$$\oplus D_1^1 D_1^2 \mathrm{cr}_4 F(D_1 G(v), D_1^1 \mathrm{cr}_2 G(v, x), \mathrm{cr}_1 G(x), G(0)) \tag{52}$$

$$\oplus D_1^1 D_1^2 \mathrm{cr}_4 F(D_1^1 \mathrm{cr}_2 G(v, x), D_1 G(v), \mathrm{cr}_1 G(x), G(0)) \tag{53}$$

$$\oplus D_1^1 D_1^2 \mathrm{cr}_4 F(D_1^1 \mathrm{cr}_2 G(v, x), D_1^1 \mathrm{cr}_2 G(v, x), \mathrm{cr}_1 G(x), G(0)) \tag{54}$$

$$\oplus D_1^1 \mathrm{cr}_2 F(D_1 G(w), G(0)) \tag{55}$$

$$\oplus D_1^1 \mathrm{cr}_2 F(D_1^1 \mathrm{cr}_2 G(w, x), G(0)) \tag{56}$$

$$\oplus D_1^1 \mathrm{cr}_2 F(D_1^1 D_1^2 \mathrm{cr}_2 G(v, v), G(0)) \tag{57}$$

$$\oplus D_1^1 \mathrm{cr}_2 F(D_1^1 D_1^2 \mathrm{cr}_3 G(v, v, x), G(0)) \tag{58}$$

$$\oplus D_1^1 \mathrm{cr}_2 F(D_1 G(w), \mathrm{cr}_1 G(x)) \tag{59}$$

$$\oplus D_1^1 \mathrm{cr}_2 F(D_1^1 \mathrm{cr}_2 G(w, x), \mathrm{cr}_1 G(x)) \tag{60}$$

$$\oplus D_1^1 \mathrm{cr}_2 F(D_1^1 D_1^2 \mathrm{cr}_2 G(v, v), \mathrm{cr}_1 G(x)) \tag{61}$$

$$\oplus D_1^1 \mathrm{cr}_2 F(D_1^1 D_1^2 \mathrm{cr}_3 G(v, v, x), \mathrm{cr}_1 G(x)) \tag{62}$$

$$\oplus D_1^1 \mathrm{cr}_3 F(D_1 G(w), \mathrm{cr}_1 G(x), G(0)) \tag{63}$$

$$\oplus D_1^1 \mathrm{cr}_3 F(D_1^1 \mathrm{cr}_2 G(w, x), \mathrm{cr}_1 G(x), G(0)) \tag{64}$$

$$\oplus D_1^1 \mathrm{cr}_3 F(D_1^1 D_1^2 \mathrm{cr}_2 G(v, v), \mathrm{cr}_1 G(x), G(0)) \tag{65}$$

$$\oplus D_1^1 \mathrm{cr}_3 F(D_1^1 D_1^2 \mathrm{cr}_3 G(v, v, x), \mathrm{cr}_1 G(x), G(0)). \tag{66}$$

5 Proof of the Chain Rule for the Second Directional Derivative

All of the key pieces to prove Theorem 2.11 are built. Specifically, we have expanded the left-hand side of Theorem 2.11 in Sect. 3, and we have expanded the right-hand side in Sect. 4.

In the proof, we will use two cases of the chain rule for abelian functors. First, there is a chain rule for D_1 if the interior functor is reduced.

Lemma 5.1 *[1, Proposition 5.7] If* $G : \mathrm{A} \rightsquigarrow \mathrm{B}$ *and* $F : \mathrm{B} \rightsquigarrow \mathrm{C}$ *are composable functors of abelian categories and* G *is a reduced functor, then there is a chain homotopy equivalence*

$$D_1(F \circ G)(x) \simeq D_1 F \circ D_1 G(x).$$

There is also a chain rule for D_1 if the interior functor is not reduced, but an additional correction term is required.

Lemma 5.2 *[1, Proposition 5.10] If $G : \mathrm{A} \rightsquigarrow \mathrm{B}$ and $F : \mathrm{B} \rightsquigarrow \mathrm{C}$ are composable functors of abelian categories, then there is a chain homotopy equivalence*

$$D_1(F \circ G)(x) \simeq D_1 F \circ D_1 G(x) \oplus D_1^x \mathrm{cr}_2 F(\mathrm{cr}_1 G(x), G(0)).$$

We make a few observations concerning D_1.

Observation 5.3 *Let F be a functor of abelian categories. The linearization of $\mathrm{cr}_1 F$ is chain homotopic to the linearization of F. In other words,*

$$D_1 \mathrm{cr}_1 F(x) \simeq D_1 F(x).$$

Proof Recall that $\mathrm{cr}_1 F(0) \cong 0$ because cross effects are strictly multi-reduced. In order to compute $\mathrm{cr}_1(\mathrm{cr}_1 F(-))(x)$, we consider the definition of the first cross effect of the functor $cr_1 F$:

$$\mathrm{cr}_1(\mathrm{cr}_1 F)(x) \oplus \mathrm{cr}_1 F(0) \cong \mathrm{cr}_1 F(x),$$

and thus $\mathrm{cr}_1 \mathrm{cr}_1 F(x) \cong \mathrm{cr}_1 F(x)$. This further implies that

$$\mathrm{cr}_2(\mathrm{cr}_1 F(-))(x, y) \cong \mathrm{cr}_2 F(x, y).$$

Recall from the definition of the linearization of F,

$$(D_1 F)_k := \begin{cases} C_2^{\times k} F & k \geq 1 \\ \mathrm{cr}_1 F & k = 0 \\ 0 & \text{otherwise} \end{cases}$$

Since $C_2^{\times k} F \cong C_2^{\times k} \mathrm{cr}_1 F$ and $\mathrm{cr}_1 \mathrm{cr}_1 F(x) \cong \mathrm{cr}_1 F(x)$, when we linearize $F(x)$ and $\mathrm{cr}_1 F(x)$, we construct equivalent complexes.

Observation 5.4 *Let F be a functor between abelian categories. The linearization of $D_1 F$ is chain homotopy equivalent to the linearization of F. In other words,*

$$D_1 D_1 F(x) \simeq D_1 F(x).$$

Proof Recall that $D_1 F$ is reduced and degree 1. It follows that $\mathrm{cr}_1 D_1 F(x) \simeq D_1 F(x)$ and $\mathrm{cr}_2 D_1 F(x, y) \simeq 0$. If we linearize $D_1 F$, then we have

$$(D_1 D_1 F)_k := \begin{cases} 0 & k \geq 1 \\ D_1 F & k = 0 \\ 0 & \text{otherwise,} \end{cases}$$

which is equivalent to $D_1 F$.

Now, we proceed with the proof of the main theorem.

Theorem 2.11 *Given two composable functors of abelian categories $G : A \rightsquigarrow B$ and $F : B \rightsquigarrow C$ with object x, v, and w in A, there is a chain homotopy equivalence*

$$\Delta_2(F \circ G)(w, v; x) \simeq \Delta_2 F(\Delta_2 G(w, v; x), \Delta_1 G(v; x); G(x)).$$

Proof We will show homotopy equivalence by matching the summands on the left-hand side (terms (2) through (32)) with their homotopy equivalents on the right-hand side (terms (35) through (66)). The justifications for equivalence between these terms are very similar. With this in mind, we will prove just one case of each type and list the remainder of the pairs of terms.

Type 1: (4) $\simeq$ (56). Only Lemma 5.1 is needed.

$$\begin{aligned}
(4) &= D_1^w \mathrm{cr}_2 F(\mathrm{cr}_2 G(w, x), G(0)) \\
&= D_1(\mathrm{cr}_2 F(-, G(0)) \circ \mathrm{cr}_2 G(-, x))(w) \\
&\simeq D_1^1 \mathrm{cr}_2 F(D_1^1 \mathrm{cr}_2 G(w, x), G(0)) \\
&= (56).
\end{aligned}$$

The proofs of (6) $\simeq$ (60) and (8) $\simeq$ (64) are similar.

Type 2: (3) $\simeq$ (38). Lemma 5.1 is used, followed by Observation 5.3.

$$\begin{aligned}
(3) &= D_1^w \mathrm{cr}_1 F(\mathrm{cr}_2 G(w, x)) \\
&= D_1(\mathrm{cr}_1 F(-) \circ \mathrm{cr}_2 G(-, x))(w) \\
&\simeq D_1 \mathrm{cr}_1 F(-) \circ D_1 \mathrm{cr}_2 G(-, x)(w) \\
&= D_1 \mathrm{cr}_1 F(D_1^1 \mathrm{cr}_2 G(w, x)) \\
&\simeq D_1 F(D_1^1 \mathrm{cr}_2 G(w, x)) \\
&= (38).
\end{aligned}$$

The proofs of (5) $\simeq$ (59) and (7) $\simeq$ (63) are similar.

Type 3: (2) $\simeq$ (35) $\oplus$ (55). Lemma 5.2 is applied, followed by Lemma 5.1 and Observation 5.3.

$$\begin{aligned}
(2) &= D_1(F \circ G)(w) \\
&\simeq D_1 \circ D_1 G(w) \oplus D_1^w \mathrm{cr}_2 F(\mathrm{cr}_1 G(w), G(0)) \\
&= D_1 F(D_1 G(w)) \oplus D_1(\mathrm{cr}_2 F(-, G(0)) \circ \mathrm{cr}_1 G(-))(w) \\
&\simeq D_1 F(D_1 G(w)) \oplus D_1^1 \mathrm{cr}_2 F(D_1 \mathrm{cr}_1 G(w), G(0)) \\
&\simeq D_1 F(D_1 G(w)) \oplus D_1^2 \mathrm{cr}_2 F(D_1 G(w), G(0)) \\
&= (35) \oplus (55).
\end{aligned}$$

Type 4: (10) $\simeq$ (57). Lemma 5.2 is applied twice, as well as Observation 5.4:

$$\begin{aligned}
(10) &= D_1^v D_1^{\bar{v}} \mathrm{cr}_2 F(\mathrm{cr}_2 G(v, \bar{v}), G(0)) \\
&= D_1^v [D_1(\mathrm{cr}_2 F(-, G(0)) \circ \mathrm{cr}_2 G(v, -))(\bar{v})] \\
&\simeq D_1^v [D_1^1 \mathrm{cr}_2 F(D_1^2 \mathrm{cr}_2 G(v, \bar{v}), G(0))] \\
&= D_1 [D_1^1 \mathrm{cr}_2 F(-, G(0)) \circ D_1^2 \mathrm{cr}_2 G(-, \bar{v})](v) \\
&\simeq D_1^1 D_1^1 \mathrm{cr}_2 F(D_1^1 D_1^2 \mathrm{cr}_2 G(v, \bar{v}), G(0)) \\
&\simeq D_1^1 \mathrm{cr}_2 F(D_1^1 D_1^2 \mathrm{cr}_2 G(v, \bar{v}), G(0)) \\
&= (57).
\end{aligned}$$

The proofs of (14) $\simeq$ (58), (15) $\simeq$ (61), (16) $\simeq$ (62), (19) $\simeq$ (42), (20) $\simeq$ (65), (21) $\simeq$ (66), (24) $\simeq$ (46), (28) $\simeq$ (50), and (32) $\simeq$ (54) are similar.

Type 5: (9) $\simeq$ (36). Lemma 5.2 is applied twice, as well as Observations 5.4 and 5.3:

$$\begin{aligned}
(9) &= D_1^v D_1^{\bar{v}} \mathrm{cr}_1 F(\mathrm{cr}_2 G(v, \bar{v}) \\
&= D_1^v [D_1(\mathrm{cr}_1 F \circ \mathrm{cr}_2 G(v, -))(\bar{v})] \\
&\simeq D_1^v [D_1 \mathrm{cr}_1 F(D_1^2 \mathrm{cr}_2 G(v, \bar{v}))] \\
&= D_1 [D_1 \mathrm{cr}_1 F \circ D_1^2 \mathrm{cr}_2 G(-, \bar{v})](v) \\
&\simeq D_1 D_1 \mathrm{cr}_1 F(D_1^1 D_1^2 \mathrm{cr}_2 G(v, \bar{v})) \\
&\simeq D_1 \mathrm{cr}_1 F(D_1^1 D_1^2 \mathrm{cr}_2 G(v, \bar{v})) \\
&\simeq D_1 F(D_1^1 D_1^2 \mathrm{cr}_2 G(v, \bar{v})) \\
&= (36).
\end{aligned}$$

The proofs of (11) $\simeq$ (39), (12) $\simeq$ (43), (13) $\simeq$ (37), (17) $\simeq$ (40), (18) $\simeq$ (41), (22) $\simeq$ (44), (23) $\simeq$ (45), (25) $\simeq$ (47), (26) $\simeq$ (48), (27) $\simeq$ (49), (29) $\simeq$ (51), (30) $\simeq$ (52), and (31) $\simeq$ (53) are similar.

6 Conclusion

We proved the chain rule formula for the second higher order directional derivative using primarily properties of linearization and cross effects. This result gave the authors of [1] hope that their definition of higher order directional derivatives of functors would produce a higher order directional derivative chain rule,

$$\begin{aligned}
\Delta_n(F \circ G)(v_n, \ldots, v_1; x_0) \simeq \Delta_n F\big(&\Delta_n G(v_n, \ldots, v_1; x_0), \ldots, \\
&\Delta_1 G(v_1; x_0); G(x_0)\big),
\end{aligned}$$

which mirrors the analogous result for functions (see [3, Theorem 3]). The proof strategy used in this paper does not provide a clear inductive procedure that could lead to the more general result.

Thus, more sophisticated machinery was developed to prove the higher order directional derivative chain rule for functors of abelian categories [1, Theorem 8.1].

Acknowledgements The authors would like to thank the Banff International Research Station (host of the second Women in Topology workshop) and the Pacific Institute for Mathematical Sciences for providing us with the opportunity to collaborate along with the other authors of [1]. We would also like to express our gratitude to Kristine Bauer, Brenda Johnson, and Emily Riehl for their guidance and support throughout this project.

References

1. K. Bauer, B. Johnson, C. Osborne, E. Riehl, A. Tebbe, Directional derivatives and higher order chain rules for abelian functor calculus. Topology Appl. **235**, 375–427 (2018)
2. S. Eilenberg, S. Mac Lane, On the groups $H(\Pi, n)$. II. Methods of computation. Ann. Math. (2) **60**, 49–139 (1954)
3. H.-N. Huang, S.A.M. Marcantognini, N.J. Young, Chain rules for higher derivatives Math. Intell. **28**(2), 61–69 (2006)
4. B. Johnson, R. McCarthy, Deriving calculus with cotriples. Trans. Am. Math. Soc. **356**(2), 757–803 (2004)

DNA Topology Review

Garrett Jones and Candice Reneé Price

Abstract DNA holds the instructions for an organism's development, reproduction, and, ultimately, death. It encodes much of the information a cell needs to survive and reproduce. It is important for inheritance and coding for proteins, and contains the genetic instruction guide for life and its processes. But also, DNA of an organism has a complex and interesting topology. For information retrieval and cell viability, some geometric and topological features of DNA must be introduced, and others quickly removed. Proteins perform these amazing feats of topology at the molecular level; thus, the description and quantization of these protein actions require the language and computational machinery of topology. The use of tangle algebra to model the biological processes that give rise to knotting in DNA provides an excellent example of the application of topological algebra to biology. The tangle algebra approach to knotting in DNA began with the study of the site-specific recombinase $Tn3$ resolvase. This chapter is a summary of some basic knot theory and biology. We then describe the tangle model developed by Ernst and Sumners using the $Tn3$ resolvase as an example. We conclude with applications of the tangle model to other biological problems.

Subject Classification 2010 57M25

1 Introduction

The DNA of any organism has a complex and interesting topology. One can take the view of it as two very long strands; as closed curves that are intertwined millions of times, perhaps linked to other closed curves or tied into knots; and,

G. Jones
Department of Mathematical Sciences, University of Wisconsin-Stevens Point, Stevens Point, WI, USA

C. R. Price (✉)
Department of Mathematics, University of San Diego, San Diego, CA, USA
e-mail: cprice@sandiego.edu

A. Deines et al. (eds.), *Advances in the Mathematical Sciences*, Association for Women in Mathematics Series 15, https://doi.org/10.1007/978-3-319-98684-5_8

subjected to supercoiling in order to convert it into a compact form for information storage. For information retrieval and cell viability, some geometric and topological features must be introduced, and others quickly removed. Some proteins preserve the topology by passing one strand of DNA through another via a protein-bridged transient break in the DNA. This protein action plays a crucial role in cell metabolism, transcription, and replication. Other proteins break the DNA and recombine the ends by exchanging them to help regulate the expression of specific genes, mediate viral insertion into and deletion from the host genome, mediate transposition and repair of DNA, and generate antibody and genetic diversity. These proteins are performing important and incredible feats of topology at the molecular level; thus, the description and quantization of these protein actions requires the language and computational machinery of topology.

The topological approach to enzymology is an indirect method in which the descriptive and analytical powers of topology are employed in an effort to infer the structure of active protein–DNA complexes *in vitro* and *in vivo*. In the topological approach to enzymology experimental protocol, molecular biologists react circular DNA substrate with protein and capture protein signature in the form of changes in the geometry (supercoiling) and topology (knotting and linking) of the circular substrate. The mathematical problem is then to deduce protein mechanism and synaptic complex structure from these observations. The mathematics of topological objects, such as knots and tangles, are then used to solve these problems.

This chapter will discuss the background information of DNA topology by providing the definitions needed from knots and tangles. It will describe the tangle model: a developed sets of experimentally observable topological parameters with which to describe and compute protein mechanism and the structure of the active protein–DNA complex. Because, one of the important unsolved problems in biology is the three-dimensional structure of proteins, DNA, and active protein–DNA complexes in solution (in the cell), and the relationship between structure and function, this model utilizes the mathematics of knots and tangle to provide some solutions. It is the 3-dimensional shape in solution which is biologically important, but difficult to determine. The chapter will conclude with a brief discussion of some of the results utilizing the tangle model.

2 Knots and Links

Although knots have been used since the dawn of humanity, the mathematical study of knots is just under 300 years old. Not only has knot theory grown theoretically in that time, the fields of physics, chemistry, and molecular biology have provided many applications of mathematical knots.

A **knot** is defined as a closed, nonintersecting curve in $\mathbb{R}^3$. Formally, it is the embedding of a circle in three dimensions (Fig. 1). Intuitively, a knot can be simply thought of as a loop of rope with no end and no beginning.

A **link** (defined as a **catenane** by biologists) is a finite union of knots properly embedded in three-dimensional space. Each of these knots, which may be trivial,

Fig. 1 Examples of simple alternating knots

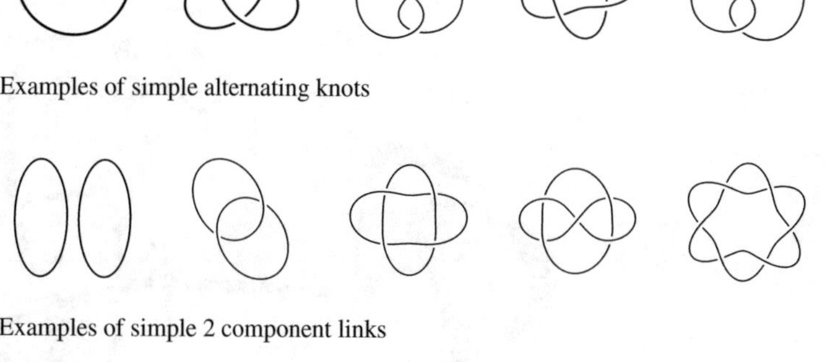

Fig. 2 Examples of simple 2 component links

Fig. 3 Ambiguous and problematic intersections not allowed in knot diagrams

Fig. 4 A polygonal projection and a smooth projection of the knot with 3 crossings has eight possible knot diagrams, two are shown here

is known as a **component** of the link. We can view a knot as a 1-component link. From here, when discussing a property of the class of links of 1 or more component, we will use the terminology "link." When discussing a 1-component link property only, we will refer to the object as a "knot" (Fig. 2).

A **link projection** is the two-dimensional image of the three-dimensional link projected onto a plane. At each double point in the projection (a crossing involving only two line segments), it is not clear which portion of the link crosses over and which crosses under. To show this, gaps are left in the projection. At a crossing, the strand of the knot at the top of the crossing, represented by a solid line segment, is called the **overcrossing**. The strand that is at the bottom of the crossing is called the **undercrossing**, represented by a broken line segment.

It is known that problematic intersections (see Fig. 3) can be avoided so that all intersections correspond to double points. A link projection drawn with these criteria is called a **link diagram**. Knots and links are studied through their diagrams. Links that have diagrams that can be drawn using a finite number of polygonal circuits (i.e., closed paths) in three-dimensional space are called **tame** (Fig. 4). All other links are known as **wild** (Fig. 5). Most applications of knot theory concern only tame links, so we will only focus on this class of links.

Fig. 5 Diagrams of wild knot. Courtesy of [35]

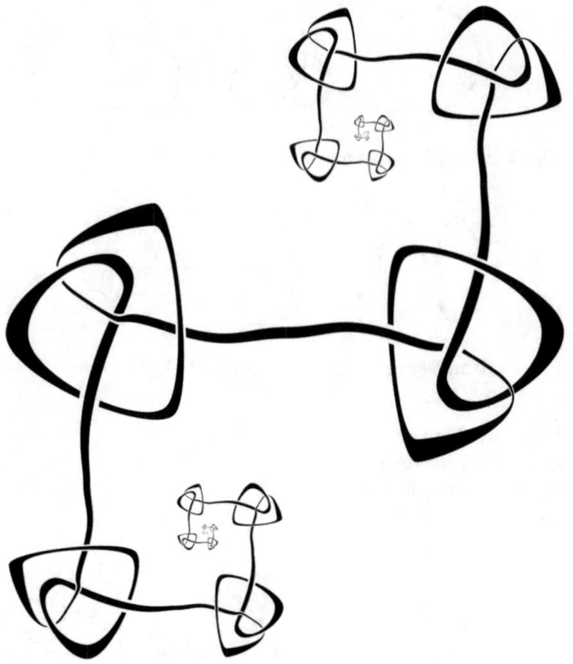

We say two links, K_1 and K_2, are **equivalent** if there is an ambient isotopy between them. An **ambient isotopy** can be described as a continuous deformation from one link diagram (K_1) to the other (K_2). It allows us to stretch, bend, and twist the link however we would like; we just cannot cut it. Mathematically, two links, K_1 and K_2, are **ambient isotopic** if there is an isotopy $h : \mathbb{R}^3 \times [0, 1] \to \mathbb{R}^3$ such that $h(s, i) = h_i(s)$ is a homeomorphism for all i where $h_0(K_1) = K_1$ and $h_1(K_1) = K_2$ [12]. If two knots are equivalent, we refer to these knots as knots of the same **knot type**, K, where K is the equivalence class under this equivalence relation.

In 1926, Kurt Reidemeister proved that if we have two distinct diagrams of K, we can go from one diagram to the other using **Reidemeister moves**, as described in Theorem 2.1.

Theorem 2.1 (Reidemeister [33]) *Two link diagrams K_1 and K_2 are equivalent if and only if they can be obtained from one another by a finite sequence of planar isotopies and the three moves: twist, poke, and slide (Fig. 6).*

Given two knots, K_1 and K_2, a knot $K_3 = K_1 \# K_2$ can be constructed as seen in Fig. 7. This knot is known as the **connected sum** of K_1 and K_2. A knot that cannot be constructed in this manner using nontrivial knots is called **prime**. All prime knots will be referred to as they are given in Rolfsen's table of prime knots [34].

Links can be split into two groups: alternating and nonalternating. A link is called **alternating** if it has a diagram in which, when traveling around each component of

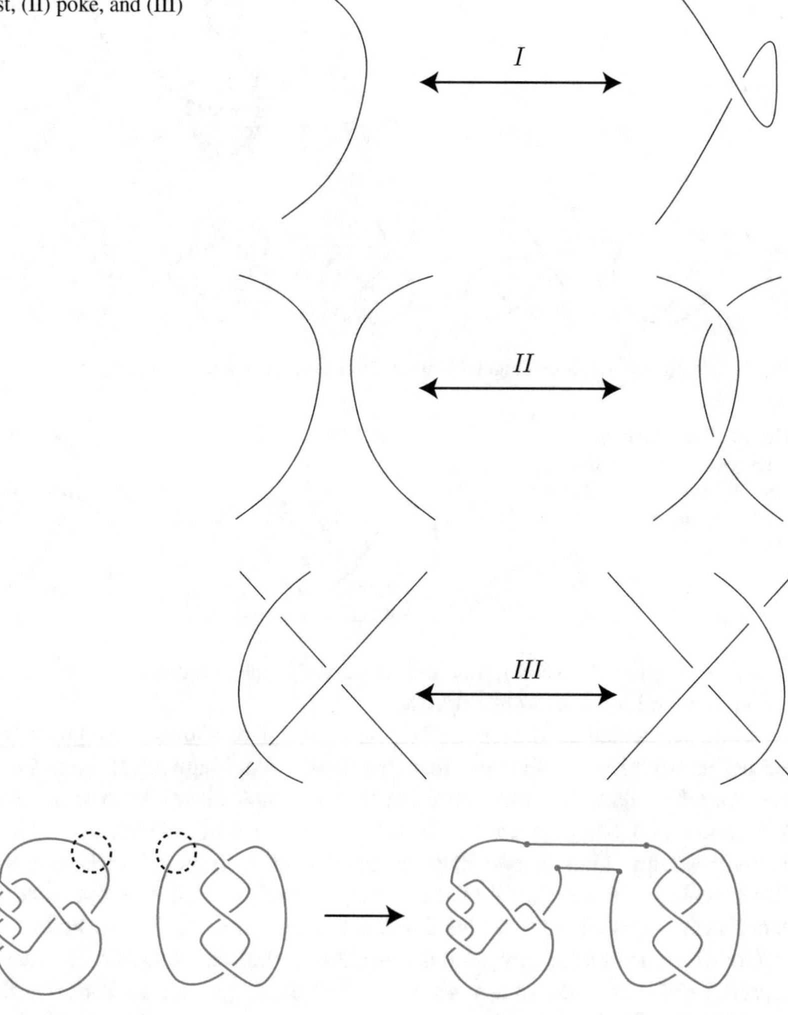

Fig. 6 Reidemeister moves: (I) twist, (II) poke, and (III) slide

Fig. 7 The connected sum of knots 5_2 and 3_1

the link, one alternates between overcrossing and undercrossings. A **nonalternating link** is one that is not alternating (i.e., every diagram has at least two overcrossings or two undercrossing in a row when traveling around the link).

An **oriented link** is a link for which each component has been given an orientation. An oriented link is **invertible** if it can be deformed to be the same link diagram with the opposite orientation [1]. The **mirror image** of a link, $\overline{L}$, is obtained by changing every overcrossing in the link to an undercrossing and vice versa. If L is equivalent to its mirror image, then we call L **amphicheiral** (or **achiral**). If L is not equivalent to $\bar{L}$, then it is **chiral**. Although not all links are achiral, most tables

Fig. 8 4_1 knot and its mirror image

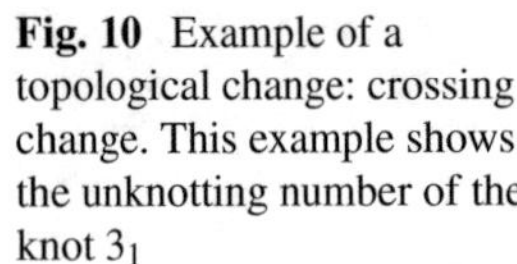

Fig. 9 Examples of minimum regular diagrams of the first five knots

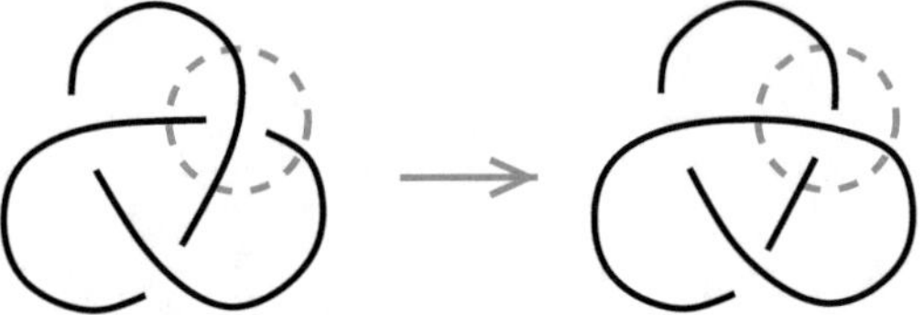

Fig. 10 Example of a topological change: crossing change. This example shows the unknotting number of the knot 3_1

do not distinguish between a link and its mirror image. One example of a knot that is amphicheiral is the 4_1 knot (Fig. 8).

While Reidemeister moves are helpful to see if two links are equivalent, they are not as useful when showing that two links are not equivalent. Link invariants are utilized to show in-equivalence between two link diagrams. A **link invariant** is a specific quality of a knot or link type that does not change its value under ambient isotopy. Thus, if two links are equivalent, then their invariants are equal. Unfortunately, for a majority of invariants, the other direction is not usually true: equal invariant values for two link diagrams do not imply equivalent links.

One example of a link invariant is the minimum crossing number. The **minimum crossing number** is the minimum number of crossings over all knot diagrams of the knot type (Fig. 9).

Some invariants keep count of the number of topological changes made to a link diagram. Looking at a knot diagram, exchange locally overcrossings and undercrossings. This type of alteration may change the knot type. The **unknotting number** is the least number of crossing changes in a diagram of a knot to get to the trivial knot, minimized over all diagrams (Fig. 10).

The **linking number** is a link invariant for links of two or more components. It is calculated using the **crossing sign convention** (Fig. 11). The linking number is calculated by taking the sum of the crossing signs of each crossing between the different components of the link and dividing by two.

While the previous invariants give numerical quantities, other invariants can associate a polynomial to a knot type: the Alexander polynomial, the Jones

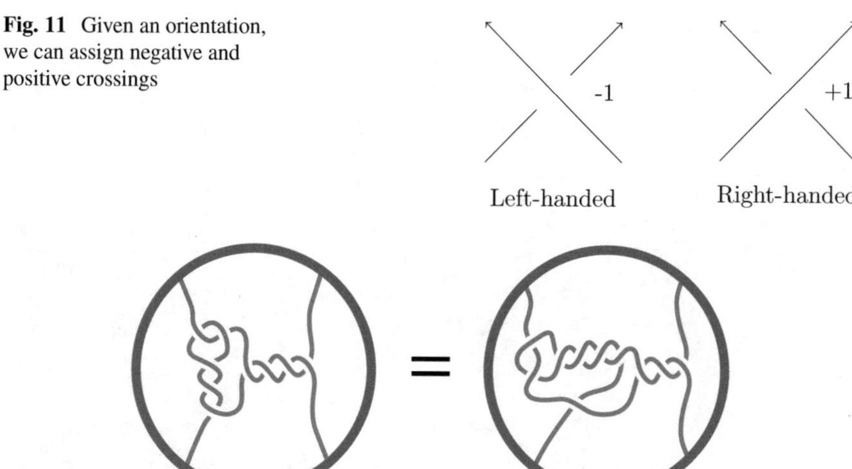

Fig. 11 Given an orientation, we can assign negative and positive crossings

Fig. 12 Equivalent tangles

polynomial, and the HOMFLY-PT polynomial [3, 19, 25] or associate to a knot diagram even more complicated algebraic structures like chain complexes of abelian groups: Khovanov Homology and Knot Floer Homology [26, 30].

3 Tangles

An n**-string tangle** is defined as a pair (B, t) of a 3-dimensional ball B and a collection of disjoint, simple, properly embedded arcs, denoted t. An n-string tangle is formed by placing $2n$ points on the boundary of B and attaching n nonintersecting curves inside B such that $\partial B \cap t = \partial t$. We consider tangles $T_1 = (B, t_1)$ and $T_2 = (B, t_2)$ to be **equivalent** if there is an ambient isotopy of one tangle to the other keeping the boundary of the ball fixed (Fig. 12).

This work will focus on **2-string tangles**. As part of the definition, we consider a 2-string tangle to be a pair (B, t) and a homeomorphism sending (B, t) to the unit ball in $\mathbb{R}^3$. We send the four endpoints of the arcs to the four equatorial points NW, NE, SE, and SW in the yz-plane described in $\mathbb{R}^3$ as the points:

$$NE : \left(0, \frac{1}{\sqrt{2}}, \frac{1}{\sqrt{2}}\right) \; NW : \left(0, -\frac{1}{\sqrt{2}}, \frac{1}{\sqrt{2}}\right)$$

$$SE : \left(0, \frac{1}{\sqrt{2}}, -\frac{1}{\sqrt{2}}\right) \; SW : \left(0, -\frac{1}{\sqrt{2}}, -\frac{1}{\sqrt{2}}\right).$$

The simplest tangles are the zero tangle, denoted (0); the (∞)-tangle, denoted (0, 0); the positive one tangle, denoted (1); and, the negative one tangle, denoted (-1) (Fig. 13).

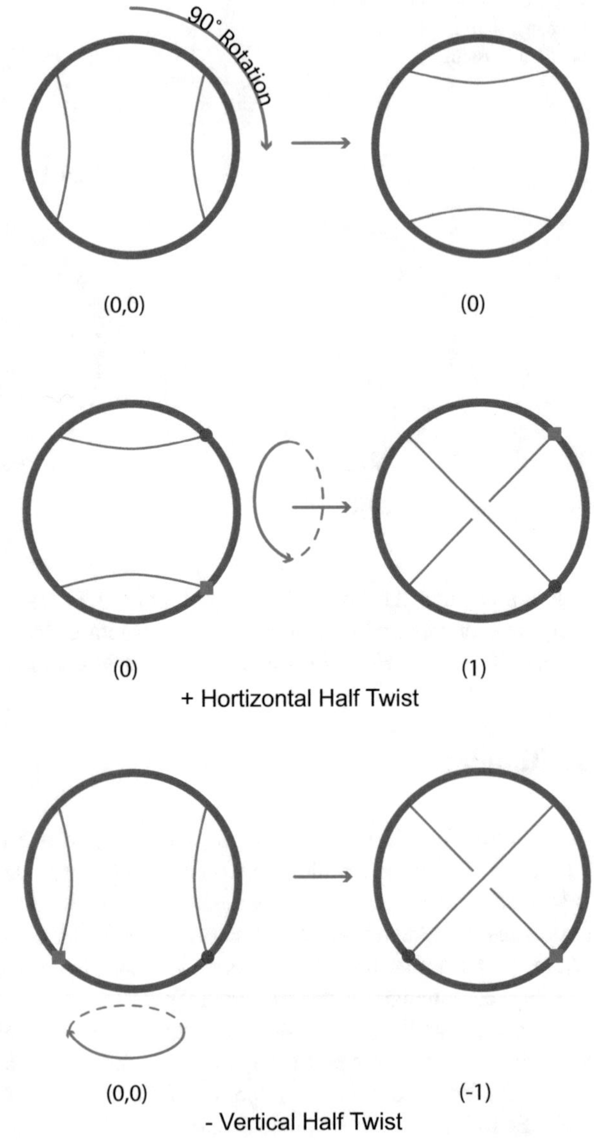

Fig. 13 Simplest 2-string tangles: the (∞)-tangle, $(0, 0)$, is a 90° rotation of the zero tangle, (0). The positive one tangle, (1) is shown here as a positive horizontal half twist added to (0). The negative one tangle, (-1) is shown as a negative vertical half twist added to $(0, 0)$

We can take the **sum** of two tangles, T_1 and T_2, creating a new tangle, $T_1 + T_2$ (Fig. 14). Another tangle operation is the **numerator closure**, which connects the northern endpoints with the shortest arc on the exterior of B and similarly the southern endpoints, resulting in a knot or link denoted $N(T)$. We can also perform this operation on a sum of tangles (Fig. 15).

This work will focus on 2-string tangle rational tangles. A 2-string tangle is **rational** if it is ambient isotopic to the zero tangle, allowing the boundary of the

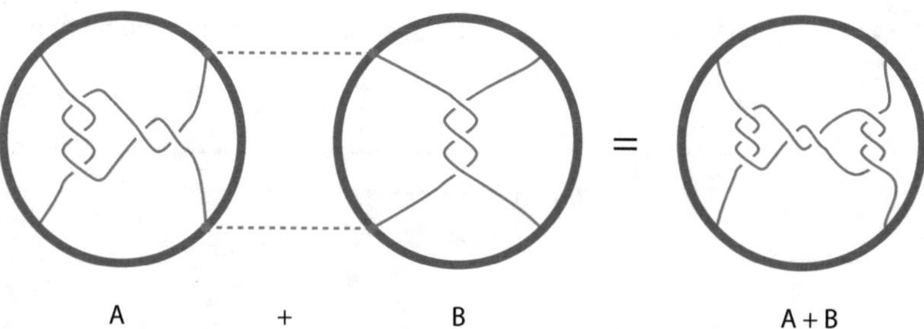

Fig. 14 Sum of two tangles

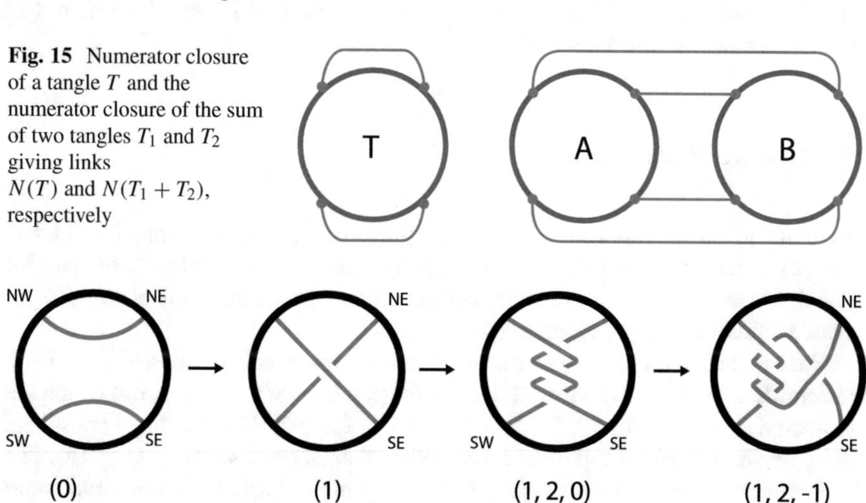

Fig. 15 Numerator closure of a tangle T and the numerator closure of the sum of two tangles T_1 and T_2 giving links $N(T)$ and $N(T_1 + T_2)$, respectively

Fig. 16 Creating a rational tangle with Conway vector $(1, 2, -1)$

3-ball to move. A rational tangle diagram is created by starting with the zero tangle and interchanging the NE and SE boundary points a finite number of times creating **horizontal half twists**. Then, continue construction by interchanging the SW and SE boundary points a finite number of times creating **vertical half twists**. Continue in this manner, alternating between adding vertical and horizontal twists (Fig. 16).

John Conway associated to each rational 2-string tangle an extended rational number, $\frac{m}{n} \in \mathbb{Q} \cup \{\infty\}$, stating that there exists a 1–1 correspondence [11]. This number can be calculated using the **Conway vector**, denoted $(a_1, a_2, \ldots, a_i)$ where we choose i to be odd. This finite sequence of integers represents the sequence of moves performed on the zero tangle to produce a rational tangle. (Note: One can start with the (∞) tangle by rotating the zero tangle by 90°.) Each integer represents the number of half twists given to the tangle, alternating between horizontal and vertical, ending with horizontal twists. The sign of the crossing follows that of Fig. 13.

If a tangle T is denoted $T(a_1, a_2, \ldots, a_i)$, then its extended rational number is calculated as:

$$\frac{m}{n} = a_i + \cfrac{1}{a_{i-1} + \cfrac{1}{a_{i-2} + \cfrac{1}{a_{i-3} + \cdots + \cfrac{1}{a_1}}}}$$

The numerator closure of a rational tangle, $\frac{m}{n}$, is referred to as a **2-bridge knot/link** denoted $N\left(\frac{m}{n}\right)$ or $< a_1, a_2, \ldots a_i >$. These links are also referred to as 4-**plats** and **rational knots/links**.

4 Biology Background

A crucial advancement in molecular biology was made when the structure of DNA was determined by James Watson and Francis Crick in 1953. Its structure revealed how DNA can be replicated and provided clues about how a molecule of DNA might encode directions for producing proteins [2].

Nucleic acids consist of a chain of linked units called **nucleotides**. Each nucleotide contains a deoxyribose, a sugar ring made of five carbon atoms which are numbered as seen in Fig. 17. This sugar ring then forms bonds to a single phosphate group between the third and fifth carbon atoms of adjacent sugar rings (Fig. 18). The backbone of a DNA strand is made from alternating phosphate groups and sugar rings. The four bases found in DNA are Adenine (A), Thymine (T), Cytosine (C), and Guanine (G). The shapes and chemical structure of these bases allow hydrogen bonds to form efficiently between A and T and between G and C. These bonds, along with base stacking interactions, hold the DNA strands together [2]. Each base is attached to the first carbon atom in the sugar ring to complete the nucleotide (Fig. 18).

The bonds between the sugars and the phosphate group give a direction to DNA strands. The asymmetric ends of the strands are called the 5′ (**five prime**) and 3′

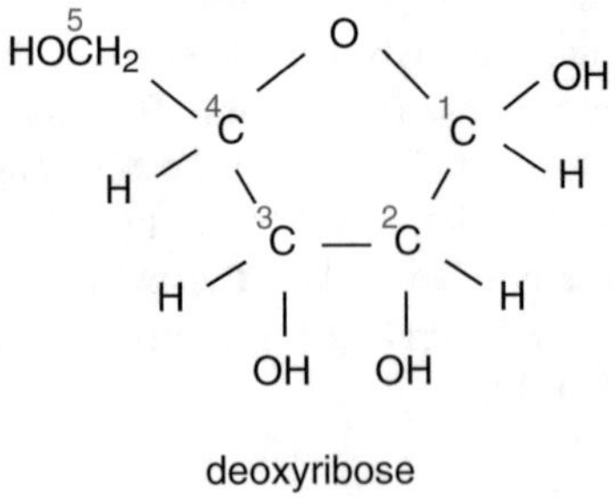

Fig. 17 Sugar ring made of five carbon atoms. Courtesy of [45]

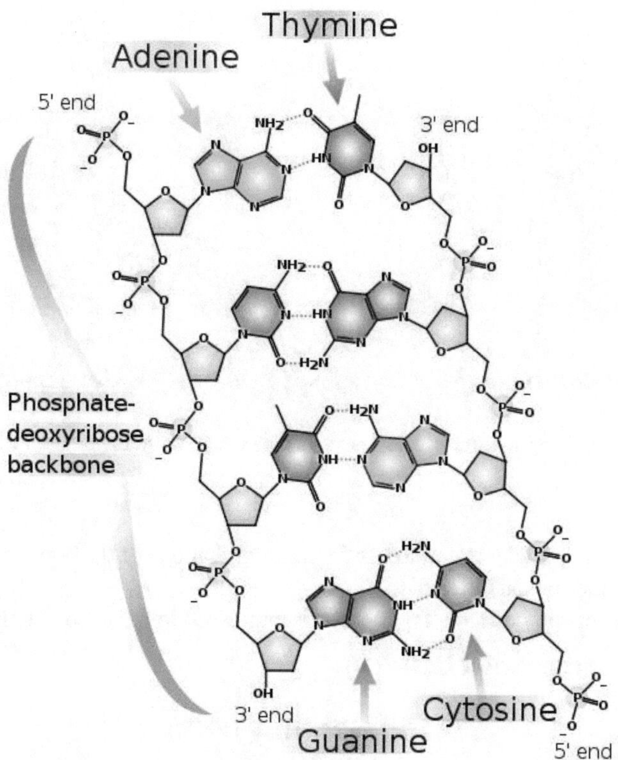

Fig. 18 Deoxyribonucleic acid. Using the direction convention given to DNA strands, we read this sequence as $ACTG$, or equivalently $CAGT$. Courtesy of [46]

(**three prime**) ends, with the $5'$ end having a phosphate group attached to the fifth carbon atom of the sugar ring and the $3'$ end with a terminal hydroxyl group attached to the third carbon atom of the sugar ring (Fig. 18). The direction of the DNA strands is read from $5'$ to $3'$. In a double helix, the direction of one strand is opposite to the direction of the other strand: the strands are **antiparallel** [2].

Besides the standard linear form, a molecule of DNA can take the form of a ring known as **circular DNA**. One way to model circular DNA mathematically is as an annulus, $\mathbf{R}$, an object that is topologically equivalent to $S^1 \times [-1, 1]$. The axis of $\mathbf{R}$ is $S^1 \times \{0\}$. With this model, we can choose an orientation for the axis of $\mathbf{R}$ and use the same orientation on $\partial\mathbf{R}$; thus, the axis and boundary curves of $\mathbf{R}$ have a parallel orientation. Note that this is a different convention than the biology/chemistry orientation. We use geometric invariants twist and writhe, denoted **Tw** and **Wr**, to describe the structure of the circular DNA molecule. **Writhe** can be determined by viewing the axis of $\mathbf{R}$ as a spatial curve and is measured as the average value of the sum of the positive and negative crossings of the axis of $\mathbf{R}$ with itself, averaged over all projections [28]. The sign convention for a crossing is

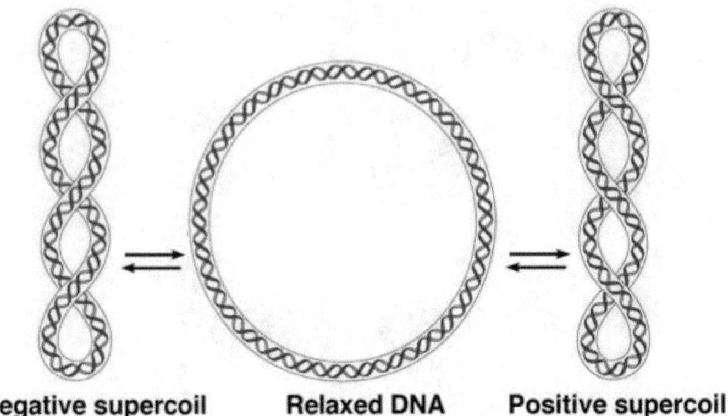

Fig. 19 Cartoon of negative, relaxed, and positive supercoiled DNA. Reproduced with permission from [22]

given in (Fig. 11). **Twist** is defined as the amount that one of the boundary curves of **R** twists around the axis of **R** [4].

One relationship between **Tw** and **Wr** is expressed in the following law:

LAW 4.1 (CONSERVATION LAW [20])

$$\mathbf{Lk(R)} = \mathbf{Tw(R)} + \mathbf{Wr(R)}$$

where **Lk(R)** is the linking number of the oriented link formed by the two boundary curves of **R** with a parallel orientation.

We say that a DNA molecule is **supercoiled** when $\mathbf{Wr} \neq 0$ (Fig. 19). Native circular DNA appears negatively supercoiled under an electron microscope, i.e., $\mathbf{Wr} < 0$ (Fig. 20) [4].

Recall that the structure of DNA is a double-stranded helix, where the four bases are paired and stored in the center of this helix. While this structure provides stability for storing the genetic code, Watson and Crick noted that the two strands of DNA would need to be untwisted in order to access the information stored for transcription and replication [2]. They foresaw that there should be some mechanism to overcome this problem.

4.1 *Transcription and Replication*

DNA can be viewed as two very long strands; as closed curves that are intertwined and perhaps linked to other closed curves or tied into knots, and supercoiled. Thus, the main topologically interesting forms that circular DNA can take: supercoiled,

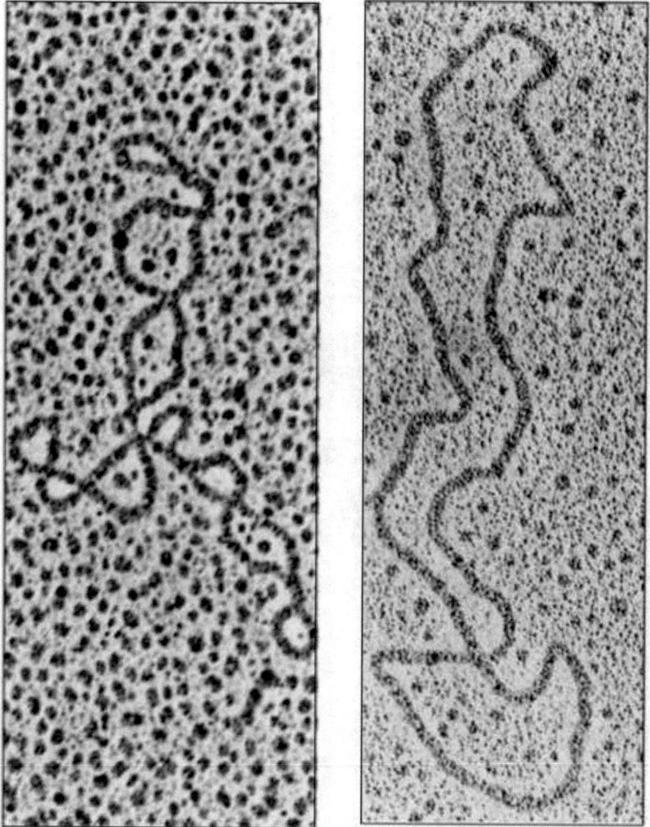

Fig. 20 Two examples of supercoiled DNA seen through an electron microscope. Reproduced with permission from [22]

knotted, linked, or a combination of these. DNA is kept as compact as possible when in the nucleus, and these three states help or hinder this cause. However, when transcription or replication occur, DNA must be accessible [41]. **Ribonucleic acid (RNA)** is a nucleic acid made up of a chain of nucleotides (Fig. 21). There are three main differences between RNA and DNA: (a) RNA contains the sugar ribose, while DNA contains a different sugar, deoxyribose; (b) RNA contains the base uracil (U) in place of the base thymine (T), which is present in DNA; and, (c) RNA molecules are single stranded, but have interesting tertiary structure. **Transcription** is the process of creating a complementary RNA copy of a sequence of DNA. Transcription begins with the unwinding of a small portion of the DNA double helix to expose the bases of each DNA strand. The two strands are then pulled apart creating an opening known as the **transcription bubble**. During this process,

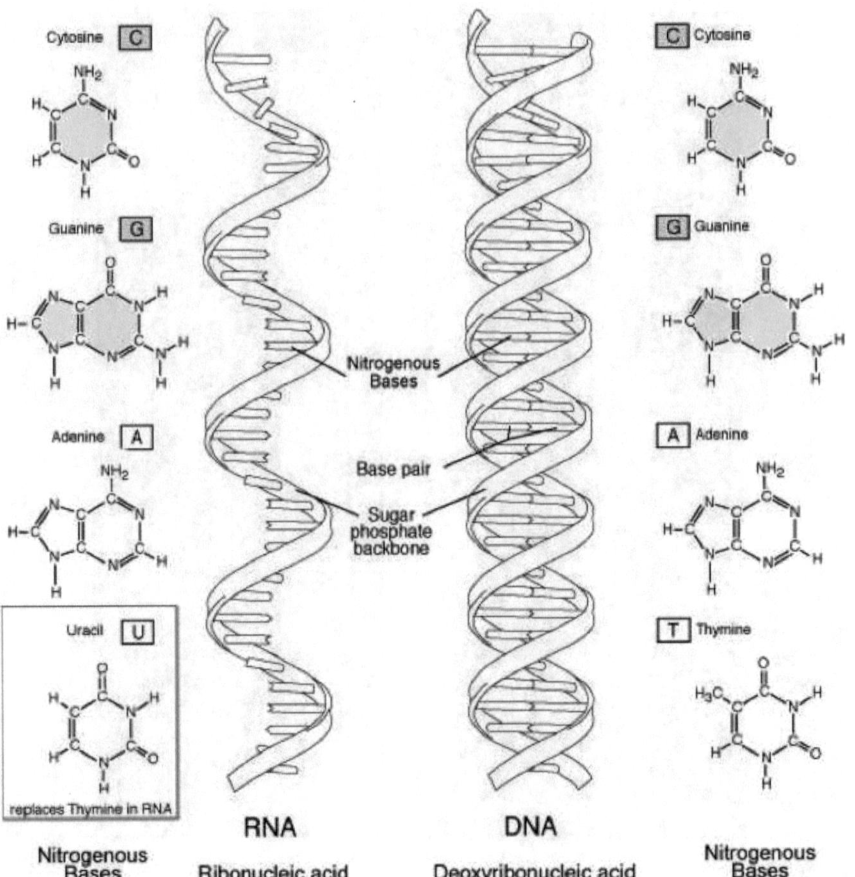

Fig. 21 Like DNA, **ribonucleic acid (RNA)** is a nucleic acid made up of a long chain of nucleotides. Courtesy of [47]

DNA ahead of the transcription bubble becomes positively supercoiled, while DNA behind the transcription bubble becomes negatively supercoiled (Fig. 22).

DNA replication is the process that starts with one DNA molecule and produces two identical copies of that molecule. During replication, the DNA molecule begins to unwind at a specific location and starts the synthesis of the new strands at this location, forming **replication forks** (Fig. 23, left). The DNA ahead of the replication fork becomes positively supercoiled, while DNA behind the replication fork becomes entangled, creating pre-catenanes, a state where the DNA molecules are beginning to form linked DNA molecules (Fig. 23, center). A topological problem occurs at the end of replication, when daughter chromosomes must be fully disentangled before mitosis occurs (Fig. 23, right) [41]. Topoisomerases play an essential role in resolving this problem.

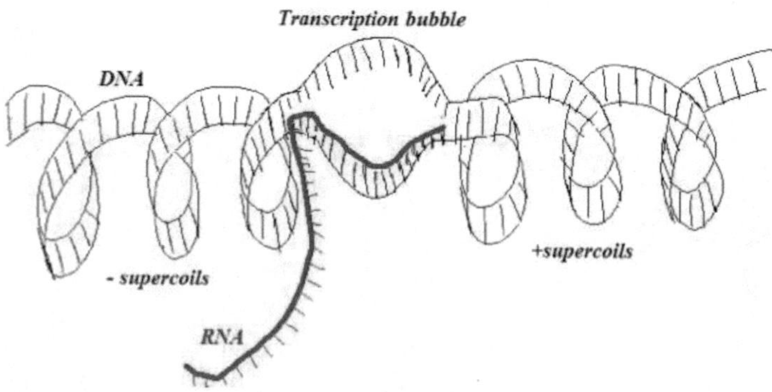

Fig. 22 Transcription-driven supercoils

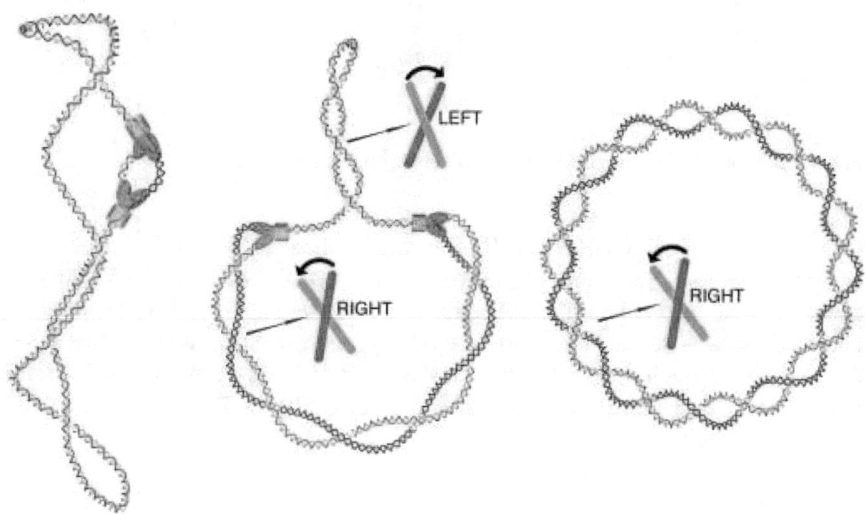

Fig. 23 Topological changes to DNA during replication of circular DNA. The process of replication begins with negatively supercoiled DNA. The replication forks are shown in purple and gold. Partially replicated DNA molecule: the replicated portions of the DNA are interwound with positive (right-handed) crossings, creating a pre-catenane, while the remaining unreplicated DNA is still negatively (left-handed) supercoiled. Completely replicated DNA shown as a DNA catenane with positive (right-handed) crossings. Used with permission from [48]

4.2 Topoisomerase

Topoisomerases are proteins that are involved in the packing of DNA in the nucleus and in the unknotting and unlinking of DNA links that can result from replication and other biological processes. These proteins bind to either single- or double-stranded DNA and cut the phosphate backbone of the DNA. A **type I topoisomerase**

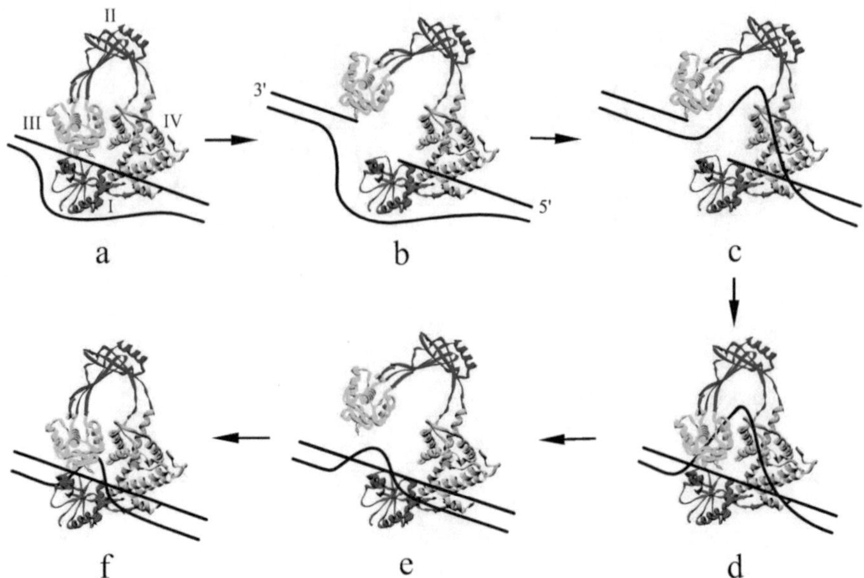

Fig. 24 Schematic of topoisomerase I action. Used with permission from [10]

cuts one strand of a DNA double helix allowing for the reduction or the introduction of stress (Fig. 24). Such stress is introduced or needed when the DNA strand is supercoiled or uncoiled during replication or transcription. **Type II topoisomerase** cuts both phosphate backbones of one DNA double helix, passes another DNA double helix through it, and then reseals the cut strands (Fig. 25). This action does not change the chemical composition and connectivity of DNA, but potentially changes its topology.

4.3 Recombinase

In various biological processes, there often is a need to integrate, excise, or invert portions of a DNA molecule. For example, gene expression is often regulated by the absence or presence of repressor or promoter sites. Inserting a promoter or repressor site can result in the expression, or lack of expression, respectively, of a particular gene. Another example is the insertion of viral DNA into its host cell. Insertion of the viral DNA into the host genome allows it to replicate and continue its life cycle. **Recombination** is a process involving the genetic exchange of DNA where DNA sequences are rearranged by proteins known as **recombinases** [2]. **Site-specific recombination** is an operation on DNA molecules where recombination proteins, **site-specific recombinases**, recognize short specific DNA sequences on the recombining DNA molecules. First, two sequences from the same or different

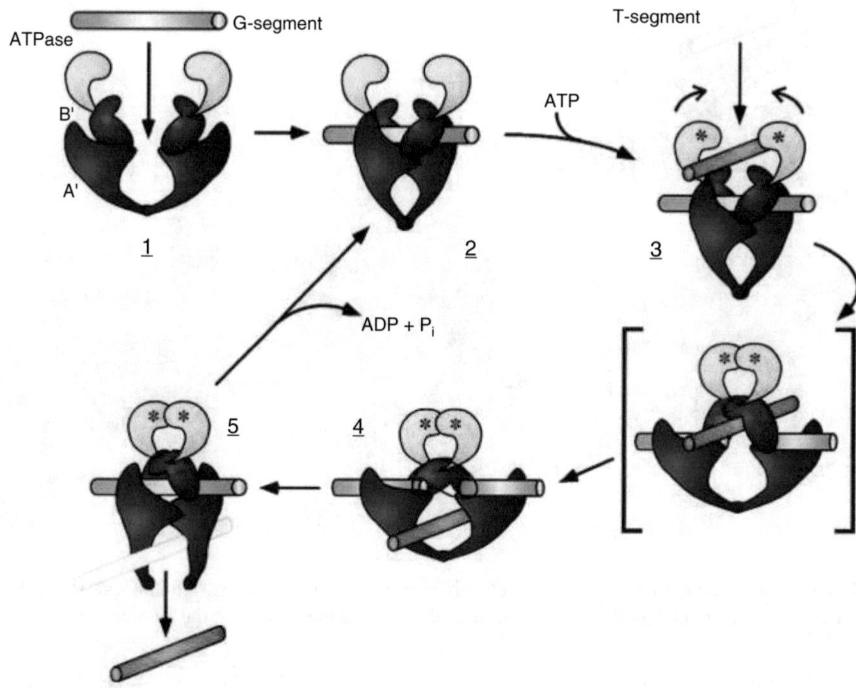

Fig. 25 Schematic of topoisomerase II action. Used with permission from [5]

DNA molecule are drawn together. The recombinase then introduces a break near a specific site, known as a **recombination site**, on the double-stranded DNA molecule. The protein then recombines the ends in some manner and seals the break (Fig. 26). We call this DNA-protein complex a **synaptic complex**. We will call the part of the synaptic complex that consists of only the protein together with the part of the substrate DNA bound to the protein, the local synaptic complex. After synapsis occurs, the recombinase then cleaves the DNA at the recombination sites and rejoins the ends by exchanging them. The specific way in which the exchange occurs is determined by the particular protein [21, 39, 43].

The DNA sequence of a recombination site can be used to give an orientation to this site. When two sites are oriented in the same direction, the sites are called **direct repeats** (Fig. 27). Recombinase action on direct repeats normally results in a change in the number of components, taking knots to links and links to knots or a link with a higher number of components (Fig. 27). If the two sites are oriented in opposite directions, the sites are called **inverted repeats** (Fig. 28). The action of a recombinase on inverted repeats normally results in no change in the number of components (Fig. 28).

There are two families of site-specific recombinases: **tyrosine recombinases** and **serine recombinases**. Tyrosine recombinases break and rejoin one pair of DNA

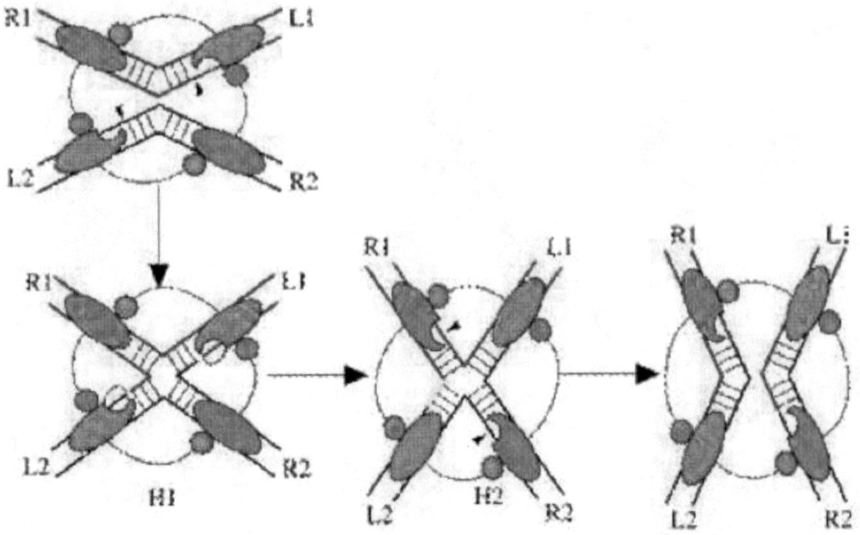

Fig. 26 An example of a site-specific recombinase mechanism where the protein makes breaks one strand of the double helix, recombines it, and then does the same with the other strand

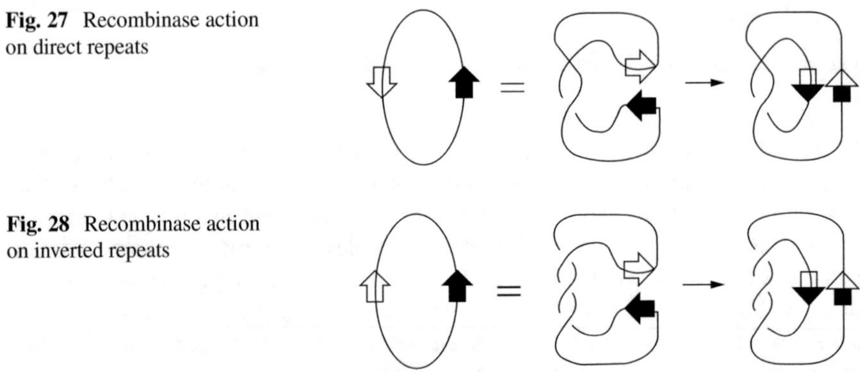

Fig. 27 Recombinase action on direct repeats

Fig. 28 Recombinase action on inverted repeats

strands at a time (Figs. 26, 29). Serine recombinases introduce double-stranded breaks in DNA and then recombines them in some manner (Fig. 30) [36].

5 Tangle Model

DNA encodes much of the information a cell needs to survive and reproduce. If we were to unwind the chromosomes from one human cell and place the DNA strands end to end, it would span approximately 2 m [9]. All of this DNA is packed

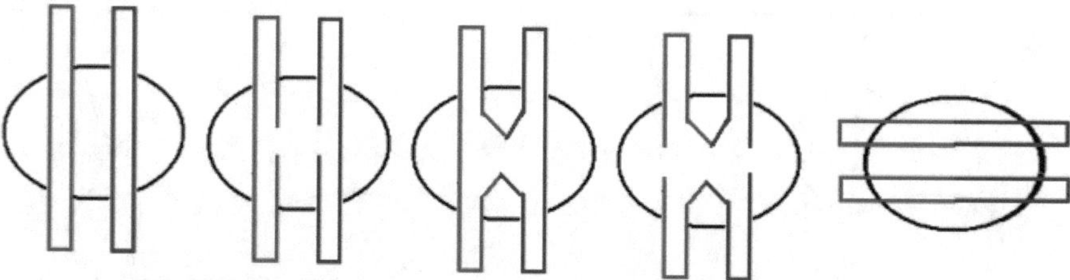

Fig. 29 Schematic of tyrosine recombinase action: single-stranded breaks. We model the tyrosine protein as a black ball, while the double-stranded DNA is modeled by red and blue rectangles

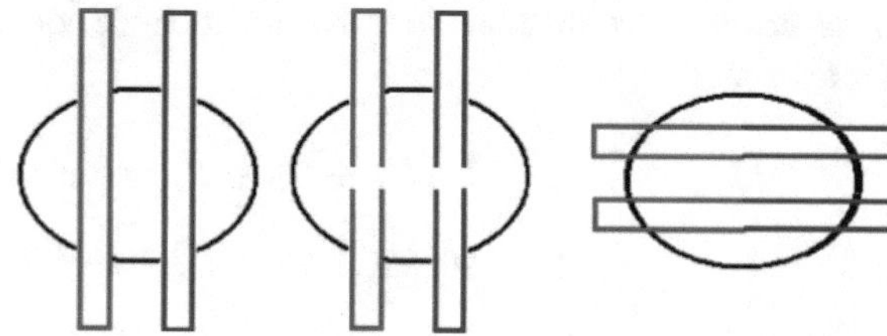

Fig. 30 Schematic of serine recombinase action: double-stranded breaks. We model the serine protein as a black ball, while the double-stranded DNA is modeled by red and blue rectangles

inside the nucleus of a cell whose diameter is measured on the scale of micrometers, that is one thousandth of a millimeter, 0.001 mm. The DNA must not only be arranged to sit inside such a small space, but it must also be organized so that the information it contains is accessible. Inside this complex environment, vital functions like transcription and replication must take place. It is no surprise, then, that various mechanisms have evolved over time to change the structure of the DNA. One mechanism is the action of proteins.

Understanding how a particular protein acts on DNA can be a difficult task. Proteins and their actions cannot be directly observed with the naked eye. Even with electron microscopy, there is not enough detail to see exactly how a particular protein binds to and acts on its substrate. We must rely on well-designed experiments to gain this knowledge. Additional use of mathematical models can help to further clarify the results obtained by experiments and this is exactly what was done to determine the action of a particular tyrosine recombinase called **Tn3 resolvase**.

In the 1990s, C. Ernst and D. Sumners developed the tangle calculus which was then successfully used to model the action of recombinases on circular DNA substrate [18]. In this model, the synaptic complex was represented by the numerator closure of a sum of 2-string tangles. A pair of 2-string tangle, O_b and P, represented the local synaptic complex, that is the protein and bound DNA. The parental tangle, P, contains the site where strand breakage and reunion takes place. The outside bound tangle, O_b, was the rest of the DNA in the local synaptic complex outside of the tangle P. Finally, another 2-string tangle, the outside free tangle, O_f, represented the DNA in the synaptic complex which is free and not bound to the protein. The action of the protein was then modeled as a tangle surgery, where the tangle, P, is replaced by a new tangle R. The knotted products which were observed

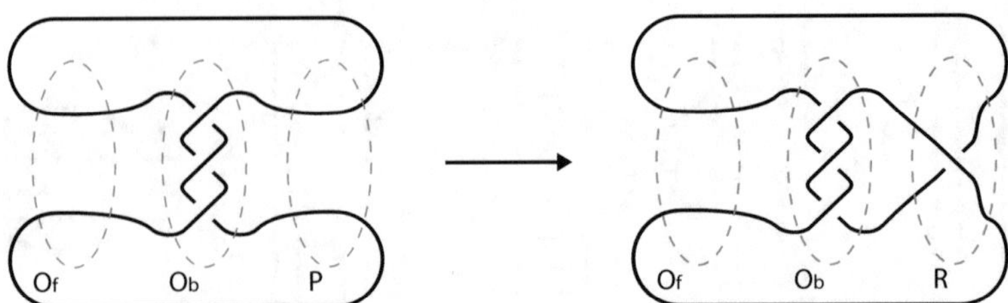

Fig. 31 A schematic of the tangle model. This particular example shows the first round of recombination for $Tn3$ resolvase

in experiments would allow for a system of tangle equations to be set up (see Fig. 31 for a visual):

$$N\left(O_f + O_b + P\right) = \text{substrate}, \tag{5.1}$$

$$N\left(O_f + O_b + R\right) = \text{product}. \tag{5.2}$$

Several assumptions had to be made for this model to work [18, 37]. One assumption was that the local synaptic complex could be modeled with a 2-string tangle that subdivided into the sum of two tangles. It was assumed that the recombination takes place entirely inside the protein ball, while the substrate configuration outside the protein ball remains fixed. The protein mechanism in a single recombination event is assumed constant, and independent of the geometry and topology of the substrate. Also, it was assumed that **processive recombination**, consecutive reactions without releasing its substrate, could be modeled with tangle addition by adding the tangle R for each additional round of recombination:

$$N\left(O_f + O_b + P\right) = \text{substrate}, \tag{5.3}$$

$$N\left(O_f + O_b + R\right) = \text{1st round product}, \tag{5.4}$$

$$N\left(O_f + O_b + R + R\right) = \text{2nd round product}, \tag{5.5}$$

$$\vdots$$

$$N\left(O_f + O_b + \underbrace{R + R + \ldots + R}_{n}\right) = \text{nth round product}. \tag{5.6}$$

Experiments with $Tn3$ resolvase acting on circular DNA substrate, which carried two copies of the recombination site, were carried out and the products of this reaction were observed. Resolvase typically mediates a single recombination event and releases the substrate. The principle product of the experiments were the Hopf link, $\langle 2 \rangle$, which was believed to be the result of this single recombination event. In about one in 20 encounters though, resolvase acts processively. Other products observed were the Fig. 8 knot, $\langle 2, 1, 1\rangle$, the result of two rounds of recombination, the Whitehead link, $\langle 1, 1, 1, 1, 1\rangle$, the result of three rounds of recombination, and the 6_2 knot, $\langle 1, 2, 1, 1, 1\rangle$, the result of four rounds of recombination [42, 44]. Using observations from electron micrographs of the synaptic complex, it was also assumed that O_f was the (0) tangle, thus we can reduce the tangle $O_f + O_b$ to one single tangle O. Using the information above, the following system of tangle equations could be set up:

$$N(O + P) = \langle 1\rangle \text{ (the unknot)}, \tag{5.7}$$

$$N(O + R) = \langle 2\rangle \text{ (the Hopf link)}, \tag{5.8}$$

$$N(O + R + R) = \langle 2, 1, 1\rangle \text{ (the Fig. 8 knot)}, \tag{5.9}$$

$$N(O + R + R + R) = \langle 1, 1, 1, 1, 1\rangle \text{ (the (+) Whitehead link)}. \tag{5.10}$$

Due to the amount of unknowns, it is not possible to explicitly solve for the tangle P; there are the infinite possibilities of a solution for any given O. However, there are biological and mathematical arguments to support the idea that $P = (0)$ [38]. Thus, with this assumption and using only the first three rounds of recombination and the tangle calculus, Ernst and Sumners were able to prove the following theorem about this system of equations:

Theorem 5.1 ([18]) *Suppose that tangles O, P, and R satisfy the following:*

$$N(O + P) = \langle 1\rangle \text{ (the unknot)}, \tag{5.11}$$

$$N(O + R) = \langle 2\rangle \text{ (the Hopf link)}, \tag{5.12}$$

$$N(O + R + R) = \langle 2, 1, 1\rangle \text{ (the Fig. 8 knot)}, \tag{5.13}$$

$$N(O + R + R + R) = \langle 1, 1, 1, 1, 1\rangle \text{ (the (+) Whitehead link)}. \tag{5.14}$$

Then, $\{O; R\} = \{(-3, 0), (1)\}$, and $N(O + R + R + R + R) = \langle 1, 2, 1, 1, 1\rangle$.

Not only did this theorem show that the tangles O and R must equal $(-3, 0)$ and (1), respectively, but it also predicted that a fourth round of recombination would result in $\langle 1, 2, 1, 1, 1\rangle$. This is exactly the product that was observed experimentally. This theorem can then be viewed as a mathematical proof that the synaptic complex structure as proposed by Wasserman et al. in [44] is the only possibility.

6 Further Applications of the Tangle Model

The use of tangle algebra to model the biological processes that give rise to knotting in DNA provides an excellent example of the application of topological algebra to biology. The tangle algebra approach to knotting in DNA began with the study of the site-specific recombinase $Tn3$ resolvase. It is assumed that this protein acts on unknotted DNA processively, producing a series of products, thus providing ample information for systematic mathematical analysis [16]. The model arising from this assumption produced testable, and verified, predictions of knot products [44] and the tangle algebra approach made it possible to write down tangle equations that reflected the progressive repeat action of the protein [18].

A similar approach was taken to study the effect of many site-specific recombinase [6, 7, 13, 14, 29, 40]. Tangles continue to be used to describe the synaptic structure during recombination. One example is extending the tangle model to include 3-string tangles [15, 17, 23], and has also been used in [6, 8] to make predictions of the possible knots that may arise under different hypotheses about the substrate arrangement. Tangle algebras are also being used to study proteins that do not change the topological structure of DNA, but bind to it in interesting ways. This experimental technique has been the focus of many papers, yielding interesting results [24, 27, 31, 32].

References

1. C.C. Adams, *The Knot Book* (American Mathematical Society, Providence, 2004), An elementary introduction to the mathematical theory of knots, Revised reprint of the 1994 original. MR MR2079925 (2005b:57009)
2. B. Alberts, D. Bray, K. Hopkins, A. Johnson, J. Lewis, M. Raff, K. Roberts, P. Walter, *Essential Cell Biology*, 2nd edn. (Garland Science/Taylor & Francis Group, New York, 2003)
3. J.W. Alexander, Topological invariants of knots and links. Trans. Am. Math. Soc. **30**(2), 275–306 (1928)
4. A.D. Bates, A. Maxwell, *Dna Topology* (Oxford University Press, Oxford, 2005)
5. J.M. Berger, S.J. Gamblin, S.C. Harrison, J.C. Wang, Structure and mechanism of DNA topoisomerase ii. Nature **379**(6562), 225–232 (1996)
6. D. Buck, E. Flapan, A topological characterization of knots and links arising from site-specific recombination. J. Phys. A **40**(41), 12377–12395 (2007). MR 2394909 (2010h:92064)
7. D. Buck, C.V. Marcotte, Tangle solutions for a family of DNA-rearranging proteins. Math. Proc. Camb. Philos. Soc. **139**(1), 59–80 (2005). MR 2155505 (2006j:57010)
8. D. Buck, K. Valencia, Characterization of knots and links arising from site-specific recombination on twist knots. J. Phys. A Math. Theor. **44**(4), 045002 (2011)
9. C.R. Calladine, H.R. Drew, B.F. Luisi, A.A. Travers, *Understanding DNA*, 3rd edn. (Elsevier Academic Press, Amsterdam, 2004)
10. J.J. Champoux, DNA topoisomerases: structure, function, and mechanism. Annu. Rev. Biochem. **70**, 369–413 (2001)
11. J.H. Conway, An enumeration of knots and links, and some of their algebraic properties, in *Computational Problems in Abstract Algebra (Proc. Conf., Oxford, 1967)* (Pergamon, Oxford, 1970), pp. 329–358. MR 0258014 (41 #2661)
12. P.R. Cromwell, *Knots and Links* (Cambridge University Press, Cambridge, 2004). MR MR2107964 (2005k:57011)
13. I.K. Darcy, Biological distances on DNA knots and links: applications to XER recombination. J. Knot Theory Ramifications **10**(2), 269–294 (2001), Knots in Hellas '98, Vol. 2 (Delphi). MR 1822492 (2002m:57008)
14. I. Darcy, J. Chang, N. Druivenga, C. McKinney, R. Medikonduri, S. Mills, J. Navarra-Madsen, A. Ponnusamy, J. Sweet, T. Thompson, Coloring the Mu transpososome. BMC Bioinf. **7**(1), 435 (2006)

15. I. Darcy, J. Luecke, M. Vazquez, Tangle analysis of difference topology experiments: applications to a mu protein-DNA complex. Algebr. Geom. Topol. **9**(4), 2247–2309 (2009)
16. P. Dröge, N.R Cozzarelli, Recombination of knotted substrates by tn3 resolvase. Proc. Natl. Acad. Sci. **86**(16), 6062–6066 (1989)
17. E. Eftekhary, *Heegaard floer homologies of pretzel knots*. arXiv preprint math/0311419.
18. C. Ernst, D. Sumners, A calculus for rational tangles: applications to DNA recombination. Math. Proc. Camb. Philos. Soc. **108**, 489–515 (1990)
19. P. Freyd, D. Yetter, J. Hoste, W.B.R. Lickorish, K. Millett, A. Ocneanu, A new polynomial invariant of knots and links. Bull. Am. Math. Soc. (N.S.) **12**(2), 239–246 (1985)
20. F.B. Fuller, Decomposition of the linking number of a closed ribbon: a problem from molecular biology. Proc. Natl. Acad. Sci. U. S. A. **75**(8), 3557–3561 (1978). MR 0490004 (58 #9367)
21. N.D.F. Grindley, K.L. Whiteson, P.A. Rice, Mechanisms of site-specific recombination. Annu. Rev. Biochem. **75**(1), 567–605 (2006), PMID: 16756503
22. J. Hardin, G.P. Bertoni, L.J. Kleinsmith, *Becker's World of the Cell*, 8th edn. (Benjamin Cummings, San Francisco, 2010)
23. H.C. Ibarra, D.A.L. Navarro, An algorithm based on 3-braids to solve tangle equations arising in the action of gin {DNA} invertase. Appl. Math. Comput. **216**(1), 95–106 (2010)
24. M. Jayaram, R.M. Harshey, The Mu transpososome through a topological lens. Crit. Rev. Biochem. Mol. Biol. **41**(6), 387–405 (2006)
25. V.F.R. Jones, A polynomial invariant for knots via von Neumann algebra. Bull. Am. Math. Soc. (N.S.) **12**, 103–111 (1985)
26. M. Khovanov, A categorification of the Jones polynomial. Duke Math. J. **101**(3), 359–426 (2000)
27. S. Kim, I.K. Darcy, *Topological Analysis of DNA-Protein Complexes*. Mathematics of DNA structure, function and interactions (Springer, Berlin, 2009), pp. 177–194
28. K. Murasugi, *Knot Theory and Its Applications* (Modern Birkhäuser Classics, Birkhäuser Boston Inc., Boston, 2008), Translated from the 1993 Japanese original by Bohdan Kurpita, Reprint of the 1996 translation [MR1391727]. MR 2347576
29. F.J. Olorunniji, D.E. Buck, S.D. Colloms, A.R. McEwan, M.C.M. Smith, W.M. Stark, S.J. Rosser, Gated rotation mechanism of site-specific recombination by *ϕc31 integrase*. Proc. Natl. Acad. Sci. **109**(48), 19661–19666 (2012)
30. P. Ozsváth, *Knot Floer Homology*. Advanced Summer School in Knot Theory, May 2009, International Center for Theoretical Physics
31. S. Pathania, M. Jayaram, R.M. Harshey, Path of DNA within the Mu transpososome. Cell **109**(4), 425–436 (2002)
32. C.R. Price, *A Biological Application for the Oriented Skein Relation* (ProQuest LLC, Ann Arbor, 2012), Thesis (Ph.D.)–The University of Iowa. MR 3078590
33. K. Reidemeister, *Knotentheorie* (Springer, Berlin, 1974), Reprint. MR MR0345089 (49 #9828)
34. D. Rolfsen, *Knots and Links*. Mathematics Lecture Series, vol. 7 (Publish or Perish Inc., Houston, 1990), Corrected reprint of the 1976 original. MR MR1277811 (95c:57018)
35. R.G. Scharein, *Interactive topological drawing*, Ph.D. thesis, Department of Computer Science, The University of British Columbia, 1998
36. M.C.M. Smith, H.M. Thorpe, Diversity in the serine recombinases. Mol. Microbiol. **44**(2), 299–307 (2002)
37. D.W. Sumners, Lifting the curtain: using topology to probe the hidden action of enzymes. Not. Am. Math. Soc. **42** (1995), 528–537.
38. D.W. Sumners, C. Ernst, S.J. Spengler, N.R. Cozzarelli, Analysis of the mechanism of DNA recombination using tangles. Q. Rev. Biophys. **28**, 253–313 (1995)
39. L.-P. Tan, G.Y.J. Chen, S.Q. Yao, *Expanding the scope of site-specific protein biotinylation strategies using small molecules*. Bioorg. Med. Chem. Lett. **14**(23), 5735–5738 (2004)
40. M. Vazquez, D.W. Sumners, Tangle analysis of Gin site-specific recombination. Math. Proc. Camb. Philos. Soc. **136**(3), 565–582 (2004). MR 2055047 (2005d:57013)
41. J.C. Wang, *Untangling the Double Helix* (Cold Spring Harbor Laboratory Press, 2009), DNA Entanglement and the Action of the DNA Topoisomerases

42. S.A. Wasserman, N.R. Cozzarelli, Determination of the stereostructure of the product of tn3 resolvase by a general method. Proc. Natl. Acad. Sci. U. S. A. **82**(4), 1079–1083 (1985). (English)
43. S.A. Wasserman, N.R. Cozzarelli, Biochemical topology: applications to DNA recombination and replication. Science **232**(4753), 951–960 (1986). (English)
44. S.A. Wasserman, J.M. Dungan, N.R. Cozzarelli, Discovery of a predicted DNA knot substantiates a model for site-specific recombination. Science **229**(4709), 171–174 (1985)
45. Wikipedia, *Deoxyribose—Wikipedia, the free encyclopedia*, 2012. Accessed 20 May 2012
46. Wikipedia, *DNA—Wikipedia, the free encyclopedia*, 2012. Accessed 20 May 2012
47. Wikipedia, *Nucleic acid—Wikipedia, the free encyclopedia*, 2012. Accessed 20 May 2012
48. G. Witz, A. Stasiak, DNA supercoiling and its role in DNA decatenation and unknotting. Nucl. Acids Res. **38**(7), 2119–2133 (2010)

Structural Identifiability Analysis of a Labeled Oral Minimal Model for Quantifying Hepatic Insulin Resistance

Jacqueline L. Simens, Melanie Cree-Green, Bryan C. Bergman, Kristen J. Nadeau, and Cecilia Diniz Behn

Abstract Insulin resistance (IR) is associated with aging, trauma, and many diseases including obesity, type 2 diabetes, polycystic ovarian syndrome, and sepsis. Determining tissue-specificity of IR in a given individual or disease state may have important implications for clinical care and requires detailed assessment of glucose–insulin dynamics. Previous work introduced a differential-equations-based model to interpret data collected under a stable isotope-based oral glucose tolerance test designed to differentiate the dynamics of exogenous and endogenous glucose. We investigated the structural identifiability of this model using the Taylor expansion method. We found that the model is structurally unidentifiable due to parameters involving the rate of appearance of exogenous glucose and the volume of distribution of the compartment that cannot be separately identified. Our analysis informs a two-step approach to model implementation that overcomes limitations in identifiability and provides a reliable methodology to estimate parameters used to quantify tissue-specific IR. This work contributes to an improved understanding of methods designed to investigate tissue-specific IR.

J. L. Simens
Department of Applied Mathematics and Statistics, Colorado School of Mines, Golden, CO, USA

M. Cree-Green · K. J. Nadeau
Division of Endocrinology, Department of Pediatrics, University of Colorado Anschutz Medical Campus, Aurora, CO, USA

B. C. Bergman
Division of Endocrinology and Metabolism, University of Colorado Anschutz Medical Campus, Aurora, CO, USA

C. Diniz Behn (✉)
Department of Applied Mathematics and Statistics, Colorado School of Mines, Golden, CO, USA

Division of Endocrinology, Department of Pediatrics, University of Colorado Anschutz Medical Campus, Aurora, CO, USA
e-mail: cdinizbe@mines.edu

A. Deines et al. (eds.), *Advances in the Mathematical Sciences*, Association for Women in Mathematics Series 15, https://doi.org/10.1007/978-3-319-98684-5_9

1 Introduction

A primary metabolic goal of the body is to provide a constant energy source of glucose for the brain [24]. Maintaining glucose homeostasis requires different mechanisms depending on the availability of exogenous glucose. In the fasted state when exogenous glucose is absent, glucose is produced from endogenous sources including glycogen degradation and gluconeogenesis with the breakdown of liver glycogen stores representing the primary endogenous source of glucose. Following a meal, exogenous glucose is readily available, so endogenous glucose sources are suppressed and excess glucose is stored for later use. Thus, metabolic systems switch between states of glucose utilization and production depending on the availability of glucose from exogenous sources. These transitions are primarily governed by insulin, a hormone that is released from the pancreas in response to a meal.

Insulin promotes the clearance of excess exogenous glucose from blood by stimulating muscle and adipose tissue. Insulin also suppresses hepatic tissue to stop glucose production and switch to storing glucose through glycogen synthesis, primarily through suppression of the hormone glucagon. Beyond these glucose effects, insulin also plays a role in several other metabolic functions such as stimulating amino acid uptake, protein synthesis, and switching overall metabolism from deriving energy from fat to utilizing glucose to supply energy.

Insulin resistance (IR) occurs when insulin does not induce expected signaling changes, and compensatory, higher than normal, concentrations of insulin are required to maintain normal glucose concentrations. When these conditions are prolonged, pancreatic insufficiency may develop, thereby compromising the regulation of the system and resulting in hyperglycemia. IR is associated with aging, trauma, and a range of diseases including obesity, type 2 diabetes (T2D), polycystic ovarian syndrome, and sepsis [4, 12, 13]. Since insulin plays several roles in glucose regulation, IR may vary in different tissues in the same person. Improved understanding of the tissue-specificity of IR in a given patient or disease condition may facilitate targeted therapeutic treatment.

Stable isotope tracers provide a methodology for probing tissue-specific IR [18, 20, 22]. Here, we focus on quantifying the effect of insulin on glucose disposal, a factor involved in the assessment of hepatic IR, a condition in which higher than normal insulin concentrations are needed to suppress glucose release from the liver. Most protocols designed to quantify hepatic IR have focused on the fasted state or the steady-state conditions of a hyperinsulinemic euglycemic clamp [24]. However, these methods cannot account for the influence of postprandial hormone and nutrient absorption dynamics, and thus answer very limited questions. Including a stable isotope tracer in a more physiologic protocol involving an oral glucose challenge allows separate assessment of the dynamics of endogenous and exogenous glucose [17]. However, since an oral glucose tolerance test (OGTT) involves time-dependent changes in concentrations of endogenous and exogenous glucose, dynamic mathematical modeling is required to optimally mine these data and to interpret the resulting implications for tissue-specific IR.

To quantify the effect of insulin on glucose disposal, Dalla Man and colleagues developed the differential-equations-based oral minimal model (OMM) and labeled oral minimal model (OMM*) to describe glucose–insulin dynamics following an oral challenge involving a stable isotope tracer [15]. Parameter values from OMM and OMM* were proposed as measures of overall insulin sensitivity, S_I, and insulin sensitivity of glucose disposal, S_I^*.

The reliable use of estimated parameters from these models as measures of an individual's insulin sensitivity requires structural and numerical identifiability of the models. Previous work has established that OMM is structurally unidentifiable [14], so this work focuses on investigating the structural identifiability of OMM*. The paper is organized as follows: in Sect. 2, we recall the derivations of OMM and OMM*; in Sect. 3, we apply a Taylor series approach to perform a formal structural identifiability analysis on OMM*; and in Sect. 4, we provide a brief summary of our results and discuss the implications of this work for future applications of OMM*.

2 Model Derivations

OMM is a differential-equations-based model that extends the original Minimal Model [7] to describe an oral challenge [8, 14]. OMM exploits known metabolic physiology to describe glucose and insulin dynamics using biologically relevant parameters. In order to further describe subject-specific glucose and insulin dynamics following an oral glucose challenge that includes a glucose tracer, Dalla Man and colleagues developed OMM* [15]. The differential equations involved in OMM* track the dynamics of the labeled glucose and allow differentiation of the labeled glucose introduced by the drink from the total glucose described by OMM. Here, we recall the derivations of OMM and OMM* in order to provide context for the models and the interpretation of parameter values in the models. In addition, this derivation highlights the interactions between OMM and OMM* that allow a robust determination of the differentiated dynamics of glucose from exogenous and endogenous sources in this OGTT protocol.

As discussed in the introduction, dynamic interactions between glucose and insulin work to maintain normal glucose homeostasis, since the availability of exogenous glucose varies widely during fasting and postprandial states. In both OMM and OMM*, glucose and insulin interactions are assumed to take place in a single compartment representing plasma (Fig. 1). Glucose concentrations $G(t)$ increase in response to glucose ingested in the meal (Ra_{meal}) and glucose released by the liver (Ra_L). Glucose concentrations decrease in response to uptake by the liver (Rd_L) and utilization of glucose by the periphery (Rd_P). The variable $\hat{I}(t)$ describes the action of insulin on glucose uptake, and we will use a convenient scaling of this variable, $X(t) = c\hat{I}(t)$ where the constant c will be specified in the derivation. These models do not account for renal excretion of glucose as may occur

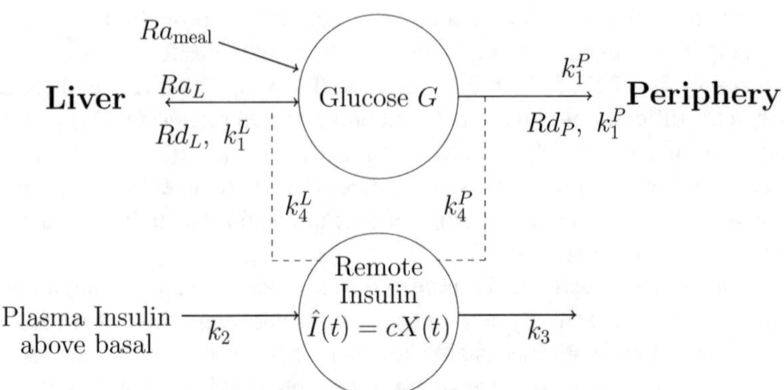

Fig. 1 Schematic summarizing the sources of production and clearance of glucose, $G(t)$, and insulin mediation of glucose uptake, $\hat{I}(t)$. $X(t)$ represents a convenient scaling of $\hat{I}(t)$ by parameter c and will be used in the final form of OMM and OMM*. Ra_L and Rd_L describe the rate of appearance and disappearance, respectively, of glucose from the liver; Ra_{meal} is the rate of appearance of exogenous glucose; k_i for $i = 1, \ldots, 4$ are rate constants, and superscripts L and P denote liver and periphery, respectively. The constants k_1 and k_4 describe insulin-independent and insulin-dependent rates of glucose uptake, respectively. The constants k_2 and k_3 describe the rates of insulin release and clearance, respectively

in hyperglycemia, so they may be most appropriate for describing glucose–insulin dynamics in people without severe hyperglycemia [19, 24].

2.1 Derivation of Oral Minimal Model (OMM)

OMM is a parametric model in which the rate of appearance into plasma of oral glucose is coupled to the classical minimal model for glucose kinetics [14]. The dynamics of insulin action on glucose are described by:

$$\frac{d\hat{I}(t)}{dt} = k_2[I(t) - I_b] - k_3\hat{I}(t), \qquad \hat{I}(0) = 0 \tag{1}$$

where $\hat{I}$ is the remote insulin acting on glucose (μU/ml); $I(t)$ is a linear interpolation of measured plasma insulin concentration (μU/ml); I_b is the basal value (μU/ml); and k_2, k_3 are constant nonnegative parameters (min^{-1}).

To derive the equation for total glucose, we start with the mass balance principle:

$$\frac{dG(t)}{dt} = \frac{Ra(t) - Rd(t)}{V}, \qquad G(0) = G_b \tag{2}$$

where G is the total plasma glucose concentration (mg/dl), G_b is the basal plasma glucose concentration (mg/dl), V is the distribution volume (dl/kg), Ra is the rate of appearance of glucose (mg/kg/min), and Rd is the rate of disappearance of glucose (mg/kg/min). Since Ra represents the rate of appearance of glucose coming from both exogenous and endogenous sources, total $Ra = Ra_{\text{meal}} + Ra_L$ where Ra_{meal} describes exogenous glucose coming from the meal and Ra_L describes endogenous glucose coming from the liver. Similarly, $Rd = Rd_L + Rd_P$ represents the rate of disappearance of glucose due to uptake by the liver (Rd_L) and the periphery (Rd_P).

The cycle of glucose release (Ra_L) and uptake (Rd_L) by the liver is called net hepatic glucose production (NHGP). Following previous work by Cobelli and colleagues [11], we have described NHGP and Rd_P as follows:

$$\text{NHGP}(t) = Ra_L - Rd_L = B_0 - k_1^L G(t) - k_4^L \hat{I}(t) G(t)$$

and

$$Rd_P(t) = k_1^P G(t) + k_4^P \hat{I}(t) G(t)$$

where B_0 is extrapolated NHGP at zero glucose (mg/kg/min), and $k_1^L, k_1^P, k_4^L, k_4^P$ are constant nonnegative parameters. Substituting these equations for NHGP and Rd_P into Eq. (2), and combining this with Eq. (1) gives the original form of OMM:

$$\left.\begin{aligned} \frac{dG(t)}{dt} &= \frac{Ra_{\text{meal}}(t) + B_0 - (k_1^L + k_1^P)G(t) - (k_4^L + k_4^P)\hat{I}(t)G(t)}{V}, \quad & G(0) = G_b \\ \frac{d\hat{I}(t)}{dt} &= k_2[I(t) - I_b] - k_3\hat{I}(t), & \hat{I}(0) = 0. \end{aligned}\right\} \tag{3}$$

Several assumptions and simplifications are needed to obtain the final form of OMM. First, following [14, 15], we define the following parameter combinations:

$$p_1 = k_1^L + k_1^P; \; p_2 = k_3; \; p_3 = \frac{k_2(k_4^L + k_4^P)}{V}; \text{ and } p_4 = B_0.$$

The additional parameter combination $S_G = p_1/V$ is the fractional glucose effectiveness, a measure of the ability of glucose to promote glucose disposal and inhibit NHGP [15]. Next, we introduce a rescaling of $\hat{I}(t)$ with $X(t) = \frac{(k_4^L + k_4^P)}{V}\hat{I}(t) = c\hat{I}(t)$ for $c = \frac{(k_4^L + k_4^P)}{V}$. $X(t)$ represents the insulin action on glucose disposal and production. Finally, we assume that, in the basal steady state, glucose is at a basal level and there is no action of insulin on excess glucose. This provides the additional constraint that $p_4 = p_1 G_b$ [11] and leads to the final form of OMM:

$$\left.\begin{aligned} \frac{dG(t)}{dt} &= -[S_G + X(t)]G(t) + S_G G_b + \frac{Ra_{\text{meal}}(t)}{V}, \quad & G(0) = G_b \\ \frac{dX(t)}{dt} &= p_3[I(t) - I_b] - p_2 X(t), & X(0) = 0. \end{aligned}\right\} \tag{4}$$

where G_b and I_b are known parameters; $I(t)$ is measured insulin concentration; and S_G, V, p_3, p_2, and the parameters associated with Ra_{meal} are the parameters to be estimated. The parameters for the final form of OMM and their relationship to the initial parameters are summarized in Table 1. The parameters in Ra_{meal} depend on the functional form chosen for Ra_{meal} and are not included in this table. These parameters will be discussed further in Sect. 3.

2.2 Derivation of Labeled Oral Minimal Model (OMM*)

In order to isolate the action of insulin on glucose disposal from its action on glucose production, we will consider the dynamics of glucose labeled with the $[1-{}^{13}\mathrm{C}]$glucose oral tracer. This tracer represents the exogenous glucose introduced into the system from the drink, denoted G_{meal}. OMM* will be used to track this labeled glucose and its interaction with insulin. We use asterisks to denote variables and parameters of OMM*. As in the derivation of OMM, we begin with the mass balance principle:

$$\frac{dG_{\mathrm{meal}}(t)}{dt} = \frac{Ra_{\mathrm{meal}}(t) - Rd_{\mathrm{meal}}(t)}{V^*}, \qquad G_{\mathrm{meal}}(0) = 0,$$

where Ra_{meal} and Rd_{meal} are the rates of appearance and disappearance, respectively, of oral glucose and V^* is the distribution volume for labeled glucose. Although the volume of distribution of labeled and unlabeled glucose are the same at steady state, the rapidly mixing pool for ingested glucose varies in the time period immediately following consumption of the drink. Therefore, V and V^* are taken to be distinct parameters. Note that by contrast with the unlabeled glucose described by OMM, the liver does not contribute to the rate of appearance of labeled glucose described by OMM*. Since the rate of disappearance is the same for labeled and unlabeled glucose, $Rd_{\mathrm{meal}}(t) = (k_1^L + k_1^P)G_{\mathrm{meal}}(t) + (k_4^L + k_4^P)\hat{I}(t)G_{\mathrm{meal}}(t)$.

Using the analogous parameter combinations previously described for OMM, we obtain the standard form for OMM*:

$$\left.\begin{aligned} \frac{dG_{\mathrm{meal}}(t)}{dt} &= -[S_G^* + X^*(t)]G_{\mathrm{meal}}(t) + \frac{Ra_{\mathrm{meal}}(t)}{V^*}, & G_{\mathrm{meal}}(0) = 0 \\ \frac{dX^*(t)}{dt} &= p_3^*[I(t) - I_b] - p_2^* X^*(t), & X^*(0) = 0 \end{aligned}\right\} \tag{5}$$

where X^* is insulin action on glucose disposal; I_b is basal insulin; $I(t)$ represents measured insulin; and the parameters to be estimated are S_G^*, V^*, p_3^*, p_2^*, and the parameters present in Ra_{meal} [15]. Combining the systems of Eqs. (4) and (5) yields the complete labeled oral minimal model for glucose–insulin dynamics following an oral challenge that includes a glucose tracer. The parameters for the final form of OMM* are summarized in Table 1.

Table 1 Summary of parameter values for OMM and OMM*

Model	Parameter	Meaning	Parameter combinations
OMM	S_G	Fractional glucose effectiveness	$S_G = p_1/V$
	G_b	Basal glucose concentration	
	V	Volume of distribution (total glucose)	
	p_1	Insulin-independent rate of glucose uptake	$p_1 = k_1^L + k_1^P$
	p_2	Rate of decay of insulin action $X(t)$	$p_2 = k_3$
	p_3	Rate of growth for $X(t)$	$p_3 = \dfrac{k_2(k_4^L + k_4^P)}{V}$
	p_4	Extrapolated NHGP at zero glucose	$p_4 = B_0$
OMM*	S_G^*	Fractional glucose effectiveness	$S_G^* = p_1^*/V^*$
	V^*	Volume of distribution (labeled glucose)	
	p_1^*	Insulin-independent rate of glucose uptake	$p_1^* = k_1^L + k_1^P$
	p_2^*	Rate of decay of insulin action $X^*(t)$	$p_2^* = k_3^*$
	p_3^*	Rate of growth for $X^*(t)$	$p_3^* = \dfrac{k_2^*(k_4^L + k_4^P)}{V^*}$
Both	I_b	Basal insulin concentration	

2.3 Model-Dependent Measures of Insulin Sensitivity

Parameters from OMM and OMM* are used to define the measures of insulin sensitivity S_I and S_I^*, respectively. Both S_I and S_I^* are defined to be a ratio of growth and decay rates of the variables representing the effect of insulin on glucose, $X(t)$ and $X^*(t)$: $S_I = \dfrac{p_3}{p_2}V$ and $S_I^* = \dfrac{p_3^*}{p_2^*}V$. $X(t)$ describes the effect of insulin on total glucose, so S_I quantifies the sensitivity to insulin of both glucose suppression and disposal; S_I^* quantifies the sensitivity to insulin of glucose disposal only, since it is based on the dynamics of the glucose coming from the meal, $X^*(t)$.

The parameters p_3 and p_3^* are the proportionality constants relating the growth of the variables $X(t)$ and $X^*(t)$ to the difference between insulin concentration at time t and basal insulin concentration, $I(t) - I_b$. The parameters p_2 and p_2^* represent the rate of decay of $X(t)$. Therefore, S_I and S_I^* provide measures of how quickly the effect of insulin changes as a function of raw insulin concentration; dividing by p_2 and p_2^* normalizes this effect by the individual's intrinsic insulin dynamics. Therefore, lower values of S_I and S_I^* represent reduced insulin sensitivity.

3 Structural Identifiability Analysis of OMM*

In order to use S_I and S_I^* as metrics for quantifying different aspects of insulin sensitivity, it is necessary to be confident that the estimated parameters represent a unique best fit to the measured data. This requires establishing the structural identifiability of the model. Structural identifiability, also known as *a priori* global identifiability or system identifiability, refers to the theoretical possibility of uniquely identifying model parameters from measured data. For establishing structural identifiability, data is assumed to be known without error. The ability to uniquely identify model parameters in practice using noisy data is known as numerical or practical identifiability. Thus, structural identifiability is a necessary but not sufficient condition for uniquely estimating model parameters.

Dalla Man and colleagues previously established that OMM is structurally unidentifiable [14]. They asserted that OMM* is also structurally unidentifiable, but they did not provide a formal analysis [15]. To address this gap, we applied the Taylor expansion method (also called the power series expansion method) [21] to analyze the structural identifiability of OMM*.

3.1 Taylor Expansion Method

The Taylor expansion method for determining structural identifiability analyzes the power series expansion of the measurement function $y(t)$ as a function of the unknown parameters [21]. The measurement function $y(t) = h(x(t), \mathbf{p})$ where $x(t)$ is the state vector, $\mathbf{p}$ is the parameter vector, and $h(x(\cdot), \mathbf{p})$ has infinitely many derivatives with respect to the state vector components and with respect to time. We let the superscript k denote the kth derivative. Since y is a unique function of time, the condition that the set of equations

$$y^{(k)}(0) = h^{(k)}(x(0), \mathbf{p}), \qquad k = 0, \ldots, \infty \tag{6}$$

have a unique solution for $\mathbf{p}$ is sufficient to define a unique Taylor expansion for $y(t)$ and, thus, to establish the structural identifiability of the system [21]. For a given parameter p_i, the set of equations (6) may be used to determine if the parameter is uniquely identifiable, identifiable (finite number of solutions for p_i), or unidentifiable (infinitely many solutions for p_i) [10].

3.2 Determining Structural Identifiability of OMM*

To apply the Taylor expansion method to investigate the structural identifiability of the final form of OMM*, we take the measurement function $y(t) = h(x(t), \mathbf{p}) = G_{meal}(t)$ where $x(t) = [G_{\text{meal}}(t), X^*(t)]$ is the state vector, $\mathbf{p}$ is the parameter vector, and $h(x(\cdot), \mathbf{p})$ has infinitely many derivatives with respect to the state vector components. By restricting to the appropriate time interval (as discussed more below), this system satisfies the assumptions on the existence of infinitely many derivatives that are necessary to apply this method.

Recall the final form of OMM*:

$$\left.\begin{aligned} \frac{dG_{\text{meal}}(t)}{dt} &= -[S_G^* + X^*(t)]G_{\text{meal}}(t) + \frac{Ra_{\text{meal}}(t;\boldsymbol{\alpha})}{V^*}, \quad & G_{\text{meal}}(0) = 0 \\ \frac{dX^*(t)}{dt} &= p_3^*[I(t) - I_b] - p_2^* X^*(t), & X^*(0) = 0 \end{aligned}\right\} \tag{7}$$

where $I(t)$ is a measured input function; I_b is a fixed parameter; and S_G^*, V^*, p_3^*, and p_2^* and the parameters present in Ra_{meal} are the parameters to be estimated.

The functional form assumed for Ra_{meal} specifies the number and type of parameters to be estimated for Ra_{meal}. Previous work has compared piecewise linear, spline, and dynamic models of Ra_{meal} to explore the effect of the functional form for Ra_{meal} on estimated insulin sensitivity [14]. For all functional forms considered, they found that estimates of S_I were strongly correlated with independent estimates of insulin sensitivity [14]. Therefore, we have adopted the piecewise linear form for $Ra_{\text{meal}}(t)$ to simplify the application of the Taylor expansion method as discussed in more detail below.

The piecewise linear form for $Ra_{\text{meal}}(t;\boldsymbol{\alpha})$ takes the following form:

$$Ra_{\text{meal}}(t;\boldsymbol{\alpha}) = \begin{cases} \alpha_{i-1} + \frac{\alpha_i - \alpha_{i-1}}{t_i - t_{i-1}}(t - t_{i-1}), & t_{i-1} \le t \le t_i, i = 1, \ldots, n \\ 0, & \text{otherwise} \end{cases}. \tag{8}$$

Clearly, $Ra_{\text{meal}}(t;\boldsymbol{\alpha})$ depends on the parameters $\boldsymbol{\alpha}$, but we will suppress this notation for clarity. When restricted to the interval $[t_{i-1}, t_i]$, $Ra_{\text{meal}}(t)$ reduces to a linear function defined by α_i and α_{i-1}. Thus, the choice of a piecewise form for Ra_{meal} guarantees that higher order derivatives will vanish on individual intervals $[t_{i-1}, t_i]$. In order to ensure differentiability of Ra_{meal}, we will restrict to individual intervals $[t_{i-1}, t_i]$ for $i = 1, \ldots, n$ when taking derivatives. By sequentially considering identifiability of the relevant subset of model parameters on each interval $[t_{i-1}, t_i]$ for $i = 1, \ldots, n$, we will be able to evaluate identifiability for all parameters $\mathbf{p} = [S_G^*, p_2^*, p_3^*, V^*, \alpha_i]^T$ for $i = 1, \ldots, n$.

For $y(t) = G_{\text{meal}}(t)$, we compute derivatives as follows:

$$\begin{aligned} y(t) =& G_{\text{meal}}(t) \\ y'(t) =& -[S_G^* + X^*(t)]G_{\text{meal}}(t) + \frac{Ra_{\text{meal}}(t)}{V^*} \\ y''(t) =& -S_G^* \frac{dG_{\text{meal}}(t)}{dt} - \frac{dX^*(t)}{dt} G_{\text{meal}}(t) - X^*(t)\frac{dG_{\text{meal}}(t)}{dt} + \\ & + \frac{1}{V^*}\frac{d}{dt} Ra_{\text{meal}}(t) \\ & \vdots \end{aligned}$$

$$y^{(n)}(t) = -S_G^* \frac{d^{(n-1)} G_{\text{meal}}(t)}{dt^{(n-1)}} + \frac{1}{V^*} \frac{d^{(n-1)}}{dt^{(n-1)}} Ra_{\text{meal}}(t) - $$
$$- \sum_{k=0}^{n-1} \binom{n-1}{k} \frac{d^{(n-1-k)} X^*(t)}{dt^{(n-1-k)}} \frac{d^{(k)} G_{\text{meal}}(t)}{dt^{(k)}}.$$

We begin by considering the system on the interval $t_0 = 0 \le t \le t_1$ as $t \to t_0^+$. In order to evaluate $y(t)$ and its derivatives as $t \to t_0$, we first evaluate $Ra_{\text{meal}}(t)$, $X^*(t)$, and their derivatives as $t \to t_0$. We have that $Ra_{\text{meal}}(t_0^+) = \alpha_0$, and $Ra_{\text{meal}}(0) = 0$ which implies that $\alpha_0 = 0$. Therefore, $\frac{d}{dt} Ra_{\text{meal}}(t_0^+) = \lim_{t \to t_0^+} \left[\frac{\alpha_1 - \alpha_0}{t_1 - t_0} \right] = \frac{\alpha_1}{t_1}$, and all higher derivatives of $Ra_{\text{meal}}(t)$ are 0. Evaluating $X^*(t)$ and its derivatives as $t \to t_0^+$ and letting $z(t) = I(t)$ gives the following:

$$X^*(t_0^+) = \lim_{t \to t_0^+} X^*(t) = 0$$

$$\frac{dX^*(t_0^+)}{dt} = \lim_{t \to t_0^+} \left[p_3^* [I(t) - I_b] - p_2^* X^*(t) \right] = 0 \text{ since } I(0) = I_b$$

$$\frac{d^2 X^*(t_0^+)}{dt^2} = \lim_{t \to t_0^+} \left[p_3^* \frac{dI(t)}{dt} - p_2^* \frac{dX^*(t)}{dt} \right] = p_3^* z'(0)$$

$$\vdots$$

$$\frac{d^{(i)} X^*(t_0^+)}{dt^{(i)}} = p_3^* \sum_{j=1}^{i-1} (-1)^{j-1} (p_2^*)^{j-1} z^{(i-j)}(0), \quad \forall i \in \mathbb{N}.$$

Finally, evaluating the measurement function $y(t)$ and its first six derivatives at t_0^+, we have

$$\begin{aligned}
y(t_0^+) &= 0 \\
y'(t_0^+) &= 0 \\
y''(t_0^+) &= \frac{1}{V^*} \frac{\alpha_1}{t_1} \\
y'''(t_0^+) &= -S_G^* y''(t_0^+) \\
y^{(4)}(t_0^+) &= -S_G^* y'''(t_0^+); \\
y^{(5)}(t_0^+) &= S_G^* y^{(4)}(t_0^+) - 6\left[p_3^* z'(0) \right] y''(t_0^+); \\
y^{(6)}(t_0^+) &= -S_G^* y^{(5)}(t_0^+) - 10\left[p_3^* z''(0) - p_2^* p_3^* z'(0) \right] y''(t_0^+) - 10\left[p_3^* z'(0) \right] y'''(t_0^+).
\end{aligned}$$

Recall that the unknown parameters for this model on this time interval are $\mathbf{p} = [S_G^*, p_2^*, p_3^*, V^*, \alpha_1]^T$. Summarizing the equations above, we represent these unknown parameters in terms of the following derivatives of $y(t)$ and $z(t)$ which are assumed to be known:

$$y''(t_0^+) = \frac{1}{V^*}\frac{\alpha_1}{t_1} \tag{9}$$

$$y'''(t_0^+) = -S_G^* y''(t_0^+) \tag{10}$$

$$y^{(5)}(t_0^+) = S_G^* y^{(4)}(t_0^+) - 6\Big[p_3^* z'(0)\Big] y''(t_0^+) \tag{11}$$

$$y^{(6)}(t_0^+) = -S_G^* y^{(5)}(t_0^+) - 10\Big[p_3^* z''(0) - p_2^* p_3^* z'(0)\Big] y''(t_0^+) - \tag{12}$$

$$- 10\Big[p_3^* z'(0)\Big] y'''(t_0^+). \tag{13}$$

Solving (10) for S_G^*, we establish that

$$S_G^* = -\frac{y'''(t_0^+)}{y''(t_0^+)} \tag{14}$$

is uniquely identifiable. Next, solving Eq. (11) for p_3^* establishes that

$$p_3^* = \frac{S_G^* y^{(4)}(t_0^+) - y^{(5)}(t_0^+)}{6z'(0)y''(t_0^+)} \tag{15}$$

is uniquely identifiable. Similarly, solving (13) for p_2^* establishes that

$$p_2^* = \frac{y^{(6)}(t_0^+) + S_G^* y^{(5)}(t_0^+) + 10p_3^* z''(0)y''(t_0^+) + 10p_3^* z'(0)y'''(t_0^+)}{10p_3^* z'(0)y''(t_0^+)} \tag{16}$$

is also uniquely identifiable. It remains to consider the identifiability of V^* and α_1. Separating (9) into knowns (right-hand side) and unknowns (left-hand side), we obtain $\frac{\alpha_1}{V^*} = t_1 y''(0)$. Thus, the parameter combination $\frac{\alpha_1}{V^*}$ is uniquely identifiable, but the individual parameters α_1 and V^* are not. Since both α_1 and V^* drop out of the y-derivatives after the second derivative, we cannot obtain any additional information about these parameters from higher derivatives. Therefore, we have established that OMM* is structurally unidentifiable on $[t_0, t_1]$.

To investigate the identifiability of the α_i parameters for $i > 1$, we consider OMM* over different time intervals. We present an approach using t_1^+ to determine α_2; this approach may be generalized to use t_i^+ to determine α_{i+1} for all $i > 1$. From $y(t)$ and its first derivative, we have $y(t_1^+) = G_{\text{meal}}(t_1)$ and $y'(t_1^+) = -[S_G^* + X^*(t_1^+)]y(t_1^+) + \frac{\alpha_1}{V^*}$. Since $X^*(t_i^+)$ is unknown for $i \neq 0$, we cannot proceed directly as in the $t = t_0^+$ case. To solve for $X^*(t_1^+)$ in terms of known quantities, we take the first, second, and third derivatives of the entire system:

$$\begin{cases} \dfrac{dG_{\text{meal}}(t_1^+)}{dt} = -[S_G^* + X^*(t_1^+)]G_{\text{meal}}(t_1^+) + \dfrac{\alpha_1}{V^*} \\ \dfrac{dX^*(t_1^+)}{dt} = p_3^*[I(t_1^+) - I_b] - p_2^* X^*(t_1^+) \end{cases} \tag{17}$$

$$\begin{cases} \dfrac{d^2G_{\text{meal}}(t_1^+)}{dt^2} = -[S_G^* + X^*(t_1^+)]\dfrac{dG_{\text{meal}}(t_1^+)}{dt} - \dfrac{dX^*(t_1^+)}{dt}G_{\text{meal}}(t_1^+) + \\ \qquad + \dfrac{1}{V^*}\dfrac{\alpha_2 - \alpha_1}{t_2 - t_1} \\ \dfrac{d^2X^*(t_1^+)}{dt^2} = p_3^*\dfrac{dI(t_1^+)}{dt} - p_2^*\dfrac{dX^*(t_1^+)}{dt} \end{cases} \tag{18}$$

and

$$\begin{cases} \dfrac{d^3G_{\text{meal}}(t_1^+)}{dt^3} = -[S_G^* + X^*(t_1^+)]\dfrac{d^2G_{\text{meal}}(t_1^+)}{dt^2} - \\ \qquad - \dfrac{d^2X^*(t_1^+)}{dt^2}G_{\text{meal}}(t_1^+) - 2\dfrac{dX^*(t_1^+)}{dt}\dfrac{dG_{\text{meal}}(t_1^+)}{dt} \\ \dfrac{d^3X^*(t_1^+)}{dt^3} = p_3^*\dfrac{d^2I(t_1^+)}{dt^2} - p_2^*\dfrac{d^2X^*(t_1^+)}{dt^2}. \end{cases} \tag{19}$$

Substituting information from the second equations in (17) and (18) into the G_{meal} derivative in (19), we obtain an expression for $\dfrac{d^3G_{\text{meal}}(t_1^+)}{dt^3} = y'''(t_1^+)$ in terms of $y(t_1^+)$, its derivatives evaluated at t_1^+, $I(t_1^+) = z(t_1^+)$, its derivatives evaluated at t_1^+, $X^*(t_1^+)$, S_G^*, p_2^*, and p_3^* that reduces to the following:

$$\begin{aligned} y'''(t_1^+) = {} & [2p_2^* y'(t_1^+) - (p_2^*)^2 y(t_1^+) - y''(t_1^+)]X^*(t_1^+) - \\ & - S_G^* y''(t_1^+) - p_3^* z'(t_1^+)y(t_1^+) + p_2^* p_3^* y(t_1^+)[z(t_1^+) - I_b] - \\ & - 2p_3^* y'(t_1^+)[z(t_1^+) - I_b]. \end{aligned} \tag{20}$$

In (20), all quantities are known except $X^*(t_1^+)$, so we solve for $X^*(t_1^+)$ to obtain an expression in terms of known quantities:

$$\begin{aligned} X^*(t_1^+) = {} & \left(\frac{1}{2p_2^* y'(t_1^+) - (p_2^*)^2 y(t_1^+) - y''(t_1^+)}\right)\Big(y'''(t_1^+) + S_G^* y''(t_1^+) + \\ & + p_3^* z'(t_1^+)y(t_1^+) - p_2^* p_3^* y(t_1^+)[z(t_1^+) - I_b] + 2p_3^* y'(t_1^+)[z(t_1^+) - I_b]\Big). \end{aligned} \tag{21}$$

Substituting this expression for $X^*(t_1^+)$ into the equation in (17) to obtain $\frac{dX^*(t_1^+)}{dt}$ in terms of known quantities, and substituting the resulting expression for $\frac{dX^*(t_1^+)}{dt}$ into the first equation in (18) gives the following:

$$\begin{aligned}\frac{d^2G_{\text{meal}}(t_1^+)}{dt^2} &= -S_G^*\frac{dG_{\text{meal}}(t_1^+)}{dt} - \frac{dX^*(t_1^+)}{dt}G_{\text{meal}}(t_1^+) - X^*(t_1^+)\frac{dG_{\text{meal}}(t_1^+)}{dt} + \\ &\quad + \frac{1}{V^*}\frac{d}{dt}Ra_{\text{meal}}(t_1^+) \\ &= -S_G^*y'(t_1^+) - \frac{dX^*(t_1^+)}{dt}y(t_1^+) - X^*(t_1^+)y'(t_1^+) + \frac{1}{V^*}\frac{\alpha_2-\alpha_1}{t_2-t_1}.\end{aligned} \tag{22}$$

Recall that S_G^*, p_2^*, p_3^* and the parameter combination α_1/V^* are uniquely identifiable from the analysis at t_0^+. The functions $y(t_1^+)$, $y'(t_1^+)$, ..., $y^{(n)}(t_1^+)$, and $z(t_1^+)$, $z'(t_1^+)$, ..., $z^{(n)}(t_1^+)$ are assumed to be known by definition, and we have established that $X^*(t_1^+)$ and $\frac{dX^*(t_1^+)}{dt}$ are known. Thus, separating the expression in (22) into knowns (right-hand side) and unknowns (left-hand side) we obtain the following expression for α_2/V^*:

$$\frac{\alpha_2}{V^*} = (t_2-t_1)\left(y''(t_1^+) + S_G^*y'(t_1^+) + \frac{dX^*(t_1^+)}{dt}y(t_1^+) + X^*(t_1^+)y'(t_1^+)\right) + t_1y''(t_0^+). \tag{23}$$

As in the analysis at t_0^+, both α_2 and V^* drop out of higher derivatives, so higher derivatives cannot provide additional information about these parameters. Thus, the parameter combination $\frac{\alpha_2}{V^*}$ is uniquely identifiable, but the individual parameters α_2 and V^* are not uniquely identifiable. A similar result is obtained for α_i for $i = 3, \ldots, n$ using an analogous approach. Thus, the parameters α_i and V^* are not uniquely identifiable on any time interval, so OMM* is structurally unidentifiable.

4 Discussion

In order to use estimated model parameters to quantify metabolic features of an individual patient, careful consideration of the modeling approach, its applicability to a given experimental protocol, and its identifiability are necessary. In this work, we applied the Taylor expansion method to examine the structural identifiability of the labeled oral minimal model, OMM*. We established that OMM* is structurally unidentifiable due to inseparable parameter combinations involving the rate of appearance of glucose from the drink (parameters $\alpha_i, i = 0, \ldots, n$) and the volume of distribution of labeled glucose, V^*. To our knowledge, this work represents the first detailed, formal analysis of structural identifiability for OMM*, and it contributes to the existing literature examining the identifiability of other models of glucose–insulin dynamics [9, 11, 14]. Future work considering the numerical identifiability of OMM and OMM* will determine the precision of estimated parameters given typical measurement error for these data and facilitate comparison between distinct groups of participants.

Although our analysis established that OMM* is not structurally identifiable, it also provided insights into relationships among parameters that may guide appropriate numerical implementations of the model. Specifically, because V^* and $\alpha_i, i = 0, \ldots, n$ appear as identifiable parameter combinations, these parameters may be reliably estimated in submodels in which either V^* or $\alpha_i, i = 0, \ldots, n$, are fixed. Dalla Man and colleagues proposed a two-step implementation of OMM* exploiting such structurally identifiable submodels [15]. In step one, they defined a reference-labeled model, RM*, in which additional tracers were used to obtain a model-independent representation of Ra_{meal} [5, 15]. Fixing this representation of Ra_{meal}, they estimated the model parameters $\{S_G^*, V^*, p_2^*$, and $p_3^*\}$. In step two, they defined a final model, FM*, in which V^* and S_G^* are fixed to the values estimated using RM*: $V^* = V^{*\text{ref}}$ and $S_G^* = S_G^{*\text{ref}}$ where $V^{*\text{ref}}$ and $S_G^{*\text{ref}}$ are the parameter values obtained for RM*. Then, they estimated the parameters p_2^*, p_3^*, and α_i for $i = 1, \ldots, n$, and they used these parameters to compute S_I^*.

Based on our analysis, both submodels of OMM*, RM* and FM*, are structurally identifiable. In each submodel, certain subsets of the parameters of OMM* are fixed, thereby enabling identification of the remaining parameters. Note that fixing V^* is sufficient to establish identifiability of FM*; it is not necessary to additionally fix S_G^*, but this likely improves the numerical identifiability of FM*. Although the submodels fail to represent the full physiology of the system in which the rate of appearance of labeled glucose from the drink may interact with a possibly dynamic volume of distribution, the two-step approach resulted in meaningful estimates of insulin sensitivity. The estimates of S_I^* from FM* were validated against estimates of S_I^* computed using the model-independent estimate of Ra_{meal} as well as against estimates of insulin sensitivity measured in the same individuals during a labeled intravenous glucose tolerance test [15]. The high correlation in these measures suggests that the parameters estimated using the two-step approach reliably quantify insulin sensitivity.

OGTT protocols are necessary for evaluating glucose–insulin dynamics in a physiologic context, and the inclusion of stable isotopes in these protocols represents a powerful methodology for differentially tracking the dynamics of exogenous and endogenous glucose in plasma following ingestion of the drink. Mathematical modeling of these data allows quantification of exogenous and endogenous glucose dynamics and their interactions with insulin, thereby providing meaningful metrics for tissue-specific IR. However, models represent limited physiology, they must be tailored to the protocol employed, and the choice of a particular model introduces model dependence into the analysis. Several models to describe OGTT data have been proposed [1, 2, 16, 23], and some of these models include more detailed physiology such as gut absorption dynamics or the role of glucagon in glucose–insulin interactions compared to OMM and OMM* [14, 15]. Here, we focused on OMM and OMM*, because these models have been successfully used to describe the dynamics of labeled and unlabeled glucose during an OGTT. However, most models for OGTT data, including OMM and OMM*, were developed using data from healthy adult participants or adult participants with T2D. Additional work is needed to determine the utility of these models for representing physiology in populations with other demographics or disease conditions.

Finally, in this work, the Taylor expansion method was sufficient for establishing structural unidentifiability of OMM*. This occurred, in part, due to the choice to represent Ra_{meal} as a piecewise linear function; since higher order derivatives of Ra_{meal} vanished, we could identify a general form for higher order derivatives of $y(t)$ and conclude that these derivatives would not yield additional information about identifiability of model parameters. However, future work involving other approaches to investigating structural identifiability, such as the differential algebra approach, may be useful for gaining additional insights into OMM* and other metabolic models [3, 6].

Acknowledgments This work was supported by the following grants: CDB: NSF DMS 1412571; MCG and CDB: Children's Hospital Colorado/Colorado School of Mines Collaborative Pilot Award; MCG: BIRCWH K12-HD057022; NORC P30 DK048520; UL1 TR001082; Boettcher Foundation Boettcher Webb Waring award; NIDDK K23DK107871; University of Colorado, Anschutz: NIH/NCATS Colorado CTSA Grant Number UL1 TR001082. Contents are the authors' sole responsibility and do not necessarily represent official NIH views. The authors would like to thank Kai Bartlette for a critical reading of the manuscript.

References

1. M. Abdul-Ghani, B. Balas, M. Matsuda, R. DeFronzo, Muscle and liver insulin resistance indexes derived from the oral glucose tolerance test. Diabetes Care **30**, 89–94 (2007)
2. E. Ackerman, J. Rosevear, W. McGuckin, A mathematical model of the glucose-tolerance test. Phys. Med. Biol. **9**, 21–33 (1964)
3. S. Audoly, G. Bellu, L. D'Angiò, M.P. Saccomani, C. Cobelli, Global identifiability of nonlinear models of biological systems. IEEE Trans. Biomed. Eng. **48**, 55–65 (2001)
4. A. Baranova, T. Tran, A. Birerdinc, Z. Younossi, Systematic review: association of polycystic ovary syndrome with metabolic syndrome and non-alcoholic fatty liver disease. Aliment. Pharmacol. Ther. **33**, 801–814 (2011)
5. R. Basu, B.D. Camillo, G. Toffolo, A. Basu, P. Shah, A. Vella, R. Rizza, C. Cobelli, Use of a novel triple-tracer approach to assess postprandial glucose metabolism. Am. J. Physiol. Endocrinol. Metab. **284**(1), E55–E69 (2003)
6. G. Bellu, M.P. Saccomani, S. Audoly, L. D'Angiò, Daisy: a new software tool to test global identifiability of biological and physiological systems. Comput. Methods Prog. Biomed. **88**, 52–61 (2007)

7. R. Bergman, Toward physiological understanding of glucose tolerance: minimal-model approach. Diabetes **38**, 1512–1527 (1989)
8. A. Caumo, R. Bergman, C. Cobelli, Insulin sensitivity from meal tolerance tests in normal subjects: a minimal model index. J. Clin. Endocrinol. Metab. **85**(11), 4396–4402 (2000)
9. S.V. Chin, M.J. Chappell, Structural identifiability and indistinguishability analyses of the minimal model and a euglycemic hyperinsulinemic clamp model for glucose-insulin dynamics. Comput. Methods Prog. Biomed. **104**, 120–134 (2011)
10. C. Cobelli, J.J. DiStefano 3rd, Parameter and structural identifiability concepts and ambiguities: a critical review and analysis. Am. J. Physiol. **239**(1), R7–R24 (1980)
11. C. Cobelli, G. Pacini, G. Toffolo, L. Sacca, Estimation of insulin sensitivity and glucose clearance from minimal model: new insights from labeled IVGTT. Am. J. Physiol. **250**(5 Pt 1), E591–E598 (1986)
12. M. Cree-Green, R. Wolfe, Postburn trauma insulin resistance and fat metabolism. Am. J. Physiol. Endocrinol. Metab. **294**, E1–E9 (2008)
13. M. Cree-Green, T. Triolo, K. Nadeau, Etiology of insulin resistance in youth with type 2 diabetes. Curr. Diab. Rep. **13**, 81–88 (2013)
14. C. Dalla Man, A. Caumo, C. Cobelli, The oral glucose minimal model: estimation of insulin sensitivity from a meal test. IEEE Trans. Biomed. Eng. **49**, 419–429 (2002)
15. C. Dalla Man, A. Caumo, R. Basu, R. Rizza, G. Toffolo, C. Cobelli, Measurement of selective effect of insulin on glucose disposal from labeled glucose oral test minimal model. Am. J. Physiol. Endocrinol. Metab. **289**(5), E909–E914 (2005)
16. C. Dalla Man, R. Rizza, C. Cobelli, Meal simulation model of the glucose-insulin system. IEEE Trans. Biomed. Eng. **54**, 1740–1749 (2007)
17. C. Dalla Man, G. Toffolo, R. Basu, R. Rizza, C. Cobelli, Use of labeled oral minimal model to measure hepatic insulin sensitivity, Am. J. Physiol. Endocrinol. Metab. **295**(5), E1152–E1159 (2008)
18. L. George, F. Bacha, S. Lee, H. Tfayli, E. Andreatta, S. Arslanian, Surrogate estimates of insulin sensitivity in obese youth along the spectrum of glucose tolerance from normal to prediabetes to diabetes. Endocrinol. Metab. **96**, 2136–2145 (2011)
19. J. Kaneko, D. Mattheeuws, R. Rottiers, J. Van Der Stock, A. Vermeulen, The effect of urinary glucose excretion on the plasma glucose clearances and plasma insulin responses to intravenous glucose loads in unanaesthesized dogs. Acta Endocrinol. (Copenh) **87**, 133–138 (1978)
20. M. Matsuda, R. DeFonzo, Insulin sensitivity indices obtained from oral glucose tolerance testing. Diabetes Care **22**, 1462–1470 (1999)
21. H. Pohjanpalo, System identifiability based on the power series expansion of the solution. Math. Biosci. **41**(1), 21–33 (1978)
22. M. Stumvoll, A. Mitrakou, W. Pimenta, T. Jenssen, H. Yki-Jarvinen, T.V. Haeften, W. Renn, J. Gerich, Use of the oral glucose tolerance test to assess insulin release and insulin sensitivity. Diabetes Care **23**, 295–301 (2000)
23. R.G.F.K.O. Vahidi, K. Kwok, A comprehensive compartmental model of blood glucose regulation for healthy and type 2 diabetic subjects. Med. Biol. Eng. Comput. **54**, 1383–1398 (2016)
24. R. Wolfe, D. Chinkes, *Isotope Tracers in Metabolic Research: Principles and Practice of Kinetic Analysis* (Wiley, Hoboken, 2005)

Spike-Field Coherence and Firing Rate Profiles of CA1 Interneurons During an Associative Memory Task

Pamela D. Rivière and Lara M. Rangel

Abstract Flexible, dynamic activity in the brain is essential to information processing. Neurons in the hippocampus are capable of conveying information about the continually evolving world through changes in their spiking activity. This information can be expressed through changes in firing rate and through the reorganization of spike timing in unique rhythmic profiles. Locally projecting interneurons of the hippocampus are in an ideal position to coordinate task-relevant changes in the spiking activity of the network, as their inhibitory influence allows them to constrain communication between neurons to rhythmic, optimal windows and facilitates selective responses to afferent input. During a context-guided odor–reward association task, interneurons and principal cells in the CA1 subregion of the rat hippocampus demonstrate distinct oscillatory profiles that correspond to correct and incorrect performance, despite similar firing rates during correct and incorrect trials (Rangel et al., eLife 5:e09849, 2016). Principal cells additionally contained information in their firing rates about task dimensions, reflective of highly selective responses to features such as single positions and odors. It remains to be determined whether interneurons also contain information about task dimensions in their firing rates. To address this question, we evaluated the information content for task dimensions in the firing rates of inhibitory neurons. Interneurons contained low, but significant information for task dimensions in their firing rates, with increases in information over the course of a trial that reflected the evolving availability of task dimensions. These results suggest that interneurons are capable of manifesting distinct rhythmic profiles and changes in firing rate that reflect task-relevant processing.

P. D. Rivière · L. M. Rangel (✉)
Department of Cognitive Science, University of California San Diego, La Jolla, CA, USA
e-mail: pdriviere@ucsd.edu; lrangel@ucsd.edu

A. Deines et al. (eds.), *Advances in the Mathematical Sciences*, Association for Women in Mathematics Series 15, https://doi.org/10.1007/978-3-319-98684-5_10

1 Introduction

Successful information processing in the brain is characterized by dynamic patterns of neuronal activity that reflect information about the changing world. In the hippocampus, a brain region important for learning and memory, these changes often manifest as selective responses to features of the environment [1]. During associative memory processing, neurons in the CA1 region of the rat hippocampus have been shown to exhibit highly selective and reliable changes in firing rate in response to specific spatial locations within an environment, stimuli such as odors that are relevant for performance [2], and the conjunction of specific odors in particular spatial locations [3–5]. Metrics that exploit these selective increases in firing rate have been devised to quantify neuronal information content for particular task features [3, 6–8]. These analyses have revealed subsets of neurons in the hippocampus whose transient changes in firing rate reflect the availability of behaviorally relevant features in the environment, leading to the hypothesis that the hippocampus dynamically recruits appropriate neuronal ensembles in the service of memory [9–13].

Associative memory processing is additionally accompanied by dynamic shifts in the oscillatory profile of the local field potential (LFP) in the CA1 region of the rat hippocampus [3]. In a context-guided odor–reward association task, rats must learn that odors are differentially rewarded depending upon the context in which they are encountered. During intervals in which rats correctly associated odors with the contexts in which they were rewarded, we observed large amplitude changes in the theta (4–12 Hz), beta (15–35 Hz), low gamma (35–55 Hz), and high gamma (65–90 Hz) frequency ranges. A surprising indicator of performance was the reorganization of spike timing with respect to the ongoing oscillations, such that both interneurons and principal cells exhibited distinct profiles of engagement in each of the four rhythms depending upon the trial outcome (correct or incorrect). For example, while the largest proportion of interneurons exhibited consistent spike timing relationships to each of the four frequency ranges during correct performance, many exhibited spike timing relationships to only theta during incorrect performance despite similar overall firing rates between correct and incorrect trials. This finding revealed a previously underexplored task-relevant selectivity in the rhythmic domain, and highlighted the dynamic patterns of activity in the interneuron population as a hallmark of successful processing.

Having established that the selective reorganization of interneuron spike timing is related to successful performance, we wished to additionally evaluate the extent to which interneurons manifest selectivity for task dimensions in a manner similar to principal cells of the hippocampus. Specifically, we tested whether interneurons modulate their firing rates for different task dimensions. We have previously reported that principal cells in the CA1 region of the hippocampus with distinct profiles of engagement in theta, beta, low gamma, and high gamma differentially represent task dimensions in their firing rates [3]. As the largest proportion of interneurons exhibited engagement in all four rhythms during correct performance, we examined the information content in the firing rates of this subpopulation for three key task dimensions: odors, positions, and odor–position combinations. Our preliminary results suggest that this interneuron population exhibited low, but significant information content for task dimensions in their firing rates, with increases in information for odors and odor–position conjunctions that reflected the availability of odor stimuli. This suggests that in addition to providing a task-relevant oscillatory framework for the hippocampal network, interneurons may convey information about task dimensions.

2 Materials and Methods

All experimental procedures were carried out as previously described in Rangel, Rueckemann, Rivière et al. 2016 [3]. They are briefly described here.

2.1 *Behavioral Paradigm*

Rats performed a context-guided odor–reward association task in which odors were differentially rewarded depending on the spatial context in which they were encountered. In this task, rats were placed in a behavioral apparatus consisting of two arms, or *contexts*, oriented at 180° with respect to each other and separated by a central chamber (Fig. 1). Each context contained two odor ports. On each trial, rats were given access to one of the contexts, which were distinct from each other on the basis of their spatial location within the recording room. To further distinguish each context, vinyl or plastic contextual wraps of different colors and textures were used to cover each arm and face plate containing the odor ports. Rats learned to sample odors presented in each port by poking their snouts in an odor port. LED sensors within each odor port registered nose pokes (*poke onset*). Odors were delivered into the odor port 250 ms after the initiation of a nose poke (*odor onset*). Odors were presented in pairs that were consistent throughout the experiment. One odor of a pair was assigned as correct in the first context, with the opposite odor assigned as correct in the second context. Odor positions were pseudorandomized and counterbalanced such that the rat had no prior knowledge of which odor port would contain the correct odor of a pair. *Correct trial:* In order to receive a water reward, rats were required to maintain a nose poke for 1500 ms in the odor port containing the correct odor of a pair. If an odor port contained the incorrect odor, the rat needed to remove his nose from the port before 1500 ms had elapsed from poke onset. Following an exit from the incorrect odor port, the rat was allowed to move to the adjacent port containing the correct odor, where he had to maintain a nose poke for 1500 ms to receive his reward. Upon completion of a nose poke in the correct odor port, a water reward was immediately delivered to an indentation in a tray positioned directly under the odor port. *Incorrect trial:* Holding a nose poke for 1500 ms in the odor port containing the unrewarded odor resulted in a white-noise buzz and no water reward on that trial. Rats encountered pairs of odors in blocks, with two blocks of odor pairs per half-session (four odors in a half-session, eight odors in a full session, 96 trials total). Contextual wraps with unique colors and textures covering each arm were switched between half-sessions. Data was analyzed during 1500-ms sampling intervals in correct odor ports (correct trials). Each half-session was analyzed separately.

During behavioral training, rats completed 80 trials each day until they could perform at 75% accuracy. Recordings began after rats achieved this performance criterion. Only those sessions in which the rat achieved at least 75% accuracy were included in the data analysis. Data during 1500-ms sampling intervals in incorrect odor ports (incorrect trials) were excluded from analysis due to an insufficient number of trials for each task dimension.

2.2 Neural Recordings and Interneuron Identification

We performed high-density extracellular tetrode recordings from the CA1 region of the awake-behaving rat in order to obtain single cell and local field potential (LFP) activity. Signals were amplified 4000–8000× and digitized at 40 kHz by an Omniplex Neural Acquisition system (Plexon). Local field potentials (LFPs) were digitally isolated with a band-pass filter from 1 to 400 Hz and spikes were isolated with a band-pass filter from 400 to 8000 Hz. Putative interneurons were isolated according to firing rates and waveform characteristics. Waveform features, such as peak and valley voltage amplitudes and total peak-to-valley distance, were compared across tetrode wires in OfflineSorter (Plexon). Interneurons clustered according to mean firing rate ($\geq$5 Hz), mean width at half of the maximum amplitude of the waveform ($<$ 150 μs), and mean temporal offset from peak to trough ($<$350 μs) [14–16].

2.3 Local Field Potential and Spike-Phase Coherence Analyses

A third-order Butterworth filter was used to band-pass filter the LFP in the theta (4–12 Hz), beta (15–35 Hz), low gamma (35–55 Hz), and high gamma (65–90 Hz) frequency ranges. The instantaneous phase was then calculated by taking the arctangent of the complex Hilbert transform of the filtered signal. For interneurons demonstrating spike-phase coherence to all four frequency ranges ($\text{theta}_{4\text{–}12\,\text{Hz}}$, $\text{beta}_{15\text{–}35\,\text{Hz}}$, low $\text{gamma}_{35\text{–}55\,\text{Hz}}$, and high $\text{gamma}_{65\text{–}90\,\text{Hz}}$) during correct trials (32 out of 67 total interneurons recorded, 6 rats, 53 half-sessions), the phase of the filtered LFP at the time of each spike was recorded for the 1.5 s odor sampling intervals leading up to reward delivery. Spike-phase relationships were assessed using a Rayleigh statistic, and interneurons were categorized as significantly phase coherent to a rhythm if exhibiting a $p < 0.05$. This criterion was previously used to identify interneurons with significant spike-phase coherence that, upon further characterization, also exhibited differences in the magnitude and phase of coherency across correct and incorrect trials [3].

2.4 *Quantifying Information*

For interneurons demonstrating spike-phase coherence to all four frequency ranges ($\text{theta}_{4\text{–}12\,\text{Hz}}$, $\text{beta}_{15\text{–}35\,\text{Hz}}$, low $\text{gamma}_{35\text{–}55\,\text{Hz}}$, and high $\text{gamma}_{65\text{–}90\,\text{Hz}}$) during correct trials, the information contained in the firing rate of the interneuron for a task dimension (odors, positions, or odor–position conjunctions) was calculated according to the following equation:

$$I = \sum_{i=1}^{n} P_i \left(\frac{F_i}{F}\right) \log_2 \left(\frac{F_i}{F}\right) \tag{1}$$

Where i designates a variant of the task dimension (one of four possible odor port positions, four possible odors, or eight possible odor–position combinations as each odor is rewarded in more than one position), n is the total number of variants, P_i is the probability of the occurrence of variant i, F_i is the mean firing rate during the occurrence of the variant i, and F is the overall mean firing rate of the cell. To determine whether calculated scores could be acquired by chance from the spiking behavior of a given interneuron, task conditions were randomly shuffled 1000 times and the observed information was considered significant if greater than the 95% confidence interval of the condition-shuffled scores.

For the interneurons with significant information for a given task dimension, additional information scores were calculated for three non-overlapping 500 ms intervals that directly preceded and spanned the 1.5-s nose poke interval. These intervals included a 750–250-ms interval prior to nose poke onset (*before*), the 500 ms interval after odor onset (*odor*), and the last 500 ms of the nose poke (*end*). Differences in the median information across the three time intervals were assessed using a Friedman's test, with post hoc pairwise comparisons performed using a Tukey's Honest Significant Difference test.

3 Results

All interneurons with significant spike-phase coherence to $\text{theta}_{4\text{–}12\,\text{Hz}}$, $\text{beta}_{15\text{–}35\,\text{Hz}}$, low $\text{gamma}_{35\text{–}55\,\text{Hz}}$, and high $\text{gamma}_{65\text{–}90\,\text{Hz}}$ during correct trials (32 out of 67 total interneurons recorded, 6 rats, 53 half-sessions, see Sect. 2) also contained significant information for one or more task dimensions (odor, position, and odor–position)

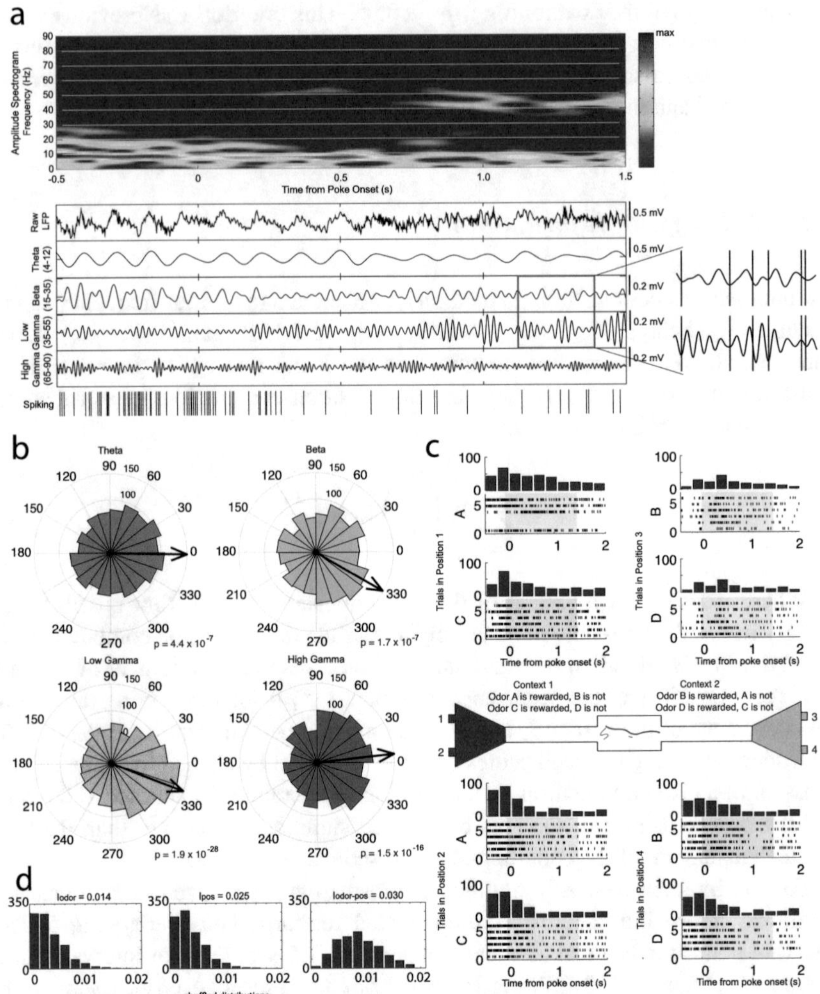

Fig. 1 Spike-field coherence and firing profiles of a single interneuron in the CA1 region of the rat hippocampus. (**a**) *Upper:* Gabor spectrogram during the odor sampling interval of a single trial. *Middle:* Corresponding raw local field potential and the band-pass filtered local field potential in the theta (4–12 Hz), beta (15–35 Hz), low gamma (35–55 Hz), and high gamma (65–90 Hz) frequency ranges. *Lower:* Spike times of a single interneuron that demonstrated significant spike-field coherence to each of the four frequency ranges. (**b**) Circular histograms indicating the phases of theta$_{4\text{–}12\,\text{Hz}}$ (orange), beta$_{15\text{–}35\,\text{Hz}}$ (blue), low gamma$_{35\text{–}55\,\text{Hz}}$ (yellow), and high gamma$_{65\text{–}90\,\text{Hz}}$ (purple) at the time of each spike, with the direction of the mean resultant length vector (R) shown by the arrow in black. *P*-values were calculated using a Rayleigh statistic. (**c**) Spiking activity of the interneuron during the sampling intervals of rewarded odors on correct trials. Each row of tick marks represents spiking during a single trial. Bar graphs above tick marks indicate mean firing rates every 250 ms. The overall mean firing rate of this interneuron was 24.563 Hz. *Center:* Schematic of the context-guided odor–reward association task. Pairs of odors are differentially rewarded depending upon the context in which they are presented. (**d**) The spiking activity of this interneuron contains significant information (see Sect. 2) for odors, positions, and odor–position combinations. Information scores are indicated above distributions of information scores from shuffled data

in their firing rates. An example of an interneuron with significant spike-phase coherence to all four rhythms and significant information for all three task dimensions is shown in Fig. 1. The median information for odors was 0.015 bits/spike (N = 21, interquartile range = 0.0194), the median information for position was 0.038 bits/spike (N = 47, interquartile range = 0.0720), and the median information for odor–position was 0.041 bits/spike (N = 51, interquartile range = 0.0609). The median of the average firing rates exhibited by this group of interneurons during correct trials was 24.57 Hz (interquartile range 12.45 Hz).

Although there were interneurons with significant spike-phase coherence to combinations of the four rhythms examined (e.g., theta only, theta and high gamma, etc.), few of these interneurons also exhibited significant information for task dimensions. For example, the population of interneurons with significant spike-phase coherence during correct trials to only $\text{theta}_{4\text{–}12\,\text{Hz}}$ (14 interneurons, 6 rats, 18 half-sessions) produced just three instances of significant information for each task dimension. Given these low numbers, analyses were restricted to the population of interneurons coherent to all four of the rhythms examined during correct trials.

We then tested whether the information for a specific task dimension changed over the course of a trial (Fig. 2). We compared the information content during the *before*, *odor*, and *end* intervals. The interneurons coherent to all four rhythms exhibited an increase in odor information across the three intervals examined (Friedman's test: d.f. = 2, $\chi^2 = 12.67$, $p = 0.0018$). Post hoc comparisons

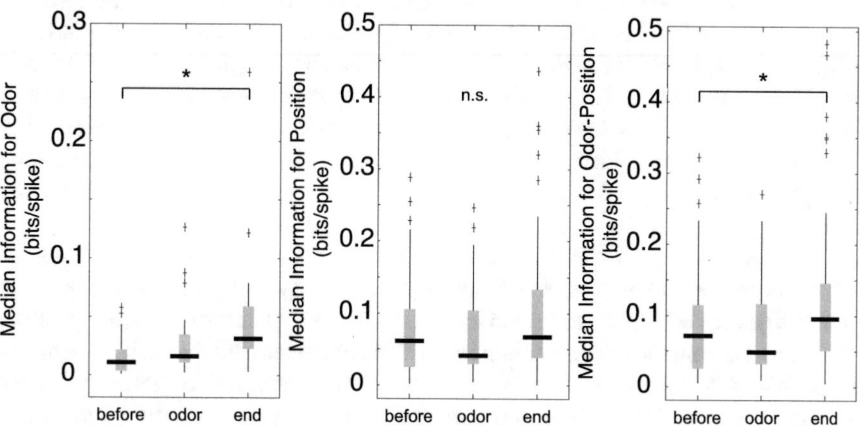

Fig. 2 *Left:* Median information (bits/spike) for odors for interneurons exhibiting significant spike-phase coherence relationships to all four rhythms examined ($\text{theta}_{4\text{–}12\,\text{Hz}}$, $\text{beta}_{15\text{–}35\,\text{Hz}}$, low $\text{gamma}_{35\text{–}55\,\text{Hz}}$, and high $\text{gamma}_{65\text{–}90\,\text{Hz}}$) during a 500-ms interval prior to nose poke (*before*), a 500-ms interval directly after odor delivery (*odor*), and 500 ms prior to the end of the nose poke (*end*). Vertical gray bars indicate the interquartile range. The top vertical line indicates $q3 + 1.5 \times (q3 - q1)$ and the bottom vertical line indicates $q1 - 1.5 \times (q3 - q1)$, where $q1$ and $q3$ are the 25th and 75th percentiles, respectively. Asterisks (*) indicate a significant pairwise comparison using a Tukey's Honest Significant Difference test, $p < 0.05$. *Middle:* Same as in A, for position information. *Right:* Same as in B, for odor–position information

revealed that this increase occurred at the end of the nose poke, but not during the interval immediately after odor delivery (Tukey's Honest Significant Difference test, $p < 0.05$ for comparisons of the before interval to odor and end intervals). These interneurons also exhibited a similar increase in odor–position information across the three intervals examined (Friedman's test: d.f. $= 2$, $\chi^2 = 11.76$, $p = 0.0028$) that occurred only at the end of the nose poke (Tukey's Honest Significant Difference test, $p < 0.05$). This interneuron group did not exhibit increases in position information across the three intervals examined (Friedman's test: d.f. $= 2$, $\chi^2 = 3.87$, $p = 0.1443$).

4 Discussion

Our preliminary results suggest that interneurons may convey information about task dimensions in their firing rates. The low, but significant information for task dimensions in the interneuron population indicates that they have potentially subtle, but reliable changes in firing rate that account for changing task conditions. Notably, the median information for each dimension is roughly an order of magnitude lower than previously reported medians derived from the firing rates of the pyramidal cell population [3]. To illustrate the types of firing rate changes that might occur to produce information scores of this magnitude, we provide the following extreme case: an interneuron with a change in firing rate at only one of four positions, a mean firing rate of 24.5741 Hz (the median for this group of interneurons), and a position information value of 0.038 bits/spike (the median for this group of interneurons), would need to exhibit a firing rate increase of approximately 14 Hz. The extent to which this degree of selectivity impacts the hippocampal network remains to be determined. Although the interneurons generally have low information values, more analyses must be applied to better characterize the range and reliability of the firing rate changes driving the significant information scores in this study [8, 17].

Interneurons demonstrated an increase in firing rate selectivity over the course of the nose poke interval that reflected the availability of information about task dimensions. Specifically, the relatively stable degree of information for position before and during the odor sampling interval is consistent with the early availability of position information prior to the nose poke. In contrast, information about odors became available 250 ms after the initiation of a nose poke, and the interneurons demonstrate increases in information for these dimensions only at the end of the odor sampling interval. These results suggest that the selectivity of interneurons' firing rates for these dimensions is a product of task-relevant engagement.

A number of future analyses could enhance the interpretive power of these preliminary results. For instance, future characterizations of this data could employ information measures that are more sensitive to the direction of changes (increases or decreases) in firing rate demonstrated by interneurons over the course of a trial. In addition, it would be informative to quantify information during a number of additional intervals during the task, including incorrect trials, correct rejection trials, and reward consumption intervals. Quantifying information during these additional intervals would help assess the degree to which the observed increases in information are related to a reward. Further exploration will facilitate the development of more nuanced and comprehensive hypotheses as to the mechanisms through which interneurons contribute to successful information processing in the hippocampus.

Leading hypotheses in the field have proposed that interneuron activity crucially shapes the oscillatory profile of the hippocampus, providing a temporal scaffolding within which pyramidal neurons can receive [18, 19], process, and successfully transmit behaviorally relevant information [20, 21]. Support for this view stems in part from the organization of the hippocampal network, where the locally projecting interneuron population densely innervates large numbers of principal cells [22, 23]. This widespread innervation, compounded with high firing rates [24], amplifies the impact of interneurons' inhibitory currents and places interneurons in a unique position to coordinate the simultaneous activity of large ensembles of pyramidal cells. In particular, periodically occurring inhibitory currents create alternating windows of suppression and relative excitability in the principal cell population, ensuring that only precisely timed inputs are able to elicit principal cell responses [21]. In this way, interneuron activity has the ability to simultaneously sculpt the oscillatory profile of the hippocampus while constraining the subset of inputs to which principal cells can respond [25]. The observed interneuron selectivity for task dimensions suggests that interneurons are capable of shaping the overall information content of the network through both the temporal coordination of principal cell excitability and reliable changes in firing rate.

The mechanisms whereby interneurons acquire the observed selectivity for task dimensions remain obscure. Here, we briefly consider two hypotheses for the emergence of this information content. On the one hand, interneurons might inherit information directly from the afferents that simultaneously recruit subsets of principal cells in a feedforward inhibitory manner. It is also possible that the observed firing rate selectivity for task dimensions in the interneuron population emerges directly from their interactions with various ensembles of pyramidal cells, each containing highly selective information for distinct task dimensions. Our finding contributes to a growing body of evidence that interneurons are active facilitators of task-relevant processing through at least two forms of selective engagement: reorganization of spike timing into multiple rhythms and the conveyance of information about task dimensions through changes in firing rates [14, 17, 26, 27].

Acknowledgments Neural recordings were collected in the laboratory of Dr. Howard Eichenbaum at the Boston University Center for Memory and Brain. This work was partially supported by the NSF DMS-1042134 and MH052090. We would like to thank Jon Rueckemann, Katherine Keefe, Blake Porter, Ian Heimbuch, Carl Budlong, Jeremiah Rosen, Khushboo Chawla, Brian Ferreri, Catherine Mikkelsen, and Rapeechai Navawongse for technical assistance.

References

1. J. O'Keefe, J. Dostrovsky, The hippocampus as a spatial map. Preliminary evidence from unit activity in the freely-moving rat. Brain Res. **34**(1), 171–175 (1971)
2. E.R. Wood, P.A. Dudchenko, H. Eichenbaum, The global record of memory in hippocampal neuronal activity. Nature **397**(6720), 613–616 (1999)
3. L.M. Rangel, J.W. Rueckemann, P.D. Rivière, K.R. Keefe, B.S. Porter, I.S. Heimbuch, C.H. Budlong, H. Eichenbaum, Rhythmic coordination of hippocampal neurons during associative memory processing. eLife **5**, e09849 (2016)
4. R.W. Komorowski, J.R. Manns, H. Eichenbaum, Robust conjunctive item-place coding by hippocampal neurons parallels learning what happens where. J. Neurosci. **29**(31), 9918–9929 (2009)
5. K.M. Igarashi, L. Lu, L.L. Colgin, M.-B. Moser, E.I. Moser, Coordination of entorhinal-hippocampal ensemble activity during associative learning. Nature **510**(7503), 143–147 (2014)
6. W. Skaggs, B. McNaughton, K. Gothard, E.J. Markus, An information-theoretic approach to deciphering the hippocampal code. Proc. IEEE **1990**, 1030–1037 (1993)
7. E.J. Markus, C.A. Barnes, B.L. McNaughton, V.L. Gladden, W.E. Skaggs, Spatial information content and reliability of hippocampal CA1 neurons: effects of visual input. Hippocampus **4**(4), 410–421 (1994)
8. A.V. Olypher, P. Lánský, R.U. Muller, A.A. Fenton, Quantifying location-specific information in the discharge of rat hippocampal place cells. J. Neurosci. Methods **127**(2), 123–135 (2003)
9. M.A. Wilson, B.L. McNaughton, Dynamics of the hippocampal ensemble code for space. Science **261**(5124), 1055–1058 (1993)
10. K. Harris, J. Csicsvari, H. Hirase, G. Dragoi, G. Buzsáki, Organization of cell assemblies in the hippocampus. Nature **424**(July), 552–556 (2003)
11. J.F. Guzowski, J.J. Knierim, E.I. Moser, Ensemble dynamics of hippocampal regions CA3 and CA1. Neuron **44**(4), 581–584 (2004)
12. G. Buzsáki, Neural syntax : cell assemblies, synapsembles, and readers. Neuron **68**, 362–385 (2010)
13. S. McKenzie, A.J. Frank, N.R. Kinsky, B. Porter, P.D. Rivière, H. Eichenbaum, Hippocampal representation of related and opposing memories develop within distinct, hierarchically organized neural schemas. Neuron **83**(1), 202–215 (2014)
14. E. Stark, R. Eichler, L. Roux, S. Fujisawa, H.G. Rotstein, G. Buzsáki, Inhibition-induced theta resonance in cortical circuits. Neuron **80**(5), 1263–1276 (2013)
15. J. Csicsvari, B. Jamieson, K.D. Wise, G. Buzsáki, Mechanisms of gamma oscillations in the hippocampus of the behaving rat. Neuron **37**(2), 311–322 (2003)
16. P. Barthó, H. Hirase, L. Monconduit, M. Zugaro, K.D. Harris, G. Buzsáki, Characterization of neocortical principal cells and interneurons by network interactions and extracellular features. J. Neurophysiol. **92**(1), 600–608 (2004)
17. W.B. Wilent, D.A. Nitz, Discrete place fields of hippocampal formation interneurons. J. Neurophysiol. **97**(6), 4152–4161 (2007)
18. F. Pouille, M. Scanziani, Enforcement of temporal fidelity in pyramidal cells by somatic feed-forward inhibition. Science **293**(5532), 1159–1163 (2001)
19. F. Pouille, M. Scanziani, Routing of spike series by dynamic circuits in the hippocampus. Nature **429**(6993), 717–723 (2004)
20. G. Buzsáki, J.J. Chrobak, Temporal structure in spatially organized neuronal ensembles: a role for interneuronal networks. Curr. Opin. Neurobiol. **5**(4), 504–510 (1995)
21. J. Cannon, M.M. McCarthy, S. Lee, J. Lee, C. Borgers, M.A. Whittington, N. Kopell, Neurosystems: brain rhythms and cognitive processing. Eur. J. Neurol. **39**(5), 705–719 (2014)
22. T.F. Freund, G. Buzsáki, Interneurons of the hippocampus. Hippocampus **6**(4), 347–470 (1996)
23. A. Sik, M. Penttonen, A. Ylinen, G. Buzsáki, Hippocampal CA1 interneurons: an in vivo intracellular labeling study. J. Neurosci. Off. J. Soc. Neurosci. **15**, 6651–6665 (1995)

24. S. Fox, J. Ranck, Electrophysiological characteristics of hippocampal complex-spike cells and theta cells. Exp. Brain Res. **41**(3), 399–410 (1981)
25. M.A. Whittington, R.D. Traub, N. Kopell, B. Ermentrout, E.H. Buhl, Inhibition-based rhythms: experimental and mathematical observations on network dynamics. Int. J. Psychophysiol. **38**(3), 315–336 (2000)
26. D. Nitz, B. McNaughton, Differential modulation of CA1 and dentate gyrus interneurons during exploration of novel environments. J. Neurophysiol. **91**, 863–872 (2004)
27. S. Royer, B.V. Zemelman, A. Losonczy, J. Kim, F. Chance, J.C. Magee, G. Buzsáki, Control of timing, rate and bursts of hippocampal place cells by dendritic and somatic inhibition. Nat. Neurosci. **15**(5), 769–775 (2012)

Learning-Induced Sequence Reactivation During Sharp-Wave Ripples: A Computational Study

Paola Malerba, Katya Tsimring, and Maxim Bazhenov

Abstract During sleep, memories formed during the day are consolidated in a dialogue between cortex and hippocampus. The reactivation of specific neural activity patterns—replay—during sleep has been observed in both structures and is hypothesized to represent a neuronal substrate of consolidation. In the hippocampus, replay happens during sharp-wave ripple complexes (SWR), when short bouts of excitatory activity in area CA3 induce high-frequency oscillations in area CA1. In particular, recordings of hippocampal cells which spike at a specific location ("place cells") show that recently learned trajectories are reactivated during CA1 ripples in the following sleep period. Despite the importance of sleep replay, its underlying neural mechanisms are still poorly understood.

We used a previously developed model of sharp-wave ripples activity, to study the effects of learning-induced synaptic changes on spontaneous sequence reactivation during CA3 sharp waves. In this study, we implemented a paradigm including three epochs: Pre-sleep, learning, and Post-sleep activity. We first tested the effects of learning on the hippocampal network activity through changes in a minimal number of synapses connecting selected pyramidal cells. We then introduced an explicit trajectory-learning task to the learning portion of the paradigm, to obtain behavior-induced synaptic changes. Our analysis revealed that recently learned trajectories were reactivated during sleep more often than other trajectories in the training field. This study predicts that the gain of reactivation rate during sleep following vs sleep preceding learning for a trained sequence of pyramidal cells depends on Pre-sleep activation of the same sequence, and on the amount of trajectory repetitions included in the training phase.

P. Malerba (✉) · M. Bazhenov
Department of Medicine, University of California San Diego, San Diego, CA, USA
e-mail: pmalerba@ucsd.edu

K. Tsimring
University of California San Diego, San Diego, CA, USA

A. Deines et al. (eds.), *Advances in the Mathematical Sciences*, Association for Women in Mathematics Series 15, https://doi.org/10.1007/978-3-319-98684-5_11

1 Introduction

Memories are composed of three stages: acquisition, consolidation, and retrieval. The consolidation phase provides memory resilience to interference, and is influenced by sleep. Indeed, memory performance benefits from sleep [1–5], and specific sleep features have been linked to increased memory performance [4, 6–8]. Sleep is a stage in which the brain can be described as self-organizing, because it is not processing external inputs. During sleep, the local average activity of brain regions can be measured to find characteristic oscillations which vary across frequency, brain regions, and timescales. Furthermore, experimental results show that the coordination of brain rhythms during sleep promotes (and possibly mediates) memory consolidation both in humans and animals [9, 10]. In particular, sharp-wave ripples (SWR) are rhythms present in wake and sleep in the hippocampus, a brain region which is crucial in encoding and retrieving recently formed memories [11–13]. An SWR complex is formed by the combination of two events, both occurring simultaneously in different layers of the local field potential (LFP) of hippocampal area CA1. (1) In stratum radiatum, a large deflection (the sharp wave) lasting 100–200 ms is caused by a barrage of excitatory inputs coming from area CA3, where a large spiking event among the highly interconnected pyramidal cells generates intense drive for area CA1 cells (both pyramidal and interneurons). (2) In stratum pyramidale, high-frequency (>150 Hz) short-lived (50–80 ms) oscillations (the ripple) are seen as a result of the CA3 input. In fact, the local fast-spiking interneurons of CA1 are known to fire at high frequency for the duration of a ripple, so their spikes are hypothesized to pace the oscillation [12, 14–17]. During a ripple, CA1 pyramidal cells are receiving excitation from CA3 and inhibition at ripple frequency from CA1 interneurons; as a result, most of them are suppressed, and the few (about 10% [14, 15]) that spike do so within windows of opportunity left by the local interneurons. In the following, we will refer to CA3 activity during SWR complexes as "CA3 sharp waves" and the CA1 activity during SWR complexes as "CA1 ripples."

During learning, some hippocampal pyramidal cells (mostly located in CA1, but also in CA3) that fire in a specific location as the animal explores an environment have been labeled "place cells" [18–20]; and it has been shown that cells which are active together during learning (e.g., place cells that code for nearby locations explored during a recently learned task) also activate in the same SWR complexes during sleep [21–23]. Furthermore, if the task involves learning a specific path and the spikes of place cells along that path are recorded, the sequence of spikes among those cells is reactivated in the correct order (in a time-compressed manner) during CA1 ripples in the subsequent sleep epoch [24, 25]. A specific path learned during a task can then be seen reactivated (in a time-compressed manner) during both awake and sleep CA1 ripples [26]. Sequence reactivation during ripples in sleep directly affects memory: both the spike sequences within ripples and the number of ripples during sleep correlate with memory performance, and suppression of sleep ripples impairs memory consolidation [27, 28]. Hence, it is crucial to explain how spiking activity related to learning enables the reactivation of the correct spike sequences in ripples during sleep.

In general, the influence of learning over sleep reactivation content is performed by comparing spiking in the nights before (Pre-sleep) and after (Post-sleep) a learning paradigm. When place cells which code for locations that are relevant to the learned task reactivate more strongly in Post-sleep compared to Pre-sleep, it is hypothesized that learning the task has influenced the spiking content of SWR complexes reactivation in both CA3 and CA1. The specific mechanisms mediating such influences are yet to be determined. Common hypotheses involve synaptic plasticity and neuromodulators, and the overall theory states that during learning (and possibly during awake SWR complexes) synapses in the hippocampus are changed, and those changes induce the enhanced representation of the learned spiking sequences in SWR complexes during Post-sleep [13, 22, 25].

In vivo studies have shown that during sleep, pyramidal cells can be divided into those involved in many CA1 ripples and those mostly not spiking in ripples, and this separation seems to persist across days [11, 14–16]. Furthermore, in a Pre-sleep/experience/Post-sleep paradigm, Grosmark et al. [29] have shown that CA1 pyramidal cells can be separated into "rigid" and "plastic." The rigid group has a higher firing rate and low spatial specificity, and shows very little change across the sleep/experience/sleep paradigm. The plastic group has a lower firing rate, but shows increasing spatial specificity and increasing ripple reactivation during the experience phase of the paradigm. The plastic cells are those which go on to show increased bursting and co-activation during CA1 ripples in the Post-sleep phase of the paradigm. In this work, we investigated how synaptic plasticity can influence spontaneous reactivation of spike sequences during sleep in our previously developed biophysical model of SWR activity, by focusing on CA3 activity in a Pre-sleep/learning/Post-sleep paradigm. The effect of learning on cell reactivation was first studied in a simplified representation, where we manually changed very few synapses, and then in a learning paradigm that was explicitly modeled to represent a "virtual rat" exploring an environment, within which a specific trajectory was rewarded (and hence learned). The spike times obtained for the training phase were used to induce offline spike-timing-dependent plasticity (STDP) [30–32] in synapses among cells in the network, leading to different spiking profiles in Pre- and Post-sleep simulations. We find that learning-dependent plasticity is able to enhance the representation of cell sequences (and hence space trajectories) during spontaneous CA3 sharp waves. Our model predicts that this enhancement depends on the co-activation in the Pre-sleep epoch, the timing (within sharp waves) of the first spike of the spike sequence, and the amount of training included in the learning phase of the experiment.

2 Results

2.1 Changes in the Minimal Number of Synapses can Promote Reactivation of Cell Triplet in CA3

In this study, we started from our previously developed computational model of stochastic, randomly emerging SWR complexes [33, 34]. This model includes two hippocampal regions, CA3 and CA1 (Fig. 1a), each represented with pyramidal cells (excitatory) and basket cells (inhibitory interneurons). Area CA3 has highly recurrent excitatory connectivity [35] and sends excitatory projections to area CA1 (the Schaffer Collaterals), reaching both excitatory and inhibitory neurons [35]. Within CA1, the excitatory connections between pyramidal cells are very few and sparse, as shown by experimental results [36]. Parameters for cells in our model are chosen so that: (1) pyramidal cells have physiological firing rates and show bursting [37], and (2) basket cells are fast-spiking neurons, with very low spike-frequency adaptation, and are then capable of spiking at high frequencies in response to sustained strong inputs [37]. While the rules shaping network connectivity stayed the same, multiple instances of specific connectivity matrices were used, to represent different "virtual rats" in our computational study.

Our CA3–CA1 model activity is carefully fitted to biophysical data derived from in vivo sharp-wave ripple recordings. In agreement with data [11, 14–16, 38], our model shows: (a) the bursting input response of pyramidal cells and the fast-spiking input response of interneurons [37], (b) the stochasticity of sharp-wave ripple events in time, (c) the background activity being random and not showing a specific frequency signature (called Large Irregular Activity in the electrophysiology literature), (d) the low fraction of CA1 pyramidal cells spiking during ripples and the large number of spikes from CA1 inhibitory cells during ripples, (e) the ripple frequency and duration matching physiological data, (f) the relative difference between pyramidal cell activity in CA3 sharp waves vs CA1 ripples (as shown in [11, 38, 39]) (see also [40]).

In our model, every cell was in a noise-driven spiking regime (as opposed to a limit cycle periodic spiking regime), a feature which introduced random fluctuations in the background network activity. Recurrence within excitatory neurons in CA3 was the gateway for occasional large excitatory spiking events (the sharp waves, SPW) which were projected to CA1 interneurons and pyramidal cells (Fig. 1b). Concurrently, the drive from CA3 sharp-wave spiking imposed high-frequency firing on the basket cells of area CA1, which formed the structure for the ripple, and determined the ripple frequency. All the while, CA1 pyramidal cells received excitation from CA3 and inhibition from CA1 basket cells. Since CA1 pyramidal cells lack recurrent excitatory synaptic connections to organize their firing, the competition between these inhibitory and excitatory inputs only allowed a small percentage of pyramidal cells in CA1 to spike during a ripple, with timing controlled by windows of opportunity left by the ongoing inhibitory ripple oscillations (Fig. 1b). Note that in our model, sharp waves in CA3 did not invade the totality of

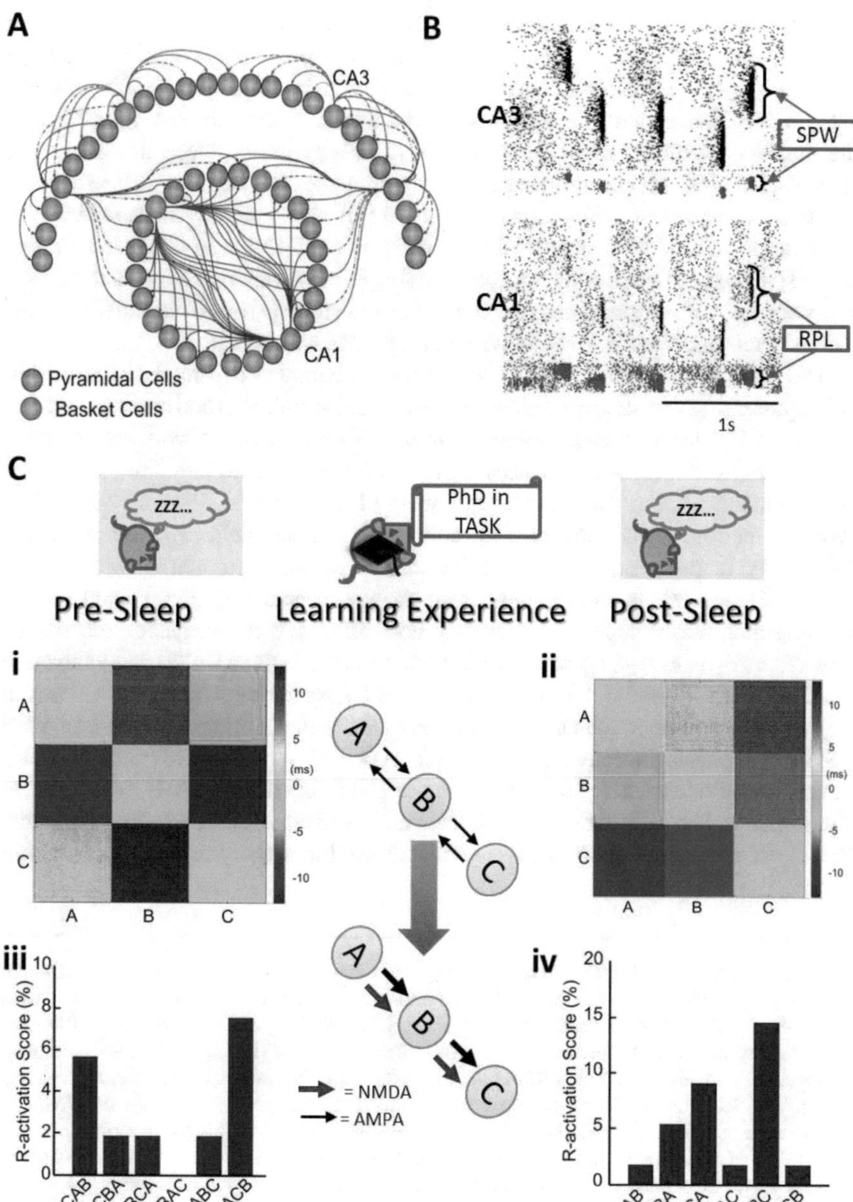

Fig. 1 Changes in the minimal number of synapses can promote reactivation of cell triplet. (**a**) Model schematic representation. The model includes 1200 pyramidal cells and 240 interneurons in CA3, and 800 pyramidal cells and 160 interneurons in CA1. Connections within CA3 include high recurrence among pyramidal cells, and a topological preference for neighbors within a radius of about one third of the network. CA3 pyramidal cells project to CA1 cells (both pyramidal cells and interneurons). Within CA1, the all-to-all connectivity is not effective between CA1 pyramidal

the network, but were instead localized. This is consistent with in vivo recordings showing that different simultaneous channels capture SWR complexes activity differently [41]. Localized sharp-wave activity in CA3 drove localized ripple activity in CA1, as can be seen in Fig. 1b. In this work, we took advantage of the biophysical model of SWR spontaneous activity during sleep to analyze the role of changing excitatory synapses in shaping the activation of cell sequences across sharp waves in CA3. In previous studies of our model, we have shown how reactivation in CA1 ripples crucially depends on reactivation in CA3 sharp waves [33, 34]; hence, to understand how learning shapes reactivation in SWR complexes it is necessary to start from reactivation within CA3 sharp waves. This study centered on CA3 cells and their spiking activity during SWR complexes.

The main setup of this study (Fig. 1c) was to compare two model simulations, one representing Pre-sleep (the sleep epoch preceding training) and one representing Post-sleep (the sleep event following training). If the specific network connectivity, parameters, and input-noise traces were kept the same in the Pre- and Post-sleep simulations, the two sleep events would look identical. We imposed offline synaptic modifications between Pre- and Post-sleep simulations to represent the effects of a learning experience and then run a Post-sleep simulation where every input and connections were copied from the Pre-sleep simulation, except for the learning-modified synapses. To quantify how often any cell sequence reactivated spontaneously across sharp waves in a simulation, we defined its "Ripple-activation score" (R-activation score) as the percentage of SWRs during which all CA3 cells in the sequence spiked in the correct order. For example, if during a given simulation 8 SWR complexes are found, and the triplet ABC spiked in this order in all 8 sharp waves, it would have an R-activation score of 100%, but if triplet ABC was found in order in only 4 sharp waves, it would have an R-activation score of 50%. Note that even if we concentrated our analysis on CA3 spiking activity and synapses within

Fig. 1 (continued) cells, which are weakly and sparsely connected. (**b**) Example of model network SWR activity. Noise-driven spiking in CA3 occasionally triggers an excitatory cascade of spikes in pyramidal cells and interneurons (the sharp wave, SPW). Spiking of CA3 pyramidal cells drives CA1 interneurons to spike in short-lived high-frequency ripples (RPL), while CA1 pyramidal cells receive competing excitation from CA3 and later inhibition from CA1 interneurons. A few CA1 pyramidal cells are driven to spike within windows of opportunity left by the rhythmic local inhibition. (**c**) Representation of the main question addressed in this work: Comparing the reactivation of cell sequences in two sleep epochs, and study the role of learning in shaping the difference. In our model, we represent learning by altering synaptic connections as shown in the middle plot: AMPA synapses promoting the spiking order "ABC" are increased, and the AMPA synapses promoting the opposite spike order are removed. In addition, NMDA synapses in the direction favoring "ABC" spiking are introduced. Data from one example simulation is used to show how spiking changes in Post- vs Pre-sleep simulations. Matrices show an example of spike-time differences between three selected pyramidal cells in the CA3 network in a Pre-sleep simulation (**c** i) and a Post-sleep simulation (**c** ii). The bar plots (**c** iii–iv) show the R-activation score (% of ripples in which a given triplet spiked) for the ordered triplet "ABC" and all its permutations. Note that the "ABC" order R-activation is larger for Post-sleep than it is for Pre-sleep

CA3, we used the outcome of full CA3–CA1 network simulations to detect ripple events in CA1, and only sharp waves in CA3 which induced ripples in CA1 were used for the CA3 spiking analysis (in fact, we have shown that only CA3 sharp waves which are large enough and synchronous enough will induce a ripple in CA1 [40]). This motivates the label of R-activation score in our analysis.

As for representing the effects of learning on synaptic changes, there are many possible factors to consider (for example, inhibitory mechanisms [42]). However, synaptic plasticity at AMPA synapses, mediated by changes in NMDA connections, has been shown to occur between hippocampal excitatory cells in spike-dependent synaptic plasticity paradigms [43, 44], where the relative timing of spikes of a pre- and a postsynaptic cell regulates the resulting change in synaptic strength. In particular, the mechanism of long-term potentiation (LTP), in which synapses are strengthened and remain stronger for a long time after the spiking of pre- and postsynaptic cells causes them to change, affecting AMPA synapses through changes in NMDA receptors, is still considered the main mechanism through which the hippocampus carries out learning of declarative memories [45]. In hippocampus, LTP results in increased AMPA receptors, counterbalanced by decreased presence of AMPA receptors at synapses where spike timing shows postsynaptic cells spiking consistently before the presynaptic cells (known as Long-Term Synaptic Depression [46], and possible increase of NMDA receptors (hence we introduced NMDA in our connections). In the real brain, many of the detailed possible variants to this mechanism are present, but the 3 steps we implemented in this test are (at least in principle) known to happen together, which is why we included them all in the picture.

In a first simplified setting, we chose three CA3 pyramidal cells (cells A, B, and C) to form an ordered triplet (our most simple cell sequence) and found their R-activation score during the Pre-sleep simulation (in Fig. 1c iii, one example of ABC R-activation score is shown together with the R-activation scores of all its permutations). We then altered the excitatory synaptic connections between the ABC cells as follows (Fig. 1c, center panel): we found the strengths of AMPA (short-lived) excitatory synapses which connect A to B and B to C, and replaced their synaptic strengths with the maximum value across all the CA3–CA3 pyramidal cell synapses in the network. We also introduced NMDA (long-lasting) excitatory synapses between A–B and B–C cells. Finally, we removed (if present) the excitatory AMPA synapses from B to A and from C to B. It is to note that in Pre-sleep there are no NMDA synapses in the network, so we did not need to remove any of them. In the example of Fig. 1c, this artificial change in very few synapses in the network led to a strong increase in the R-activation score of ABC in the Post-sleep simulation (Fig. 1c iv). The ability of the new synapses to improve the reactivation of the ABC spiking sequence was further supported by the comparison between the average spike-time differences between cells of the triplet in Pre- (Fig. 1c i) and Post-sleep (Fig. 1c ii).

2.2 Pre-Sleep Activation Modulates Learning-Induced Gain in Post-Sleep Reactivation

Using the setup we developed in the previous section, we could represent activity in the Pre- and Post-sleep phases and quantify the change in sequence reactivation between the two sleep stages due to synaptic changes introduced during learning. Across multiple Pre-sleep simulations (n = 17), we found an inherent range of R-activation scores for ordered cell triplets (Fig. 2a) which, surprisingly, was not uniformly distributed among triplets. In fact, we could fit a log-normal curve to the distribution, in agreement with numerous experimental observations [11, 12], which also emphasizes that this type of variance in cells firing rates and R-activations could optimize coding strategies for the hippocampal system. We chose to study the effects of changing few synapses over a pool of ABC triplets spanning 0–25% R-activation scores in the Pre-sleep simulation to cover most of the range of R-activations found in the distribution (see Computational Methods for details on how the cells composing each triplet were chosen). Each simulation/triplet pair (which can be thought of as representing a subject/task pair in an experimental study) then underwent the synaptic modifications introduced in Fig. 1c: maximized AMPA synapses favoring the triplet order, removing AMPA synapses opposing the triplet order and introducing NMDA synapses along the maximized AMPA ones. For each simulation/triplet case, we then compared Pre- and Post-sleep R-activations of the ABC triplet. We call the difference between the Post-sleep score and the Pre-sleep score for a triplet its "Score Gain." In Fig. 2b, we show that the Score Gain of a triplet correlated with its Pre-sleep R-activation, and hence when this simplified learning was applied to a triplet with too high R-activation during Pre-sleep it could not increase its R-activation, or might even have decreased it. This is intuitively coherent with some kind of "ceiling effect," in which Post-sleep R-activation could only reach a preset amount (possibly limited by the overall spontaneous network activity, which is imposed by average network properties) and hence a Pre-sleep R-activation too high limited the available range for Gain.

In this simplified representation of a learning effect on synapses, it is important to note that the first cell in the triplet (cell "A") does not receive any change in its input in Post-sleep compared to Pre-sleep. In some sense, this procedure could be overlooking the very beginning of the sequence to be reactivated, and that could be the reason why there was such a hard ceiling effect on the R-activation Gain for triplets in Fig. 2b. To address this limitation, we first "prolonged" our test set, introducing a fourth cell to the sequence undergoing artificial synaptic manipulation. The rules for changing synapses stayed the same: maximize AMPA connections in the direction of the ordered sequence (A to B, B to C, and C to D), remove AMPA connections opposing the ordered sequence (B to A, C to B, and D to C), and introduce NMDA connections where AMPA connections were maximized. In this case, we could compare the R-activation Score Gain of two triplets within each test: ABC and BCD. The first triplet, just like in Fig. 2b, would not receive any enhancement to its first cell, but for the second triplet (BCD), its first cell would be

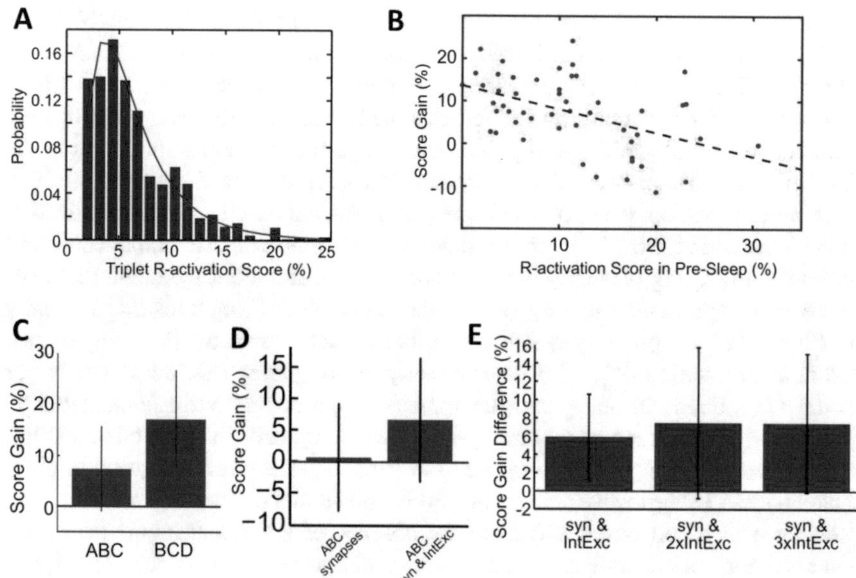

Fig. 2 Pre-sleep activation modulates learning-induced gain in Post-sleep reactivation. (**a**) Distribution of triplet reactivation scores across 17 simulations lasting 50 s. The red line shows a log-normal distribution fit to the data (mean = 1.6921 with 95% CI 1.6621–1.722, standard deviation = 0.6125 with 95% CI 0.5921–0.6345). (**b**) Gain in post-sleep reactivation is higher for triplets with lower pre-sleep activation. Each dot is a separate simulation pair, in which a triplet "ABC" R-activation is found in the Pre-sleep (x-axis value) and in the Post-sleep. The y-axis shows the difference between Post- and Pre-sleep R-activation score (i.e., Score Gain). The line shows a linear fit which emphasizes the statistically significant negative correlation ($r = -0.476$, $p = 0.0004$). (**c**) Gain in Post-sleep R-activation increases if the first cell of the triplet receives extra synaptic input. Comparing triplet "ABC" and "BCD" in simulations where the changes in synapses were applied along the word "ABCD." Hence, triplet "BCD" has all cells receiving extra synaptic input, while in triplet "ABC" only the last two cells do. Plot shows the average Post–Pre R-activation score gain across 15 simulation tests, lasting 50 s, and error bars mark standard deviations. Having synaptic input on the first cell of the sequence gives an extra gain to the R-activation in Post-sleep (paired student's t-test $p = 2.3545 \times 10^{-8}$). (**d**) Gain on Post-sleep reactivation can increase with increased intrinsic excitability of first cell in triplet. The bar plot shows the gain in R-activation score (Post-sleep minus Pre-sleep) for triplet "ABC" when only synapses are manipulated (left bar) compared with the case in which synapses and intrinsic excitability of cell "A" are manipulated to represent learning. Error bars mark standard deviation. Note that on average the gain is higher for triplets with enhanced intrinsic excitability in the first cell (paired student's t-test $p = 0.003$). In this case, we could only use triplets in which the first cell did not have a high intrinsic excitability in Pre-sleep (artificially increasing intrinsic excitability can lead to the cell spiking behavior changing from noise-driven to DC-current driven, and hence the cell spikes continuously and decoupled from network activity) (10 simulations used). (**e**) If the increase in intrinsic excitability is too large, no further gain is introduced. Bar plot shows the difference introduced by additional intrinsic excitability to cell "A" compared to Post-sleep with only synaptic manipulation. Error bars mark standard deviations. Each triplet received a fixed (equal) amount of increased excitability and such amount could be doubled (2xInt.Exc) or tripled (3xInt.Exc). Note that while the first bar shows that additional excitability to cell "A" can lead to increased gain (as compared to a case where only synapses are manipulated) the range of efficacy of such increase is small. In fact, additional increase in excitability does not reflect in additional gain in Post-sleep R-activation. Statistically, the first bar is different from zero ($p = 0.003$), while the other two are not (2xInt.Exc $p = 0.0191$, 3xInt.Exc $p = 0.0129$, student's t-test)

receiving enhanced excitatory synaptic input, both fast (AMPA) and slow (NMDA). In Fig. 2c, we show that extra excitatory synaptic input to the first cell did provide an advantage in Score Gain by comparing the Score Gains of the two different triplet types across 15 simulations. It is to note that the two distributions were significantly different (paired Student's t-test $p = 1.9 \times 10^{-6}$), which implies that input to the first cell in the triplet is capable of driving Post-sleep enhanced activation. In line with this observation, we next tested whether enhancing the changes of R-activation of the first cell in the triplet by means other than synaptic excitatory input could still introduce a favoring bias in Score Gain for a triplet. In fact, it is known that cells of the same type in the same region still show a range of firing rates (log-normally distributed [47]), a property which is considered useful for network coding of past and new information [29]. This heterogeneity in firing rates was introduced in our model via a direct current parameter representing a cell's "intrinsic excitability" (I_{DC} in the equation for membrane voltage, see Computational Methods), and has been shown to arise from heterogeneity in synaptic strength distribution in other models of CA3 sharp-wave activity [48]. It is known that different mechanisms can possibly alter a cell baseline (i.e., in the absence of additional input) firing rate and behavior, such as effects of the neuromodulators on conductances [49–51] or the balance of ionic concentrations [52]. To test if a learning-induced change in intrinsic excitability of the first cell in the triplet could compensate for the lack of additional synaptic input, we compared 10 simulations in which the same triplet underwent only synaptic changes (same paradigm as for Fig. 2b) or received a small increase of the intrinsic excitability in addition to the synaptic changes. In Fig. 2d, we show that a small increase in synaptic excitability of the first cell in the sequence led to enhanced Score Gain from Pre- to Post-sleep R-activation of a triplet. When we increased the intrinsic excitability amount added to cell "A," we found (Fig. 2e) that the effect was quickly saturated such as doubling or even tripling the increase could not significantly increase the Score Gain from Pre- to Post-sleep. This suggests that while increased excitability in the first cell could contribute to enhancing Score Gain, the overall Post-sleep R-activation was not fully dominated by any one given factor, such as synaptic excitatory inputs or intrinsic excitability, but rather established by their interaction with the overall network activity (and hence the Pre-sleep R-activation).

2.3 Simplified Learning Extends Word Length Reactivation

Within the same simplified approach of representing the effect of learning over hippocampal network connectivity by manually changing a small set of hand-selected synapses, we extended our analysis to sequences of spikes longer than a triplet. We chose a length of seven cell spikes to build our sequence, as SWR complexes are short-lived events and fitting longer sequences within a single event would be hard. In fact, there is an ongoing hypothesis that memories of paths which would require a long sequence of many place cells are reactivated across multiple

SWR complexes which happen in a very quick sequence (SWR packets [11]). While our computational model shows SWRs happening at times in groups interspersed by long pauses (data not shown), in this work we focus on characterizing how changes in synapses affect changes in R-activation within CA3 across many SWR complexes, and longer sequences would introduce an ulterior complexity, possibly masking relevant effects.

We represent a sequence of 7 cells with the 7-letter "word" ABCDEFG. In Fig. 3a, we show one example of the changes introduced in synaptic connectivity by our artificial learning representation. In comparing the matrices of synaptic connection weights between the cells representing ABCDEFG in the network, we maximized the connections favoring the direction of ordered reactivation, which resulted in a high synaptic strength in the upper diagonal of the Post-sleep synaptic connections matrix in Fig. 3a, and removed the connections favoring the opposite reactivation order, which resulted in all zeros in the lower diagonal of the Post-sleep synaptic connections. Note that all other connections between cells in the sequence were left unchanged. Furthermore, the middle schematic in Fig. 3a shows that NMDA connections were also attributed to the synapses which had maximized AMPA connection weights (further details on the composition of the sequence of 7 cells are reported in Computational Methods).

To quantify the effect of these synaptic changes on the R-activation of cells in the sequence, we considered progressively longer sub-words within the total word length, always starting from cell "A." Hence, we looked at the word "A" (length 1), "AB" (length 2), "ABC" (length 3), "ABCD" (length 4), etc. all the way to the full "ABCDEFG" word (length 7). For each word considered, across the different lengths, we found the percentage of SWR events in which the whole word was reactivated in the correct order, and called that the R-activation score of each word. We could then compare the R-activation scores of words of length 1–7 in Pre- and Post-sleep. It is important to study words of different lengths, because in principle the sequence could activate only in part, representing an incomplete memory reactivation, and estimating how much of the memory (i.e., the sequence) is reactivated in CA3 in the presence of spontaneous sharp-wave activity, and how learning influences the completeness of reactivation, is the objective of this study. In Fig. 3b, we show one example of the outcome of one simulation (Pre- and Post-sleep): as it is intuitively necessary, the R-activation scores are lower for longer words, as they include the shorter ones. In this case, our artificial synaptic manipulation resulted in larger R-activation during Post-sleep of word of all lengths but the full one, meaning the last cell never spiked at the end of the whole sequence. It is to note that this measure is strict in the sense that it does not count the reactivation of partial words in the sequence unless it represents a chunk from the beginning (i.e., reactivation of BCDE in an SWR is not counted). One reason why long sequences have very low probability to activate in our model is introduced by the structure within the sequence of spikes imposed by learning (i.e., the synaptic changes). In fact, while learning does enhance synapses to promote the "next spike" in a sequence, no synaptic strength is raised so high that it would cause a spike for sure, as this would be nonphysiological. Hence, the pattern completion with a

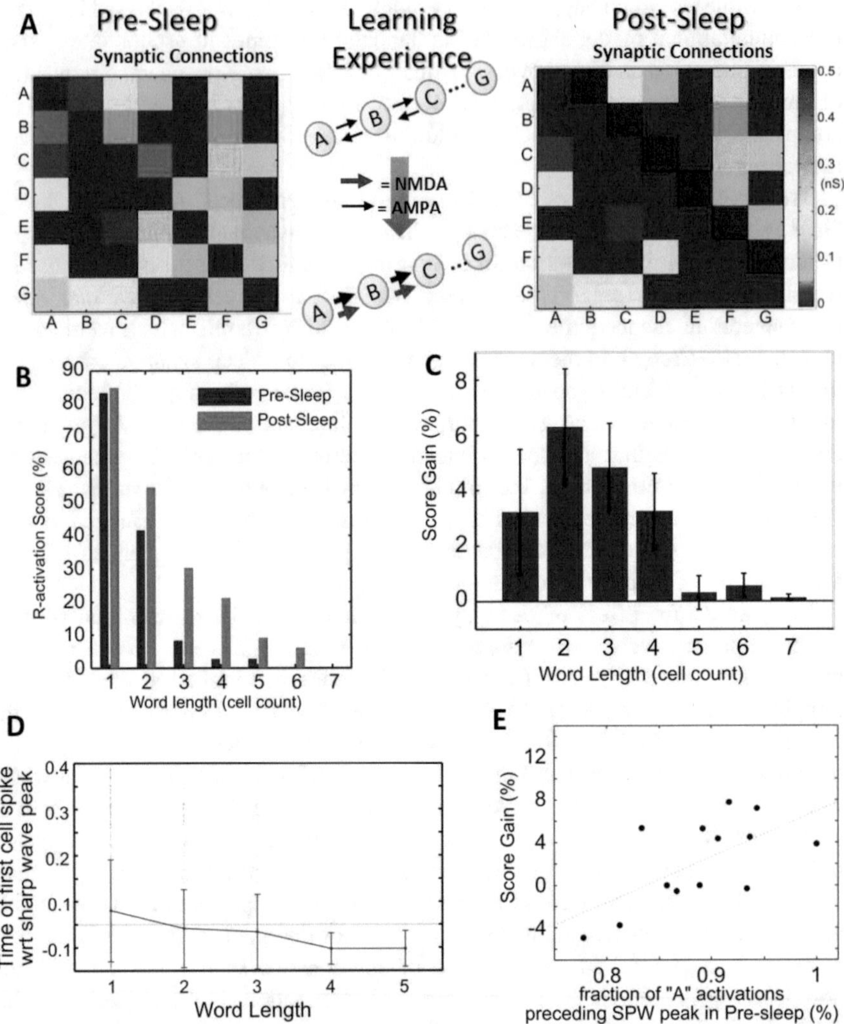

Fig. 3 Learning extends word length reactivation. (**a**) Changes in synaptic connections introduced in the paradigm. Drawing in the middle shows the manipulation we introduce to synaptic connectivity. AMPA synapses favoring the correct reactivation order were strengthened, while NMDA synapses favoring the word order were introduced. AMPA synapses promoting spiking in order opposite to the word order were removed. Matrices show the spike-time difference among cells which constitute the ABCDEFG word in one example simulation, both in Pre-sleep and in Post-sleep. (**b**) Learning increases word R-activation at many lengths in one example simulation. Plot shows the R-activation scores for words of increasing lengths ("A" = 1, "AB" = 2, "ABC" = 3, etc.) both in Pre-sleep (blue bars) and Post-sleep (green bars). (**c**) Across 14 simulations, the plot shows the differences between Post-sleep R-activation score and Pre-sleep R-activation score of words of increasing lengths, averaged. Error bars show standard error. (**d**) Length of word reactivation is related to the timing of word initiation during the sharp wave. For each word activating during ripples, the time of spike of the first cell "A" (with respect to

7-letter word would be very hard to implement for the network activity. Assuming independence of events, we can estimate that if the probability of postsynaptic spike occurring after presynaptic spike within a sequence is $P < 1$, then the probability of having all 7 letters would be P^7. Indeed, one can see in Fig. 3 that with each additional letter length in our reactivating word sequence, the ability of the network activity to pattern complete is reduced. Using this measure, we could subtract the Pre-sleep R-activation from the Post-sleep R-activation scores to obtain a Score Gain for the R-activation of words of increasing lengths. Across 14 simulations (the ones where the ABCDEFG sequence had an activation score below 23%), we found that artificially manipulating a small number of synapses promoted the R-activation of the sequence at all word lengths during Post-sleep (Fig. 3c). When studying the R-activation of triplets of cells (Fig. 2), we had found that inputs to the first cell in the triplet could influence the synaptic-induced Score Gain. For a longer sequence of spikes, the role of the first cell in the sequence was likely to still be relevant. In particular, we reasoned that the timing of the first spike in the sequence would affect the possible word length (within the sequence) which could reactivate in a given SWR complex: if the first spike happened too late in the CA3 sharp wave, only a small portion of the sequence could reactivate within the event duration. We found the time of the first spike (referenced to the time of the sharp-wave peak) for all words reactivating in a simulation, and plot them grouped by word length (Fig. 3d, $n = 14$ simulations, all Pre-sleep), and noticed that while the times were spread in a range (from -100 ms to $+400$ ms) the average time moment of the first spike was advanced for longer words compared to the shorter ones. In fact, for words of length 1 (single spikes), the average time of the first (and only spike) was after the sharp-wave peak, for words of length 2 ("AB") it was slightly before the peak, and length 3–5 showed an average first spike timing preceding the peak progressively, up to about 100 ms earlier. This property (the earliness of first spike in longer reactivations) was true for both Pre- and Post-sleep simulations (Fig. 2d), and the plots of word length vs first spike time did not look different in Pre- vs Post-sleep, meaning our plasticity manipulation did not

Fig. 3 (continued) the peak of the sharp wave) is found and plotted against the length of the word, as a red dot. $N = 14$ simulations, all Pre-sleep (i.e., before learning changes were introduced). The blue error bar identifies the mean and standard deviation of the all the dots corresponding to the words of the same length. The dotted horizontal line marks the peak of a sharp wave. Note that for longer activating words, the average timing of the first spike in a sharp wave is progressively advancing. (**e**) Learning-induced gain of R-activation of long words depends on timing of first cell in Pre-sleep. For each simulation, two values are computed: (1) the fraction of sharp waves in which cell "A" spiked at least 12 ms before the sharp-wave peak and (2) the difference between Post- and Pre-sleep R-activation scores (i.e., the score gain) for words of length 4 or above (summing the values of panel C for lengths 4–7). These two values are plotted against each other in the scattergram, where the fraction of sharp waves with early A spikes is the x-coordinate, and the total score gain is the y-coordinate. The linear fit shows a positive correlation (linear correlation $r = 0.6327$ with significance $p = 0.0203$). Note that in this figure we have discarded the data point (0.9667, 30.81) deemed an outlier because of its extremely high Score Gain

result in a generalized advancement in time of the first cell spike. Hence, whenever our plasticity manipulation induced a large Score Gain for long sub-words, it had to prolong the length of a reactivated short sub-word which started early enough within a sharp wave. Based on this idea, we compared for each simulation the total Score Gain for words of length 4 and above with the fraction of the total cell A spikes during Pre-sleep which happened early in a sharp wave. Figure 3e shows the scatterplot of such comparison, confirming a statistically significant positive correlation.

2.4 Experience-Related Learning: From Rat Trajectory to New Synaptic Connections

The next step in our modeling effort was to introduce an explicit relationship between a spatial learning experience and changes in synaptic connections, so that to create a virtual machinery in which a "rat" alternates spatial experiences (which can trigger learning among the network synapses) and sleeping experiences (which will show spontaneous reactivation during SWR activity). While modeling efforts to relate trajectory exploration to specific spiking of cells in a biophysical model have been developed [53–55], the specific mechanisms by which spatial perception results in place cell activity are currently under investigation and are not the main subject of our study. Hence, we chose to not design a network spiking model of the awake state, but instead to introduce a phenomenological model of place cell spiking activity as our awake/learning state.

To model the learning phaseof the "virtual rat," we follow a paradigm used to study episodic memory processing [56, 57]. We started by defining a 2-dimensional virtual enclosure (20 × 20 cm in size) and distributing 81 identical "place fields" along the surface: 2-D Gaussian bumps of integral 1 with peaks evenly spaced, as shown in Fig. 4a. In this virtual exploration enclosure, we introduced 8 locations of interest, marked with blue diamonds above the place fields in Fig. 4a. These locations could be of interest to the virtual rat because there could be feeders or other types of reward placed at any or all of these locations. In this virtual enclosure, we represented a learning task consistent with the experiments which drove this model [56, 57]: we assigned to 3 of the feeders a special relevance (circled in yellow in Fig. 4a) and we made the virtual rat explore 3 of these feeders in a fixed order, defining a learning trajectory (yellow arrows in Fig. 4a). A learning experience was then constructed by a string of locations to be reached in time, where the 3 ordered learning targets were interspersed with 3 randomly chosen (varying each time) locations among the remaining 5. This produced a space exploration path on the virtual enclosure (x_t, y_t), where t represents time, and x and y are the spatial coordinates on the 2-dimensional virtual enclosure. As the virtual animal traveled along its path within the enclosure, it traversed a number of place fields, triggering spikes with probabilities drawn from a Poisson process with rate given by the

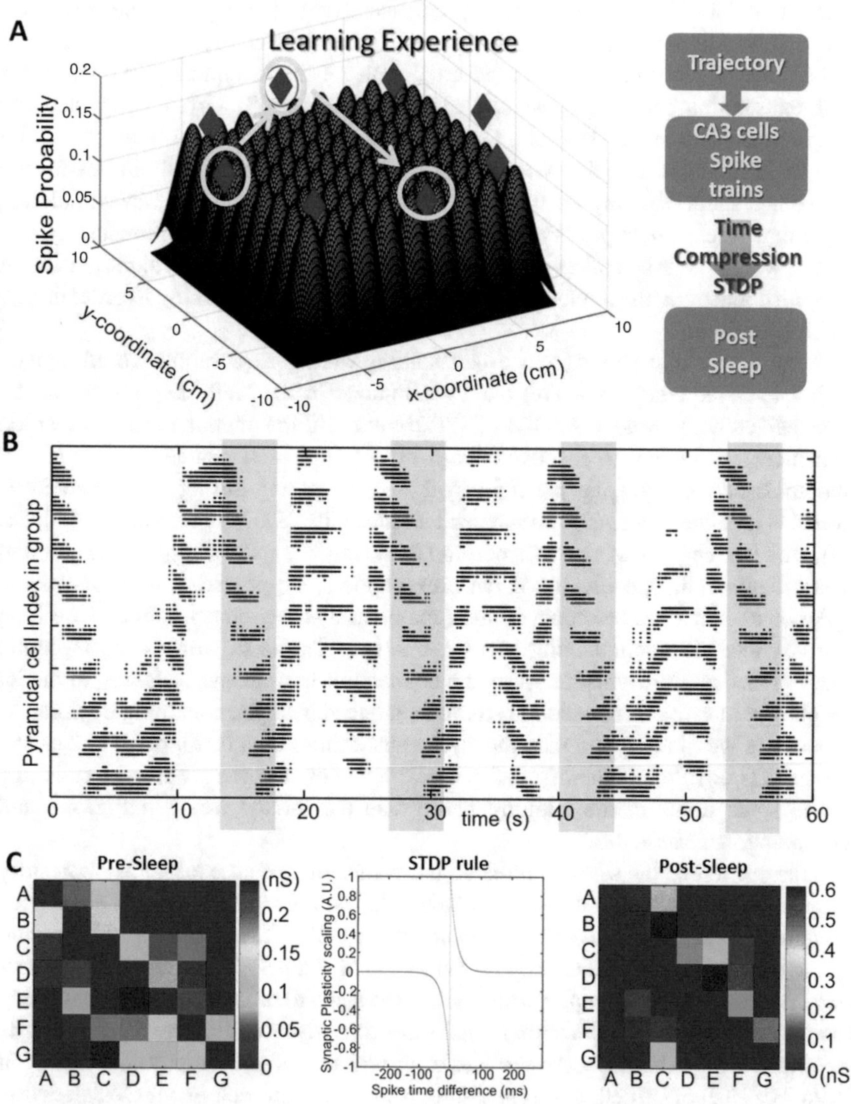

Fig. 4 Experience-related learning: from rat trajectory to new synaptic connections. (**a**) Representation of how a trajectory in space creates artificial spike times for selected CA3 cells. The available space is tiled by regions over which place fields are defined. Each field is assigned to a specific cell. The field defines the relationship between the rat coordinates in space and the probability of a cell to spike. The diagram on the right shows the procedure connecting a trajectory experience to a change in synaptic connections. (**b**) Example of spiking of cells during learning experience. In the rastergram, each dot represents a spike time (x-axis) of a given cell (y-axis). In red are the spikes of cells subsequently used to identify a spiking sequence ABCDEFG for replay analysis ("trajectory cells"). In the shaded times, the "virtual rat" is exploring the sequence of locations to be learned ("trajectory"). In between times, the animal is exploring randomly selected locations. (**c**) Synapses between trajectory cells (red in **b**) before and after applying trajectory-driven plasticity (matrices on either side). Middle plot: representation of the function mediating the STDP rule used to connect spike times (as in **b**), after time compression, to synaptic changes

place field spiking probability. This is consistent with the experimental measures which have shown that place fields can be reconstructed from the locations of spike times of place cells and can be fitted with 2-D exponentials [20]. Hence, for each exploration path (x_t, y_t), we considered spikes from 81 virtual cells encoding the place fields assigned to them. One example of such trajectory-induced spiking activity is shown in Fig. 4b, where the spike times of a subset of all place cells are shown and shaded in gray are the times in which the virtual animal was completing the assigned relevant trajectory across the 3 rewarding feeders (in yellow in Fig. 4a). In red, we marked the spikes of cells with place fields along the straight portion of the path connecting the 3 relevant feeders, which can be seen spiking in order during the shaded times.

In an actual experiment, place cell spiking during exploration would happen within low-frequency (3–8 Hz) theta oscillations in the LFP [12] and would be interspersed with awake CA3–CA1 SWR events, within which the recently learned sequences are known to reactivate, in a time-compressed manner [11, 58]. It is hypothesized that synaptic plasticity within cells spiking during the awake SWR events is mediated by spike-time-dependent plasticity (STDP) mechanisms [30, 58, 59], which strengthen synapses according to the relative spike timing of the pre- and postsynaptic cell. Specifically, if the presynaptic spike precedes the postsynaptic spike of a short time the connection is maximally strengthened, while if the time gap is wider the strengthening is a lot smaller. Vice versa, when a postsynaptic spike precedes a presynaptic spike the connection is weakened. Hence, to model the change in synaptic connection strengths induced by a given learning/exploration experience we time-compressed the awake spike times by a factor of 10 and used a classic STDP rule to quantify the synaptic changes occurring between cells due to the spike times in the learning phase (see the middle panel in Fig. 4c, and Computational Methods).

This portion of the setup completed the connection of an exploratory trajectory to spike times of a group of cells, and then to the computation of the STDP-induced rescaling of synaptic connections between those cells. This awake virtual structure had to be connected with our biophysical model of sleep SWR activity for the three-stage paradigm of Pre-sleep, learning, and Post-sleep to be completed. By assigning each place field within the virtual enclosure to a given cell in the CA3 network, we could bridge the gap between awake virtual model and biophysical model of sleep. For each place cell in the enclosure, we selected one cell in our CA3 network, which was randomly assigned within a range of cell indexes. The range of CA3 pyramidal cells was selected to have a high likelihood of co-activation during SWR; since SWR are localized within the network topology, the 81 indexes were assigned to a range of cells of 100 indexes (detailed in Computational Methods). Since the cells belonged to the CA3 network, a specific synaptic strength value was already known from the Pre-sleep SWR activity simulation (see in the left panel of Fig. 4c one example). The effect of STDP on virtual synaptic connections which resulted from the awake computation was then assigned offline to the synaptic connection strengths between the corresponding CA3 cells, hence completing the flow diagram shown in Fig. 4a. One example of the resulting difference in synaptic connections between Pre- and Post-sleep that was induced by this procedure is given in Fig. 4c. Note that for each simulation we identified a subgroup of 7 cells among the 81 CA3 cells selected, to be the ones that host the spike sequence which is "recorded" during our virtual experiment, and hence whose change in R-activation between Pre- and Post-sleep was the subject of our analysis.

2.5 *Experience Learning Increases Trajectory Reactivation in Post-Sleep*

Once we had this explicit model of the Pre-sleep, learning, and Post-sleep paradigm, we could observe the different R-activation of cells with place fields along the target trajectory in the two sleep epochs. In in vivo experiments, it is known that awake learning promotes reactivation in the subsequent sleep [21, 23–25], so we first tested if our model satisfied this requirement. Since once again it was a list of 7 cells, we used the same measure of R-activation developed in the simplified learning case (when we only manually changed some synapses, Fig. 3). Figure 5a shows one example of the R-activation of the 7 "trajectory" cells in Pre- and Post-sleep, for each word of increasing length, all starting from the first cell ("A," "AB," "ABC," etc.). For each word length, we computed its Score Gain as the difference between Post- and Pre-sleep R-activation scores and Fig. 5b shows that across 14 simulations we had an increase in R-activation at many lengths, although the longest sub-words proved difficult to obtain.

In the virtual rat setup, we proceeded to study the role of different learning experiences on the R-activation changes between Pre- and Post-sleep. For each simulation in Fig. 5b, we had a spatial experience trajectory (performed during the learning phase). We repeated each simulation experiment using only half of the spatial experience trajectory. Since the exploration of the regions of interest leading to sequence learning was uniformly distributed along the spatial experience, this paradigm can be thought of as roughly halving the amount of training that the virtual rat received on the relevant feeders. We labeled the two training conditions "long learning experience" and "short learning experience." Even with reduced learning, we still expected the plasticity induced by the spatial exploration experience to promote the R-activation of the cell sequence in Post-sleep. In Fig. 5c, we show one example of Post- and Pre-sleep R-activation scores in the short learning experience, confirming this expectation. However, reducing the learning time affected the Post-sleep R-activation: across all our simulations, the Score Gain of the short learning experience was smaller than the Gain for the long training (Fig. 5d).

In our learning experience, the virtual rat visited a sequence of 3 specific targets followed by 3 random targets, and then again the 3 specific targets, followed by 3 new random ones, etc. Hence, the path covered by the spatial learning experience involved in general spiking from any or all of the 81 CA3 cells with place fields tiling the spatial enclosure. Thus, while the goal of the experiment was to learn a specific selected sequence, in principle, many other trajectories could have also been learned—at least partially. If the learning of selected sequence was effective,

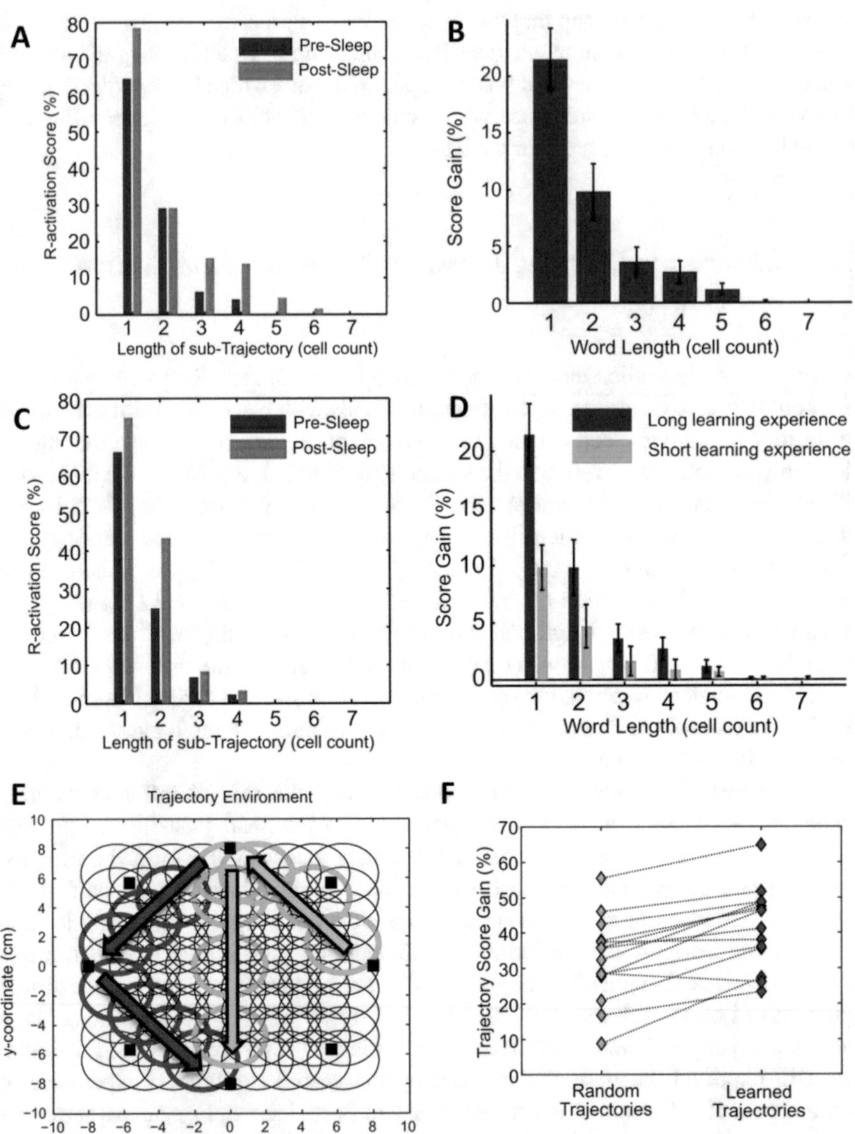

Fig. 5 Experience learning increases trajectory reactivation in Post-sleep. (**a**) Trajectory learning increases sequence reactivation during sleep. The R-activation score of sequences of trajectory cells of increasing lengths (e.g., 1 = "A," 4 = "ABCD") in one example of Pre- and Post-sleep simulations. (**b**) Score Gain (Post-sleep score minus Pre-sleep score) for the cell sequences of increasing lengths. Bar shows mean value across 14 simulations, error bars mark the standard error. (**c**) Shorter learning experience results in reduced gain of sequence R-activation. Example of R-activation score of sequences in a Pre- and Post-sleep simulation with trajectory-learning time reduced to one half. (**d**) Score Gain across sequence length in the case of long vs short learning experience. Note that at every length the increase in R-activation Post-sleep compared to pre-sleep is higher for stronger learning. Error bars mark the standard error of the mean across

not only that sequence would reactivate more in Post-sleep compared to Pre-sleep, it would also reactivate more in comparison to other nonselected paths. To verify if the learning experience was effective at separating the learned selected trajectory from occasional random trajectories, we compared the R-activation of a number of other possible trajectories, connecting any 3 feeders/targets in the enclosure, to the R-activation of the learned selected trajectory. In Fig. 5e, we show a view of the virtual enclosure from above, with circles representing the half-height of the place fields of the 81 pyramidal cells assigned to a given simulation. In red, we emphasize the path connecting the learned trajectory and the place fields of the cells which compose the spiking sequence "ABCDEFG" used in all quantifiers of R-activation scores above (Fig. 5a–d). In green, we mark one example of a different trajectory which was not the learning trajectory and of place fields of cells along its path (one Random Trajectory). We collected a 7-cell long sequence for each possible 3-target trajectory given by the 8 target enclosure (a total of 336 samples), and for each sequence we found the Pre- and Post-sleep R-activation for sub-words of increasing lengths. Since we wanted to compare learned and non-learned trajectories, we introduced a single number Trajectory R-activation Score as follows. For each length of a sub-word (from 1 to 7), we considered representative of a reactivation of a sub-word of length n all the ordered strings of letters of length n showing immediate neighbors within the main (7-letter) word. For example, for a sub-word of length 4, the representatives would be "ABCD," "BCDE," "CDEF," and "DEFG" (but "ACEF" would not qualify). The percentage of SWR in which any of the representatives of a sub-word of length 4 was spiking gave the R-activation score for length 4. A Trajectory Score in a given simulation (Pre- or Post-sleep) would then be defined as the sum of all R-activation scores across all length (1–7). The difference between the Post-sleep Trajectory Score and the Pre-sleep Trajectory Score then produced a Trajectory Gain. When the average Trajectory Score across all Random trajectories within the enclosure was compared with the Learned trajectory, the Learned trajectory scored above average in all but one simulation (Fig. 5f). Hence, our behavior-driven synaptic plasticity paradigm could enhance specific trajectory reactivation with high reliability. This type of test would have been impossible in the simplified learning setup where we artificially changed only very few synapses in the network but no cell spiking activity was explicitly considered as coding for something specific in the awake epoch.

Fig. 5 (continued) 14 simulations. (**e**) Example of the circular enclosure and two possible spatial trajectories: the trajectory which is learned in the task (in red) and one that is not learned in the task (in green). To compare the efficacy of learning on Score Gain, we tested 336 possible trajectories (green) for each simulation against the red trajectory. (**f**) For each simulation (n = 14), we compare the Pre- and Post-sleep R-activation of different trajectories. The red dots show the Trajectory Score Gain for the learned trajectory in each simulation, and the green dots show the Trajectory Score Gain averaged among the other 336 trajectories tested (Random Trajectories), which were not learned but were possibly visited in the task epoch. Lines connect dots belonging to the same simulation

3 Conclusion

In this work, we used computer models to study the impact of changing synaptic connectivity within the hippocampal CA1–CA3 network on the cell reactivation properties. We first introduced limited targeted synaptic changes between selected CA3 pyramidal cells, representing the effects of a learning experience in between two successive simulations of SWR activity, building a sequence of Pre-sleep, learning, and Post-sleep. We then expanded the learning epoch to represent explicitly a spatial exploration and path learning task, and applied STDP to trajectory-driven spiking to build a relationship between learning and connectivity, and hence study how learning can affect reactivation during sleep. For the sleep epochs, we used our previously developed model of SWR activity in CA3 and CA1 hippocampal regions. In this new work, we studied reactivation in the CA3 network component.

Specifically, for a given triplet of cells (Figs. 1 and 2) representing part of the experience that was learned, we manually strengthened excitatory (AMPA) synapses favoring the triplet R-activation order, removed synapses opposing it, and introduced some NMDA connections (again favoring the triplet order) (Fig. 1). Such manipulation in general resulted in a higher R-activation of the ABC triplet in the Post-sleep compared to the Pre-sleep (Fig. 2). We found that the R-activation gained in the Post-sleep compared to Pre-sleep depended on the Pre-sleep score: triplets with lower Pre-sleep scores reached higher gains. Note that altering only AMPA connections (through potentiation and depression) without introducing any new NMDA connections would not result in significant difference between Post- vs Pre-sleep activation (data not shown), which reinforced our idea that a combination of AMPA–NMDA effects could represent the minimal set of learning changes capable of inducing increased sleep reactivation. Furthermore, our study highlighted a role for input (synaptic or increased intrinsic excitability) to the "first" cell of the sequence in promoting Post-sleep sequence reactivation, and in particular how timing of the first cell is crucial to the length of a sequence to be reactivated during an SWR (Fig. 3).

In this work, we concentrate on the mechanisms that can influence pattern completion in CA3 during sharp waves, as our previous analysis showed that selective input to CA1 cells is effective at inducing CA1 replay during ripple [34], and we shape the analysis of learning-induced changes in the context of sleep reactivation changes. Cutsturidis and Hasselmo [60] have defined a model of activity in CA1 and medial septum, showing awake learning and sleep reactivation. In this model, which does not have CA3 spiking activity, the synaptic changes that mediate sleep sequence replay are placed at the projections from entorhinal cortex and CA3 onto CA1 pyramidal cells. Consistently with their idea, it is likely that a study that focuses on CA1 replay before and after learning would have to introduce synaptic plasticity at CA3-to-CA1 excitatory connections, and possibly address a role of input from entorhinal cortex to CA1 pyramidal cells.

The design of a full "virtual rat learning experience" model was inspired by rat behavioral paradigms used to study hippocampal ripple replay in an open enclosure

(Fig. 4). We designed a trajectory in space with a high amount of repetitions for three specific locations interspersed with varying groups of three random locations. We selected a group of CA3 pyramidal cells (81) and assigned to each of them a place field: a region in space where they were more likely to spike. Hence, for a given trajectory we could use Poisson processes to assign to each of the 81 cells spike times along the trajectory. We used these spike trains to calculate the effects of spike-time-dependent plasticity (STDP) along the trajectory, and changed AMPA and NMDA synaptic connections in the whole CA3 network according to the resulting STDP (Fig. 4). This method was again used to represent a learning experience in between two sleep simulations, and 7 cells with place fields along the trajectory were selected to represent the memory of the trajectory. When comparing R-activation in Pre- and Post-sleep (Fig. 5), we found that synaptic plasticity driven by the learning experience increased the spontaneous R-activation of the memory in CA3 during ripples, that such gain was dependent on the length of the learning experience and that while all cells involved in the learning process received changes in their synaptic connectivity the sequence representing the learned trajectory showed a larger gain than an exhaustive sample of other possible trajectories in the same enclosure.

Combining what we learned from the simplified synaptic change conditions with what we learned from our virtual rat learning experience, our study suggests the following hypothesis: that within the hippocampal system the choice of pyramidal cells that are recruited to encode for a specific task actively learned (a proxy for a human declarative memory) cannot be completely randomized. Instead, the pool of cells which are ready to be used for learning is tightly connected to the specifics of their reactivation activity in the previous night of sleep. In other words, the learning process operates in interaction with the network substrate, not independently of it. Thus, the synaptic changes mediating the selection of specific pyramidal cells for ordered co-reactivation during a Post-sleep event can be minimal, if the choice of cells used to code for the learning process is efficient. Specifically, the model predicts that a minimal number of synaptic adjustments can promote the largest increase in R-activation during Post-sleep for cell assemblies with a relatively low Pre-sleep R-activation score and with the first cell in the group showing early spiking during SWR in the Pre-sleep epoch. This hypothesis is consistent with recent data showing that across days and nights, some hippocampal cells can be classified as rigid (with not changing firing rates and participation in ripples), while other cells are able to learn and increase their involvement in SWR activity after being activated during the day [29].

We consider the model we present here a prototype which can be expanded to analyze and shape hypotheses related to a number of open questions, for example: (1) the influence of hippocampal reactivation on cortical reactivation can be studied in this model when connecting it to thalamocortical models of sleep rhythmic activity [61, 62], together with the role of cortical input in hippocampal SWR replay; (2) which synaptic changes are performed in the hippocampus when learning competing (interfering) memories, and how they can affect sleep-dependent consolidation of interfering memories [2, 63, 64]; and (3) the model can be expanded to introduce plasticity to interneurons, and physiological awake spiking activity in the whole CA3–CA1 network (characterized by theta–gamma rhythmic interaction during active exploration [65]) to study the relationship between the theta phase spiking of a sequence and its reactivation during sleep SWR [29, 38].

This study suggests an explicit mechanism for STDP-mediated learning to interact progressively with SWR across one sleep–learning–sleep cycle, and can be interpreted within the broader problem of learning progressively new things across multiple learning/consolidation (awake/sleep) cycles. One hypothesis is that sleep-dependent memory consolidation happens when reactivation of cell sequences in the hippocampus enables cortical reactivation, which in turn promotes plasticity within cortex, ultimately transferring the encoding of the memory information from the short-term storage of the hippocampus to the long-term distributed storage of cortex. It is known that across many days retrieval of a specific memory becomes progressively independent from hippocampal reactivation [4]. Our work suggests that as hippocampal sequences become less necessary to the memory consolidation process of a specific sleep cycle, they will reactivate less in SWR and therefore will become better suited to be used for coding new experiences in the following day. This evokes a scenario of cell assemblies "ranked" by their co-reactivation during Pre-sleep, where some will be used for learning, and hence increase their SWR co-activation in Post-sleep while the ones not used for learning will reactivate even less in Post-sleep. This progressive "degradation" of a memory representation in the hippocampus during sleep is generating a pool of readily available cell assemblies to use for coding the next day.

4 Computational Methods

4.1 Sleep Epochs Network Model

Sleep Model Representation of Sleep Activity

The model we use in this study to represent sleep spiking activity was derived from the CA3–CA1 network model introduced in [40, 66], which shows physiologically realistic spiking activity as follows: in the intervals between different SPW-Rs, the model shows [34, 40, 66, 67] Large Irregular Activity [11, 14, 15, 17, 38], the occurrence of sharp-wave ripple events in CA3–CA1 network is not periodic, the spiking activity of excitatory and inhibitory cells matches SWR activity in fraction of cells spiking and spike count, and the high-frequency activity in CA1 local field potential during ripples matches experimental data [11, 14, 15, 47]. Specifically, both in data and in our model, the frequency of a ripple in CA1 is $\sim$160 Hz, the length of time intervals between two sharp-wave ripples is distributed exponentially, with fitted rate $\sim$1.7 Hz, and each ripple event lasts between 50 and 80 ms.

A crucial role in the fitting of model behavior to known stochastic hippocampal activity was played by the choice of having cells in a noise-driven spiking regime, rather than oscillatory state. This was done to achieve in vivo-like spiking by addition of a noise current, which would mimic the input applied in dynamic clamp experiments when introducing in vivo-like conditions in hippocampal slice [68, 69]. The selection of parameters leading to cell-specific activity is motivated in the Rationale section, and consistent with biological data.

Rationale

We started with our previously developed [34] network of CA1 pyramidal and basket cells and constructed a network of pyramidal and basket cells to represent CA3 activity, then built Schaffer Collaterals projections from CA3 pyramidal cells to CA1 pyramidal cells and interneurons. We used equations of the adaptive exponential integrate and fire formalism [70, 71], which can show bursting activity (like CA3 and CA1 pyramidal cells [37]) or fast-spiking activity (like basket cells [37]) depending on their parameters [70]. CA3 pyramidal cells are allowed a stronger tendency to burst in response to a current step input by having a less strong spike-frequency adaptation than CA1 neurons, consistently with data [37]. For simplicity, all cells belonging to the same population had the same parameters (specified in the following section). To introduce heterogeneity, every cell receives a different direct current term (selected from a normal distribution), together with an independent Ornstein–Uhlenbeck process (OU process) [72], which can be thought of as a single-pole filtered white noise, with cutoff at 100 Hz. This noisy input is added to take into account the background activity of the cells which we did not explicitly model in the network. The standard deviation of the OU process controls the size of the standard deviation in subthreshold fluctuations of cell voltages, and is a parameter kept fixed within any cell type. Once the parameter tuning is in effect, the cells (even when disconnected from the network) show fast and noisy subthreshold voltage activity, and their spikes are nonrhythmic, driven by fluctuations in the noise input they received, which is called a noise-driven spiking regime, rather than a deterministic spiking regime, and is representative of in vivo conditions [68, 69, 73].

Cells are arranged within a one-dimensional network in CA3, and connectivity within CA3 is characterized by each cell reaching other cells within a third of the network around them, which is consistent with anatomical estimates [74]. For pyramidal to pyramidal cells connections, the probability of synaptic contact within this radius of one third was higher for neurons closer to the presynaptic cell and decayed for neurons further away. Details of all network connections are introduced in the Connectivity section. Intuitively, the highly recurrent connections between pyramidal cells in CA3 have a gradient in density that induces a convergence/divergence connectivity fairly uniform across all CA3 pyramidal cells, which represents the homogeneity of CA3 pyramidal cells arborization within the region. This connectivity represents the highly recurrent pyramidal connections in CA3 without introducing special hubs of increased excitatory recurrence in any specific location in the network.

Our CA3–CA1 network is populated by one type of hippocampal inhibitory cells: basket cells. Despite the large number of different interneuron types found in the hippocampus, basket cells are the dominant subtype of interneuron in and near the pyramidal cell layer of both CA3 and CA1. They are the most extensively studied in the context of sharp-wave ripple activity, although other inhibitory cell types have been shown to spike in relation to SWR (O-LM cells stop spiking right before the SWR, so do axo-axonic cells). Basket cells are known to dominantly spike during ripples in CA1, and no special role for different interneuron subtypes has been identified in CA3 spiking activity during sleep. Different interneurons are hypothesized to serve modulatory roles, perhaps contributing to gating the initiation of ripples (those who stop spiking right before ripples) and a theoretical big picture on this topic is proposed by Taxidis [75].

Equations and Parameters

We model SWR activity in the hippocampus using a network of 240 basket cells and 1200 pyramidal cells in CA3, and 160 basket cells and 800 pyramidal cells in CA1. The ratio of excitatory to inhibitory neurons is known to be approximately 4 [37] and since in our model we did not introduce any of the numerous hippocampal interneuron types but for basket cells, we apply that ratio to the pyramidal to basket cell network. This ratio also favored the ability of the network to support a background disorganized spiking regime, where excitatory and inhibitory currents were able to balance each other [76]. For each neuron, the equations are

$$C\,\dot{v} = -g_L\,(v - E_L) + g_L\Delta\,\exp\left(\frac{(v - V_t)}{\Delta}\right) - w + I(t)$$

$$\tau_w \dot{w} = a\,(v - E_L) - w$$

$$v(t) = V_{thr} \Rightarrow v\,(t + dt) = V_r,\, w\,(t + dt) = w(t) + b$$

$$I(t) = I_{DC} + \beta\eta_t + I_{syn}(t)$$

$$\tau\ d\eta_t = -\eta_t dt + dW_t$$

CA1 cells parameters are reported in [34], and CA3 cells parameters were as follows. Pyramidal cells parameters: C (pF) = 200; g_L (ns) = 10; E_L (mV) = −58; a = 2; b (pA) = 40; Δ (mV) = 2; τ_w (ms) = 120; V_t (mV) = −50; V_r (mV) = −46; V_{thr} (mV) = 0. Interneurons parameters: C (pF) = 200; g_L (ns) = 10; E_L (mV) = −70; a = 2; b (pA) = 10; Δ (mV) = 2; τ_w (ms) = 30; V_t (mV) = −50; V_r (mV) = −58; V_{thr} (mV) = 0.

The coefficients establishing noise size were $\beta = 80$ for pyramidal cells, $\beta = 90$ for interneurons. DC inputs were selected from Gaussian distributions with mean 24 (pA) and standard deviation 30% of the mean for pyramidal cells in CA3, mean 130 (pA) and standard deviation 30% of the mean for CA3 interneurons, mean 40 (pA) and standard deviation 10% of the mean for CA1 pyramidal cells, and mean 180 (pA) and standard deviation 10% of the mean for CA1 interneurons.

Synaptic currents were modeled with double exponential functions, for every cell n we had $Isyn\ (t) = \sum_{j=1}^{160} g^{j\to n}\, s^{j\to n}(t)\,(v_n - E_i) + \sum_{j=1}^{800} g^{j\to n} s^{j\to n}(t)\,(v_n - E_e)$, where $E_i = -80$ mV and $E_e = 0$ mV, and $s^{j\to n}(t) = \sum_{t_k} F\left(e^{H\left(\frac{t-t_k}{\tau_D}\right)} - e^{H\left(\frac{t-t_k}{\tau_R}\right)}\right)$, where t_k are all the spikes of presynaptic cell j.

In this equation, F is a normalization coefficient, set so that every spike in the double exponential within parentheses peaks at one, and $H(\bullet)$ is the Heaviside function, ensuring that the effect of each presynaptic spike affects the postsynaptic current only after the spike has happened. The timescales of rise and decay (in ms) used in the model were as follows [34, 75, 77, 78]. For AMPA connections from pyramidal cells to pyramidal cells: $\tau_R = 0.5$, $\tau_D = 3.5$. For AMPA connections from pyramidal cells to interneurons: $\tau_R = 0.5$, $\tau_D = 3$. For $GABA_A$ connections from interneurons to interneurons: $\tau_R = 0.3$, $\tau_D = 2$. For $GABA_A$ connections from interneurons to pyramidal cells: $\tau_R = 0.3$, $\tau_D = 3.5$. For NMDA synapses between CA3 pyramidal cells, which we add in the learning paradigm and are hence present in Post-sleep network activity simulations, $\tau_R = 9$, $\tau_D = 250$ [37].

Connectivity

The CA3 network was organized as a one-dimensional network. For connections from a CA3 pyramidal cell to the other CA3 pyramidal cells, we first considered a radius (of about one third of the network) around the presynaptic cell, and the probability of connection from the presynaptic cell to any cell within such radius was higher for cells with indexes nearby the presynaptic cell and reduced progressively with cell index distance [74]. Specifically, we used a cosine function to shape the probability within the radius, and parameterized how fast with index distance the probability had to decay by using a monotonic scaling of the cosine phase: if x was the index distance within the network, y = arctan(kx)/arctan(k) imposed the decay probability p(y) = Pcos(4y), where P was the peak probability and k = 2 was a parameter controlling the decay of connection probability with distance within the radius. An analogous structure underlied the probability of CA3 pyramidal cells to connect to inhibitory interneuron in CA3 and for Schaffer Collaterals to connect a CA3 pyramidal cell to CA1 pyramidal cells [74]. To balance the relationship between feed-forward excitation from pyramidal cells to interneurons and feedback inhibition from interneurons to pyramidal cells, probability of connection from a presynaptic basket cell to a cell within a radius (about 1/3 of the network size) was constant at 0.7, for $GABA_A$ connections to both CA3 pyramidal cells and interneurons. Within CA1 connectivity was all-to-all, with the caveat that synaptic weights which were sampled at or below zero caused a removal of a given synapse. As a result, most synapses between CA1 pyramidal cells were absent, consistently with experimental findings [36]. To introduce heterogeneity among synaptic connections, synaptic weights for all synapse types were sampled from Gaussian distributions with variance (σ) given by a percent of the mean (μ). Parameters used in the simulations were (we use the notation Py3 and Py1 to denote pyramidal cells in CA3 and CA1, respectively, and analogously Int3/Int1 for interneurons) as follows: Py3 $\rightarrow$ Py3: $\mu = 34$, $\sigma = 40\%\mu$; Int3 $\rightarrow$ Int3: $\mu = 54$, $\sigma = 40\%\mu$; Py3 $\rightarrow$ Int3: $\mu = 77$, $\sigma = 40\%\mu$; Int3 $\rightarrow$ Py3: $\mu = 55$, $\sigma = 40\%\mu$; Py3 $\rightarrow$ Py1: $\mu = 34$, $\sigma = 10\%\mu$; Py3 $\rightarrow$ Int1: $\mu = 320$, $\sigma = 10\%\mu$; Int1 $\rightarrow$ Int1: $\mu = 3.75$, $\sigma = 1\%\mu$; Py1 $\rightarrow$ Int1: $\mu = 6.7$, $\sigma = 1\%\mu$; Int1 $\rightarrow$ Py1: $\mu = 8.3$, $\sigma = 1\%\mu$; Py1 $\rightarrow$ Py1: $\mu = 0.67$, $\sigma = 1\%\mu$. It is to note that the mean (μ) declared was normalized by the total number of cells before the variance to the mean was introduced in the distribution. Since the CA3 and CA1 networks are of different sizes, a direct comparison of the parameter values or their magnitude across regions would not account for the effective values used in the simulations. Learning epochs models.

Introducing Targeted Synaptic Changes

"ABC" Triplets Analysis

In the CA3 network, we looked for pyramidal cells that had a chance to reactivate in a large fraction of spontaneous SWR. In our model, the probability of SPW participation among CA3 cells depends on network topology, and in particular looks bimodal with peaks about cell index 350 and 750 (data not shown). Hence, we started from an interval of accepted cell indexes between 700 and 800. Three cells within this range were then chosen uniformly (using function randperm in MATLAB, The MathWorks). For each Pre-sleep simulation (we had 15 samples, each 50 s long), 20 sets of 3 cells were first randomly chosen and then cells within the sets were permuted, generating a set of 120 ordered triplets. From this pool, three triplets were chosen with the criteria that no two triplets could be a permutation of each other and that their R-activation scores in Pre-sleep were spanning a range of available scores (meaning one triplet was chosen with a low score, one with a medium score, and one with a high score). Hence, we had a pool of 45 triplets with Pre-sleep R-activation Scores roughly spanning the available range.

For each triplet of cells "ABC," we manually modified some AMPA and NMDA synapses between the cells composing the triplet. The synaptic weights of AMPA synapses favoring the order of the triplet (A $\rightarrow$ B and B $\rightarrow$ C) were increased to the maximum AMPA synaptic strength value within the specific Pre-sleep CA3 network (remember for every simulation the synaptic connectivity matrices were generated anew). Along the maximized AMPA synapses, we introduced NMDA synaptic connections with a fixed value, identical for all simulations (1.25 ns, normalized by the total number of added NMDA synapses), which was found to be efficient at promoting co-activation during SWRs without enhancing the spiking activity of cells beyond physiological values. Finally, the weights of AMPA synapses which were favoring the reverse order of the triplet (C → B and B → A) were replaced by zeros. The new Post-sleep simulation was then run for 50 s.

The R-activation Score was found using the spike times of the cells in the triplet. The spike times of the three cells were required to be between the start and end of the SPW event in CA3. The first spike of each cell was used to identify the order in which the cells spiked relative to each other. The R-activation Score of a triplet in a simulation was the percent of the total SWR in which the ordered triplet spiked (meaning each cell of the triplet spiked at least once and in the correct order). The Score Gain was found by taking the difference between the R-activation Score in Post- and Pre-sleep. To verify whether the cells were spiking in the correct order, we also quantified the average spike-time difference (STD) across triplets.

We analyzed the input received by the triplet "ABC" by using synaptic and intrinsic excitability modifications. For the synaptic modifications, an additional cell "D" was randomly chosen from the cell index range 700–800. This cell was added to the end of each of the ordered triplets from the 15 simulations. The same synaptic modifications in the NMDA and AMPA synapses were performed between the additional cell and the last cell of the triplet. The Score Gain between the triplet "ABC" and "BCD" was then compared to analyze the effect of altering the synapses between "A" and "B" on the replay of "BCD." To confirm whether the input of the first cell influences the replay of the triplet, the intrinsic excitability of the first cell in the triplet was modified. The DC input of the first cell (I_{DC} in the membrane voltage equation) was increased by 2 pA (a small value, chosen to maintain the cells in a noise-driven spiking regime without driving them to an oscillatory bursting spiking regime). This intrinsic excitability was then doubled and tripled.

"ABCDEFG" Sequences

We tested the effects of targeted synaptic modifications for longer cell sequences (7 cells, represented with the word "ABCDEFG"). To select 7 cells from our range of cell indexes between 700 and 800 in CA3, we proceed starting from a triplet and progressively adding one more cell at a time, as follows. The starting triplet was randomly chosen, with two requirements: (1) in Pre-sleep synapses between A → B and B → C were nonzero and (2) in Pre-sleep the R-activation Score of the triplet was between 4 and 10%. The next cell ("D") was randomly chosen among the remaining cells in the 700–800 range with the same requirements as above applied to the new sub-triplet "BCD" (i.e., a synapse C → D was present in Pre-sleep and the BCD R-activation score was in the 4–10% range). The procedure of adding cells at the end of the list was iterated until a 7-letter list of cell indexes was populated. For these tests, we used 14 simulations, 50 s long. Since we did not analyze the CA1 cells, we only ran simulations of the CA3 sub-network in this part of the study.

To quantify the R-activation of sub-words in the sequence, the time of the first spike of any cell within each SWR was used. It is important to note that we effectively composed a list of 8 cells and then used the last seven of them as the ABCDEFG of our sequence, to possibly take into account the role of input to the first cell which we learned about in the triplet analysis. Furthermore, for cells which were later in a sequence (from cell D onward), we included their first spike time even if it happened up to 0.3 s after the end of a SPW, to let the possible "tales" of a reactivating sequence be included in our analysis.

4.2 Modeling a Virtual Rat's Spatial Learning Experience

The "learning experience" comprised of a rat running across an artificial enclosure, which we tiled as a grid of place fields. The size of the enclosure was 16 cm by 16 cm and the grid was spaced with 0.1 cm gaps. Each place field was numbered from 1 to 81. Each place field represented a Poisson distribution of the cell's firing rate as function of location, as a normal probability density function with center at one of the grid's 81 locations and radius (standard deviation) 3 cm. This radius was chosen to establish a sufficient overlap between place fields for the spiking activity to lead to significant STDP-mediated synaptic rescaling.

We chose a simple trajectory for the rat to learn. The rat had to move through 3 locations in the same order. These 3 locations, as well as the direction of the trajectory, were the same for all simulations. The rat then moved to 3 random locations that were not part of the learned trajectory. This counted as one repetition for the learning task. The virtual rat moved between the locations in a straight line that was completed in about 2 s. The peak instantaneous firing rate of place cells, that were activated by the movement (because they had place fields along the trajectory), was set to 100 Hz (actual spikes in time are shown in Fig. 4b).

When assigning a place field (1–81) to a CA3 cell within the sleep model network, we started from 7 place fields along the learned trajectory, and assigned to them the same 7 CA3 cells used for the simulations with targeted synaptic changes. The remaining 74 cells were randomly assigned within the 700–800 index range (excluding the cells which already received a field). Given a "learning experience" trajectory, we had 81 spike trains generated by their respective Poisson processes place fields, and used them to modify the synapses between the network cells which were assigned those place fields. We used spike-time-dependent plasticity (STDP) as the rule for strengthening or weakening AMPA and NMDA synapses. The spike times were compressed by factor of 10, to represent their reactivation during awake SWR leading to STDP-induced synaptic plasticity [32]. Every synaptic connection strength g from cell A to cell B was then replaced by $g + \Delta g$ with $\Delta g = \sum 2AGsign\,(t_A - t_B)\,\frac{e^{|t_A - t_B|}}{20}$. Where $g_{max} = A * G$ with A = 0.001 for AMPA synapses and 0.01 for NMDA synapses a scaling factor, G the maximum value of AMPA synapses in the given CA3 network, and G = 1.25 ns for NMDA synapses. t_A and t_B are spike times for cells A and B, respectively.

Within a given "learning experience," the rat repeatedly visited the learning trajectory, followed by 3 random locations, which were newly reselected with each iteration. We established the length of a given learning experience path to represent a long learning experience or a short one depending on the change induced in the synapses by the path. In fact, since the STDP effect is cumulative and every iteration results in incremental increase in the synapses favoring a target sequence-ordered R-activation. The length of the experience path was determined by incrementally increasing the repetitions until the average strength of AMPA synapses between cells along the trajectory reached a threshold of 0.4 ns. This value is close to the maximum synapse for all simulations used (the average maximum strength of the AMPA synapses over all 14 simulations was 0.517 with a standard deviation of 0.023). We chose a threshold value below the maximum AMPA synapse for each simulation, because some synapses could significantly surpass the threshold and disrupt the spontaneous network activity.

Acknowledgments This work was supported by MURI grant (MURI: N000141612829 and N000141612415) to MB.

References

1. S. Mednick, K. Nakayama, R. Stickgold, Sleep-dependent learning: a nap is as good as a night. Nat. Neurosci. **6**(7), 697–698 (2003)
2. S.C. Mednick, D.J. Cai, T. Shuman, S. Anagnostaras, J.T. Wixted, An opportunistic theory of cellular and systems consolidation. Trends Neurosci. **34**(10), 504–514 (2011)
3. S.C. Mednick, T. Makovski, D.J. Cai, Y.V. Jiang, Sleep and rest facilitate implicit memory in a visual search task. Vis. Res **49**, 2557 (2009)
4. B. Rasch, J. Born, About sleep's role in memory. Physiol. Rev. **93**(2), 681–766 (2013)
5. I. Wilhelm, S. Diekelmann, I. Molzow, A. Ayoub, M. Mölle, J. Born, Sleep selectively enhances memory expected to be of future relevance. J. Neurosci. **31**(5), 1563–1569 (2011)
6. L. Marshall, H. Helgadottir, M. Molle, J. Born, Boosting slow oscillations during sleep potentiates memory. Nature **444**(7119), 610–613 (2006)
7. E.A. McDevitt, K.A. Duggan, S.C. Mednick, REM sleep rescues learning from interference. Neurobiol. Learn. Mem. **122**, 51 (2014)
8. S.C. Mednick, E.A. McDevitt, J.K. Walsh, E. Wamsley, M. Paulus, J.C. Kanady, S.P. Drummond, The critical role of sleep spindles in hippocampal-dependent memory: a pharmacology study. J. Neurosci. **33**(10), 4494–4504 (2013)
9. C.V. Latchoumane, H.V. Ngo, J. Born, H.S. Shin, Thalamic spindles promote memory formation during sleep through triple phase-locking of cortical, thalamic, and hippocampal rhythms. Neuron **95**, 424 (2017)
10. B.P. Staresina, T.O. Bergmann, M. Bonnefond, R. van der Meij, O. Jensen, L. Deuker, C.E. Elger, N. Axmacher, J. Fell, Hierarchical nesting of slow oscillations, spindles and ripples in the human hippocampus during sleep. Nat. Neurosci. **18**(11), 1679–1686 (2015)
11. G. Buzsaki, Hippocampal sharp wave-ripple: a cognitive biomarker for episodic memory and planning. Hippocampus **25**(10), 1073–1188 (2015)
12. G. Buzsaki, D.L. Buhl, K.D. Harris, J. Csicsvari, B. Czeh, A. Morozov, Hippocampal network patterns of activity in the mouse. Neuroscience **116**(1), 201–211 (2003)
13. C.D. Schwindel, B.L. McNaughton, Hippocampal-cortical interactions and the dynamics of memory trace reactivation. Prog. Brain Res. **193**, 163–177 (2011)

14. J. Csicsvari, H. Hirase, A. Czurko, A. Mamiya, G. Buzsaki, Fast network oscillations in the hippocampal CA1 region of the behaving rat. J. Neurosci. **19**(16), Rc20 (1999)
15. J. Csicsvari, H. Hirase, A. Czurko, A. Mamiya, G. Buzsaki, Oscillatory coupling of hippocampal pyramidal cells and interneurons in the behaving rat. J. Neurosci. **19**(1), 274–287 (1999)
16. J. Csicsvari, H. Hirase, A. Mamiya, G. Buzsáki, Ensemble patterns of hippocampal CA3-CA1 neurons during sharp wave-associated population events. Neuron **28**(2), 585–594 (2000)
17. A. Ylinen, A. Bragin, Z. Nadasdy, G. Jando, I. Szabo, A. Sik, G. Buzsaki, Sharp wave-associated high-frequency oscillation (200 Hz) in the intact hippocampus: network and intracellular mechanisms. J. Neurosci. **15**(1 Pt 1), 30–46 (1995)
18. J. O'Keefe, Place units in the hippocampus of the freely moving rat. Exp. Neurol. **51**, 78–109 (1976)
19. J.M. O'Keefe, L. Nadel, *The Hippocampus as a Cognitive Map* (Clarendon Press/Oxford University Press, Oxford/New York, 1978)
20. J. O'Neill, T. Senior, J. Csicsvari, Place-selective firing of CA1 pyramidal cells during sharp wave/ripple network patterns in exploratory behavior. Neuron **49**(1), 143–155 (2006)
21. W.E. Skaggs, B.L. McNaughton, Replay of neuronal firing sequences in rat hippocampus during sleep following spatial experience. Science **271**(5257), 1870–1873 (1996)
22. G.R. Sutherland, B. McNaughton, Memory trace reactivation in hippocampal and neocortical neuronal ensembles. Curr. Opin. Neurobiol. **10**(2), 180–186 (2000)
23. M.A. Wilson, B.L. McNaughton, Reactivation of hippocampal ensemble memories during sleep. Science **265**(5172), 676–679 (1994)
24. G. Girardeau, M. Zugaro, Hippocampal ripples and memory consolidation. Curr. Opin. Neurobiol. **21**(3), 452–459 (2011)
25. H.S. Kudrimoti, C.A. Barnes, B.L. McNaughton, Reactivation of hippocampal cell assemblies: effects of behavioral state, experience, and EEG dynamics. J. Neurosci. **19**(10), 4090–4101 (1999)
26. Z. Nádasdy, H. Hirase, A. Czurkó, J. Csicsvari, G. Buzsáki, Replay and time compression of recurring spike sequences in the hippocampus. J. Neurosci. **19**(21), 9497–9507 (1999)
27. V. Ego-Stengel, M.A. Wilson, Disruption of ripple-associated hippocampal activity during rest impairs spatial learning in the rat. Hippocampus **20**(1), 1–10 (2010)
28. G. Girardeau, K. Benchenane, S.I. Wiener, G. Buzsaki, M.B. Zugaro, Selective suppression of hippocampal ripples impairs spatial memory. Nat. Neurosci. **12**(10), 1222–1223 (2009)
29. A.D. Grosmark, G. Buzsaki, Diversity in neural firing dynamics supports both rigid and learned hippocampal sequences. Science **351**(6280), 1440–1443 (2016)
30. G. Bi, M. Poo, Synaptic modification by correlated activity: Hebb's postulate revisited. Annu. Rev. Neurosci. **24**, 139–166 (2001)
31. M.S. Rioult-Pedotti, D. Friedman, J.P. Donoghue, Learning-induced LTP in neocortex. Science **290**(5491), 533–536 (2000)
32. J.H. Sadowski, M.W. Jones, J.R. Mellor, Sharp-wave ripples orchestrate the induction of synaptic plasticity during reactivation of place cell firing patterns in the hippocampus. Cell Rep. **14**(8), 1916–1929 (2016)
33. P. Malerba, A. Fodder, M. Jones, M. Bazhenov, Modeling of coordinated sequence replay in CA3 and CA1 during sharp wave-ripples, in *Society for Neuroscience Annual Meeting San Diego, CA, 2016 Neuroscience Meeting Planner* (2016)
34. P. Malerba, G.P. Krishnan, J.M. Fellous, M. Bazhenov, Hippocampal CA1 ripples as inhibitory transients. PLoS Comput. Biol. **12**(4), e1004880 (2016)
35. G.M. Shepherd, *The Synaptic Organization of the Brain* (Oxford University Press, Oxford, 2004)
36. J. Deuchars, A.M. Thomson, CA1 pyramid-pyramid connections in rat hippocampus in vitro: dual intracellular recordings with biocytin filling. Neuroscience **74**(4), 1009–1018 (1996)
37. P. Andersen, R. Morris, D. Amaral, T. Bliss, J. O'Keefe, *The Hippocampus Book* (Oxford University Press, New York, 2006)
38. K. Mizuseki, S. Royer, K. Diba, G. Buzsáki, Activity dynamics and behavioral correlates of CA3 and CA1 hippocampal pyramidal neurons. Hippocampus **22**(8), 1659–1680 (2012)

39. D. Sullivan, J. Csicsvari, K. Mizuseki, S. Montgomery, K. Diba, G. Buzsáki, Relationships between hippocampal sharp waves, ripples, and fast gamma oscillation: influence of dentate and entorhinal cortical activity. J. Neurosci. **31**(23), 8605–8616 (2011)
40. P. Malerba, M. Bazhenov, Circuit mechanisms of hippocampal reactivation during sleep. Neurobiol. Learn. Mem. 2018. https://doi.org/10.1016/j.nlm.2018.04.018
41. J. Patel, E.W. Schomburg, A. Berenyi, S. Fujisawa, G. Buzsaki, Local generation and propagation of ripples along the septotemporal axis of the hippocampus. J. Neurosci. **33**(43), 17029–17041 (2013)
42. N. Rebola, M. Carta, C. Mulle, Operation and plasticity of hippocampal CA3 circuits: implications for memory encoding. Nat. Rev. Neurosci. **18**(4), 208–220 (2017)
43. L.F. Cobar, L. Yuan, A. Tashiro, Place cells and long-term potentiation in the hippocampus. Neurobiol. Learn. Mem. **138**, 206–214 (2017)
44. R.A. Nicoll, A brief history of long-term potentiation. Neuron **93**(2), 281–290 (2017)
45. D. Manahan-Vaughan, Learning-related hippocampal long-term potentiation and long-term depression A2, in *Learning and Memory: A Comprehensive Reference*, 2nd edn., ed. By J.H. Byrne (Academic Press, Oxford, 2017), pp. 585–609
46. C. Pinar, C.J. Fontaine, J. Trivino-Paredes, C.P. Lottenberg, J. Gil-Mohapel, B.R. Christie, Revisiting the flip side: long-term depression of synaptic efficacy in the hippocampus. Neurosci. Biobehav. Rev. **80**, 394–413 (2017)
47. K. Mizuseki, G. Buzsaki, Preconfigured, skewed distribution of firing rates in the hippocampus and entorhinal cortex. Cell Rep. **4**(5), 1010–1021 (2013)
48. Y. Omura, M.M. Carvalho, K. Inokuchi, T. Fukai, A lognormal recurrent network model for burst generation during hippocampal sharp waves. J. Neurosci. **35**(43), 14585–14601 (2015)
49. D.A. McCormick, Neurotransmitter actions in the thalamus and cerebral cortex and their role in neuromodulation of thalamocortical activity. Prog. Neurobiol. **39**, 337–388 (1992)
50. D.A. McCormick, H.C. Pape, A. Williamson, Actions of norepinephrine in the cerebral cortex and thalamus: implications for function of the central noradrenergic system. Prog. Brain Res. **88**, 293–305 (1991)
51. M. Vassalle, Contribution of the Na+/K+-pump to the membrane potential. Experientia **43**(11-12), 1135–1140 (1987)
52. F.F. Offner, Ion flow through membranes and the resting potential of cells. J. Membr. Biol. **123**(2), 171–182 (1991)
53. Y. Burak, I.R. Fiete, Accurate path integration in continuous attractor network models of grid cells. PLoS Comput. Biol. **5**(2), e1000291 (2009)
54. L.M. Giocomo, M.B. Moser, E.I. Moser, Computational models of grid cells. Neuron **71**(4), 589–603 (2011)
55. E. Kropff, A. Treves, The emergence of grid cells: intelligent design or just adaptation? Hippocampus **18**(12), 1256–1269 (2008)
56. B. Jones, E. Bukoski, L. Nadel, J.M. Fellous, Remaking memories: reconsolidation updates positively motivated spatial memory in rats. Learn. Mem. **19**(3), 91–98 (2012)
57. B.J. Jones, S.M. Pest, I.M. Vargas, E.L. Glisky, J.M. Fellous, Contextual reminders fail to trigger memory reconsolidation in aged rats and aged humans. Neurobiol. Learn. Mem. **120**, 7–15 (2015)
58. L.A. Atherton, D. Dupret, J.R. Mellor, Memory trace replay: the shaping of memory consolidation by neuromodulation. Trends Neurosci. **38**(9), 560–570 (2015)
59. M.R. Mehta, From synaptic plasticity to spatial maps and sequence learning. Hippocampus **25**(6), 756–762 (2015)
60. V. Cutsuridis, M. Hasselmo, Spatial memory sequence encoding and replay during modeled theta and ripple oscillations. Cogn. Comput. **4**(3), 554–574 (2011)
61. M. Bazhenov, I. Timofeev, M. Steriade, T.J. Sejnowski, Model of thalamocortical slow-wave sleep oscillations and transitions to activated states. J. Neurosci. **22**(19), 8691–8704 (2002)
62. G.P. Krishnan, S. Chauvette, I.S. Shamie, S. Soltani, I. Timofeev, S.S. Cash, E. Halgren, M. Bazhenov, Cellular and neurochemical basis of sleep stages in thalamocortical network. Elife **5**, e18607 (2016)

63. Y. Wei, G. Krishnan, M. Bazhenov. Synaptic mechanisms of memory consolidation during NREM sleep. Society for Neuroscience. San Diego, CA. Program No. 82.01 (2016)
64. Y. Wei, G.P. Krishnan, M. Bazhenov, Synaptic mechanisms of memory consolidation during sleep slow oscillations. J. Neurosci. **36**(15), 4231–4247 (2016)
65. N. Kopell, C. Börgers, D. Pervouchine, P. Malerba, A. Tort, Gamma and theta rhythms in biophysical models of hippocampal circuits, in *Hippocampal Microcircuits* (Springer, New York, 2010), pp. 423–457
66. P. Malerba, M. W. Jones, M. Bazhenov, Defining the synaptic mechanisms that tune CA3-CA1 reactivation during sharp-wave ripples. bioRxiv (2017). https://doi.org/10.1101/164699
67. P. Malerba, N. F. Rulkov, M. Bazhenov, Large time step discrete-time modeling of sharp wave activity in hippocampal area CA3. bioRxiv (2018). https://doi.org/10.1101/303917
68. T. Broicher, P. Malerba, A.D. Dorval, A. Borisyuk, F.R. Fernandez, J.A. White, Spike phase locking in CA1 pyramidal neurons depends on background conductance and firing rate. J. Neurosci. **32**(41), 14374–14388 (2012)
69. F.R. Fernandez, T. Broicher, A. Truong, J.A. White, Membrane voltage fluctuations reduce spike frequency adaptation and preserve output gain in CA1 pyramidal neurons in a high-conductance state. J. Neurosci. **31**(10), 3880–3893 (2011)
70. R. Brette, W. Gerstner, Adaptive exponential integrate-and-fire model as an effective description of neuronal activity. J. Neurophysiol. **94**(5), 3637–3642 (2005)
71. J. Touboul, R. Brette, Dynamics and bifurcations of the adaptive exponential integrate-and-fire model. Biol. Cybern. **99**(4-5), 319–334 (2008)
72. G.E. Uhlenbeck, L.S. Ornstein, On the theory of the Brownian motion. Phys. Rev. **36**(5), 823 (1930)
73. A. Roxin, N. Brunel, D. Hansel, G. Mongillo, C. van Vreeswijk, On the distribution of firing rates in networks of cortical neurons. J. Neurosci. **31**(45), 16217–16226 (2011)
74. X.G. Li, P. Somogyi, A. Ylinen, G. Buzsáki, The hippocampal CA3 network: an in vivo intracellular labeling study. J. Comp. Neurol. **339**(2), 181–208 (1994)
75. J. Taxidis, S. Coombes, R. Mason, M.R. Owen, Modeling sharp wave-ripple complexes through a CA3-CA1 network model with chemical synapses. Hippocampus **22**(5), 995–1017 (2012)
76. B.V. Atallah, M. Scanziani, Instantaneous modulation of gamma oscillation frequency by balancing excitation with inhibition. Neuron **62**(4), 566–577 (2009)
77. M. Bartos, I. Vida, M. Frotscher, A. Meyer, H. Monyer, J.R. Geiger, P. Jonas, Fast synaptic inhibition promotes synchronized gamma oscillations in hippocampal interneuron networks. Proc. Natl. Acad. Sci. U. S. A. **99**(20), 13222–13227 (2002)
78. V. Cutsuridis, B. Graham, S. Cobb, I. Vida, *Hippocampal Microcircuits: A Computational Modeler's Resource Book* (Springer, New York, 2010)

A DG Method for the Simulation of CO_2 Storage in Saline Aquifer

Beatrice Riviere and Xin Yang

Abstract To simulate the process of CO_2 injection into deep saline aquifers, we use the isothermal two-phase two-component model, which takes mass transfer into account. We develop a new discontinuous Galerkin method called the "partial upwind" method for space discretization, incorporated with the backward Euler scheme for time discretization and the Newton–Raphson method for linearization. Numerical simulations show that the new method is a promising candidate for the CO_2 storage problem in both homogenous and heterogenous porous media and is more robust to the standard discontinuous Galerkin method for some subsurface fluid flow problems.

1 Introduction

CO_2 sequestration in porous media, such as saline aquifers and oil and gas reservoirs, is an important venue to reduce the excessive amount of carbon dioxide in the atmosphere. Numerical simulations for CO_2 sequestration process have been studied using many simulators. The reader is referred to the work of Class et al. [4] for a benchmark study of these simulations. The focus of this paper is to develop a discontinuous Galerkin (DG) method to simulate CO_2 storage, which is a two-component two-phase type of flow. DG methods for two-phase flow without inter-mass transfer have been heavily studied in the literature (see, e.g., [1, 2, 5, 9, 12]). Similarly, two-component single-phase flow (also referred to as miscible displacement) has been numerically and successfully modeled by DG methods [6, 10, 11, 13–17]. The most commonly used model for two-phase flow is the elliptic pressure-hyperbolic saturation formulation, in that the pressure and saturation are weakly coupled together, and the problem can be solved sequentially. Bastian and Riviere [2] used nonsymmetric interior penalty DG formula for the pressure

B. Riviere (✉) · X. Yang
Department of computational and applied mathematics, Rice University, Houston, TX, USA
e-mail: riviere@rice.edu

A. Deines et al. (eds.), *Advances in the Mathematical Sciences*, Association for Women in Mathematics Series 15, https://doi.org/10.1007/978-3-319-98684-5_12

equation and for the diffusion term in the saturation equation, and used the upwind scheme for the advection term. The nonlinear coefficients were linearized by time-lagging. A variant of DG method using similar techniques along with adaptivity in time and space was numerically investigated by Klieber and Riviere [12]. Eslinger [9] numerically studied the compressible air–water two-phase problem using the local DG method for saturation equation. The numerical DG solutions of two-phase flow problem usually have spurious overshoot and undershoot phenomena resulting from large advection, which can be controlled by slope limiting. However, the slope limiters are difficult to construct for higher-dimensional problems, and theoretical analysis is limited to one-dimensional problems. Fully implicit fully coupled DG method proposed by Epshteyn and Riviere [5] can be used to stabilize the oscillations without using slope limiters, but computational cost is increased. In [1], Bastian showed the robustness and scalability of a DG method for a wetting-phase pressure and capillary pressure formulation of the incompressible two-phase flow.

Our paper solves a two-component two-phase problem, with the additional difficulty that one component moves from one phase to the other. The novelty of this work is the approach for handling transfer from one phase to the other. Discontinuous finite element methods were first applied to the two-phase two-component model with interphase mass transfer by Ern and Mozolevski [7]. Their work took into account phase disappearance and showed the potential to handle heterogeneous porous medium. In [7], Henry's law is used to express the linear correlation of density and pressure, which allows for the easy choice of liquid pressure and dissolved gas density as the primal variables. The density changes in CO_2 sequestration problem vary greatly, and hence in our study, Henry's law is not used. Rather, we use a cubic spline interpolant for the relationship between mass fraction and pressure. Other properties like CO_2 viscosity and the mass fraction of CO_2 in water also depend on the gas pressure. Therefore, we need the gas pressure to be one of the primal variables to reduce the complexity of simulation. The difficulty to correctly simulate the accumulation of the non-wetting phase due to the discontinuous capillary pressures in heterogeneous porous media was studied by Ern et al. [8]. They enforced the nonlinear interface conditions weakly and used the weighted average numerical flux and total velocity reconstruction for the DG scheme.

An outline of the paper is the following: the mathematical models are given in the next section. The numerical scheme is described in Sect. 3 and numerical results are shown in Sect. 4. Some conclusions follow.

2 Mathematical Model

In this section, we first show the mathematical model and explain the terms used in the model. Then, we transform the model to the equations that we use for discretization.

2.1 Mass Conservation Laws

We consider the isothermal two-phase two-component model for CO_2 sequestration problems. The two phases are CO_2-rich phase and water-rich phase. The CO_2-rich phase is the non-wetting phase and is denoted by n. The water-rich phase is the wetting phase and is denoted by w. The two components considered are carbon dioxide (denoted by CO_2) and water (denoted by H_2O). We use the Reynold's transport theorem for the mass conservation of CO_2 and H_2O to obtain the following equations:

$$\phi \frac{\partial}{\partial t}\left(\sum_{\alpha\in\{w,n\}} \rho_\alpha X_\alpha^{CO_2} S_\alpha\right) - \nabla\cdot\left(\sum_{\alpha\in\{w,n\}} \rho_\alpha X_\alpha^{CO_2} \mathbf{v}_\alpha\right) = q^{CO_2}, \tag{1}$$

$$\phi \frac{\partial}{\partial t}\left(\sum_{\alpha\in\{w,n\}} \rho_\alpha X_\alpha^{H_2O} S_\alpha\right) - \nabla\cdot\left(\sum_{\alpha\in\{w,n\}} \rho_\alpha X_\alpha^{H_2O} \mathbf{v}_\alpha\right) = q^{H_2O}, \tag{2}$$

where ϕ denotes the porosity, ρ_α the density of phase α, X_α^β the mass fraction of component β in phase α, S_α the saturation of phase α, $\mathbf{v}_\alpha$ the averaged velocity on the macroscopic scale for phase α, and q^β the source term of component β.

The density of the CO_2-rich phase ρ_n depends on the pressure. Figure 1 shows the cubic spline interpolation for the function of $\rho_n(p_n)$, where p_n denotes the

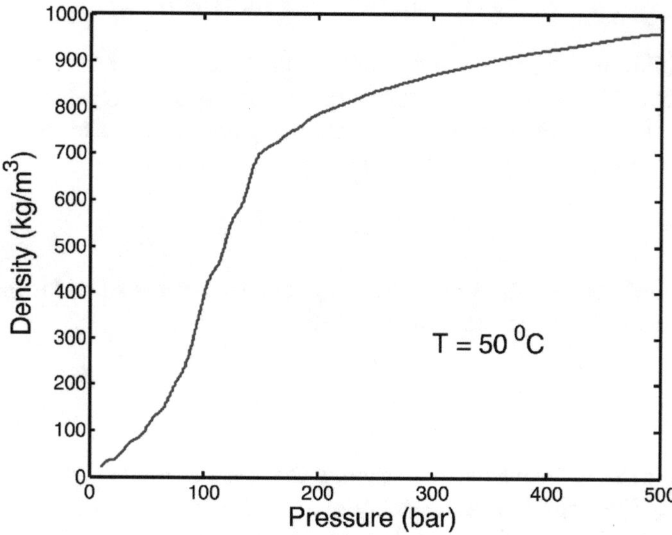

Fig. 1 Cubic spline interpolation for the correlation of the density of CO_2 and the pressure at $T = 50\,°C$

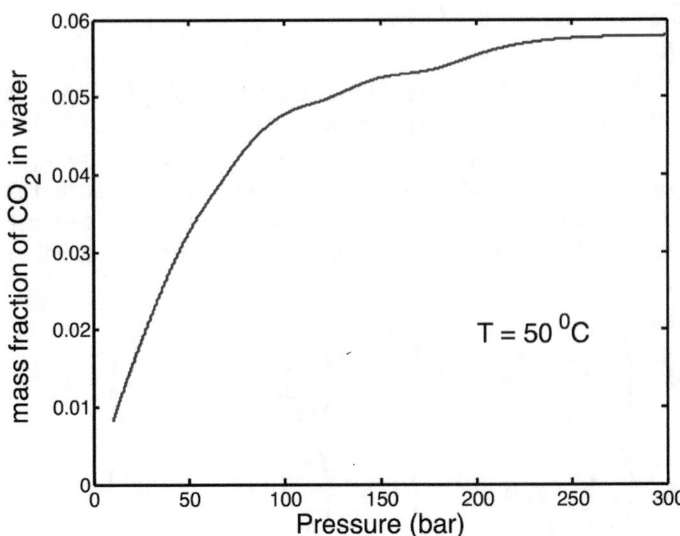

Fig. 2 Solubility of CO_2 in water using cubic spline interpolation

pressure of the non-wetting phase. The density of brine, which depends on the density of water, salinity, and the solubility of carbon dioxide, is simply assumed to be constant. The mass fractions satisfy the following equation:

$$X_\alpha^{CO_2} + X_\alpha^{H_2O} = 1, \quad \alpha \in \{w, n\}. \tag{3}$$

The solubility of CO_2 in brine depends on the pressure, and the corresponding function $X_w^{CO_2}(p_n)$ is approximated using the cubic spline interpolation as shown in Fig. 2. The solubility of H_2O in the CO_2-rich phase is approximately 100 times smaller than the solubility of CO_2 in brine, and thus we assume

$$X_n^{H_2O} = 0, \quad X_n^{CO_2} = 1.$$

The averaged velocity $\mathbf{v}_\alpha$ is given by the generalized Darcy's law for multiphase flow. If we neglect the gravity term, we have

$$\mathbf{v}_\alpha = -\frac{k_{r\alpha}}{\mu_\alpha} K \nabla p_\alpha, \tag{4}$$

where $k_{r\alpha}$ denotes the relative permeability for phase α, μ_α the dynamic viscosity of phase α, K the absolute permeability, and p_α the pressure of phase α.

We use the Brooks and Corey [3] formula for the relative permeability:

$$k_{rw} = S_e^{\frac{2+3\lambda}{\lambda}}, \tag{5}$$

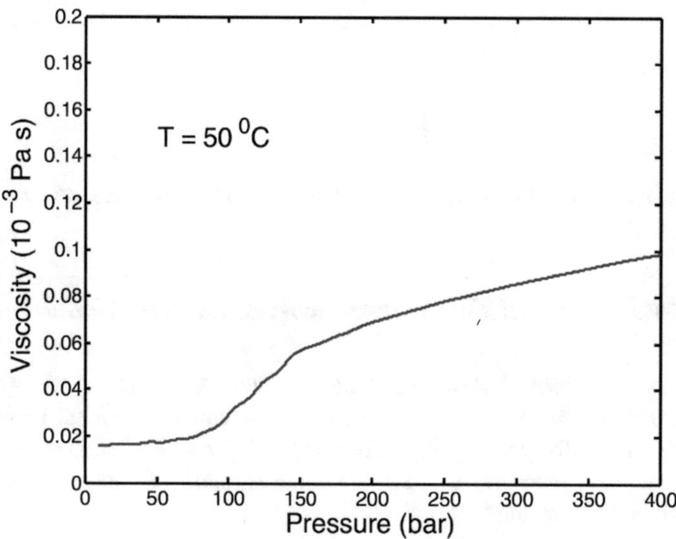

Fig. 3 Cubic spline interpolation for the correlation of CO_2 viscosity and the pressure at 50 °C

$$k_{rn} = (1 - S_e)^2 \left(1 - S_e^{\frac{2+\lambda}{\lambda}}\right), \tag{6}$$

where λ is a scalar which takes small value (e.g., $\lambda = 0.2$) for the heterogeneous material and larger value (e.g., $\lambda = 2.0$) for the homogeneous material. The notation S_e denotes the effective saturation for the wetting phase, and is defined to be

$$S_e = \frac{S_w - S_{wr}}{1 - S_{wr}}, \tag{7}$$

where S_{wr} denotes the residual saturation for the wetting phase.

The dynamic viscosity of the CO_2-rich phase μ_n is also a function of pressure. Figure 3 shows the cubic spline interpolation of the function of $\mu_n(p_n)$. The dynamic viscosity of brine μ_w is mainly dependent on the salinity while the pressure has little influence. Therefore, we assume that μ_w is constant. We now have two equations, which are Eqs. (1) and (2), and four unknowns, which are S_w, S_n, p_w, and p_n. By definition, the phase saturations sum up to one:

$$S_w + S_n = 1. \tag{8}$$

The difference between the phase pressures is the capillary pressure, p_c,

$$p_n - p_w = p_c, \tag{9}$$

and it is a function of the effective wetting phase saturation, using the Brooks and Corey formula:

$$p_c(S_e) = p_d S_e^{-\frac{1}{\lambda}}, \tag{10}$$

where p_d is the entry pressure, and λ is the same parameter in Eqs. (5) and (6).

2.2 The Isothermal Two-Phase Two-Component Model

Before numerical discretization, we manipulate the system (1)–(2) of conservation laws. We first add the two equations and obtain an equation without mass fraction terms. We choose for primary variables the wetting phase saturation, S_w, and the non-wetting phase pressure, p_n. The remaining variables, S_n and p_w, are replaced by Eqs. (8) and (9). In addition, we set

$$X_n^{H_2O} = 0, \quad X_n^{CO_2} = 1, \quad X_w^{H_2O} = 1 - X_w^{CO_2}.$$

Finally, we expand the time derivative, and obtain, after manipulation, the following equations:

$$\begin{aligned} &\phi\,(\rho_w - \rho_n)\frac{\partial S_w}{\partial t} + \phi(1 - S_w)\frac{d\rho_n}{dp_n}\frac{\partial p_n}{\partial t} \\ &-\nabla\cdot\left(\frac{k_{rw}}{\mu_w}\rho_w K(-\frac{dp_c}{dS_w})\nabla S_w\right) - \nabla\cdot\left(\frac{k_{rw}}{\mu_w}\rho_w K\nabla p_n\right) \\ &-\nabla\cdot\left(\frac{k_{rn}}{\mu_n}\rho_n K\nabla p_n\right) = q^{CO_2} + q^{H_2O}, \end{aligned} \tag{11}$$

$$\begin{aligned} &\phi\rho_w\left(1 - X_w^{CO_2}\right)\frac{\partial S_w}{\partial t} - \phi\rho_w S_w\frac{dX_w^{CO_2}}{dp_n}\frac{\partial p_n}{\partial t} \\ &-\nabla\cdot\left(\frac{k_{rw}}{\mu_w}\rho_w\left(1 - X_w^{CO_2}\right)K\left(-\frac{dp_c}{dS_w}\right)\nabla S_w\right) \\ &-\nabla\cdot\left(\frac{k_{rw}}{\mu_w}\rho_w\left(1 - X_w^{CO_2}\right)K\nabla p_n\right) = q^{H_2O}. \end{aligned} \tag{12}$$

We note that in the equations above, the functions ρ_n, $X_w^{CO_2}$, and μ_n depend on the pressure p_n and the functions k_{rw} and k_{rn} depend on the saturation S_w. Now, let us state the initial and boundary conditions. The time interval is denoted by $(0, T)$. The domain is denoted by Ω and its boundary by $\partial\Omega$. We separate $\partial\Omega$ into two parts: the outflow boundary $\Gamma^{\partial+}$ and the inflow boundary $\Gamma^{\partial-}$, satisfying

$$\overline{\Gamma^{\partial+} \cup \Gamma^{\partial-}} = \partial\Omega, \quad \Gamma^{\partial+} \cap \Gamma^{\partial-} = \emptyset.$$

The initial conditions are described below:

$$S_w(x,0) = S_0(x), \quad p_n(x,0) = p_0(x), \quad \forall x \in \Omega. \tag{13}$$

We impose Dirichlet and Neumann-type boundary conditions on different parts of the boundary:

$$S_w(x,t) = f_s(x,t), \quad \forall x \in \Gamma^{\partial -},\ t \in (0,T), \tag{14}$$

$$\nabla S_w(x,t) \cdot \mathbf{n} = 0, \quad \forall x \in \Gamma^{\partial +},\ t \in (0,T), \tag{15}$$

$$p_n(x,t) = f_p(x,t), \quad \forall x \in \Gamma^{\partial +},\ t \in (0,T), \tag{16}$$

$$\nabla p_n(x,t) \cdot \mathbf{n} = g_p(x,t), \quad \forall x \in \Gamma^{\partial -},\ t \in (0,T). \tag{17}$$

3 Numerical Method

Equations (11) and (12) are strongly coupled, and thus we use the fully coupled method to solve the problem. We also use the backward Euler method for the time discretization to avoid CFL constraints. For the space discretization, we discuss the existing discontinuous Galerkin methods and propose a new "partial upwind" method in this section.

3.1 Standard DG Discretization

The interior penalty DG methods usually use the average numerical flux with stabilization terms for the diffusion terms (elliptic operators) and use the upwind numerical flux for the advection terms (hyperbolic operators).

Both the third terms in Eqs. (11) and (12) are independent or slightly dependent on p_n, and they can be treated as the nonlinear elliptic terms in S_w and are discretized accordingly. The rest of the terms in Eqs. (11) and (12) come from the Darcy's law for phase velocities, and are elliptic terms in p_n and hyperbolic terms in S_w. They are the advection terms and are supposed to be discretized using the upwind method. However, the resulting scheme is unstable, because the fact that they are also the elliptic operator on p_n cannot be ignored. Therefore, large penalty terms for p_n are needed for stabilization, which gives inaccurate solutions. In fact, inaccuracies in ∇p_n yield large oscillations in S_w. Therefore, previous DG work for two-phase problems, such as the papers by Epshteyn and Riviere [5] and by Ern and Mozolevski [7], treat similar terms as the elliptic operator and use the usual DG discretization. This means that numerical diffusive fluxes are averaged and stabilization terms are added. In this work, we show that for the CO_2 storage problem, the average fluxes yield oscillations when advection dominates the problem.

3.2 The Partial Upwind Method

The idea behind the partial upwind method is that before applying the upwind method to the advection term, we substract an elliptic part depending on p_n from the advection term and discretize it using the usual average numerical flux with stabilization. Hence, the equation for p_n is much better stabilized. We expect the proposed method to perform well because upwinding stabilizes the numerical oscillations. A theoretical justification of the convergence of the method is challenging because the nonlinear coefficients degenerate in parts of the domain. We show in the numerical examples that the solution is unstable if we do not use partial upwinding.

We use the Brooks–Corey formula, and the fourth term in Eq. (11) becomes

$$
\begin{aligned}
-\nabla\cdot\left(\frac{\rho_w}{\mu_w}S_e^{3+\frac{2}{\lambda}}K\nabla p_n\right) &= -\nabla\cdot\left(\frac{\rho_w}{\mu_w}\left(\frac{S_w-S_{wr}}{1-S_{wr}}\right)^{3+\frac{2}{\lambda}}K\nabla p_n\right)\\
= -\nabla\cdot\left(\frac{\rho_w}{\mu_w}C(p_n,x)K\nabla p_n\right)&\\
-\nabla\cdot\left(\frac{\rho_w}{\mu_w}\left(\left(\frac{S_w-S_{wr}}{1-S_{wr}}\right)^{3+\frac{2}{\lambda}}-C(p_n,x)\right)K\nabla p_n\right),& \qquad (18)
\end{aligned}
$$

where $C(p_n, x)$ is a positive function that does not depend on S_w. Then, the first part

$$-\nabla\cdot\left(\frac{\rho_w}{\mu_w}C(p_n,x)K\nabla p_n\right),$$

is discretized as the elliptic term for p_n and the second part

$$-\nabla\cdot\left(\frac{\rho_w}{\mu_w}\left(\left(\frac{S_w-S_{wr}}{1-S_{wr}}\right)^{3+\frac{2}{\lambda}}-C(p_n,x)\right)K\nabla p_n\right),$$

is discretized using the upwind scheme. The selection of $C(p_n, x)$ depends on the value of λ. For example, when $\lambda = 2$, the best choice is

$$C(p_n,x)=\left(\frac{S_{wr}}{1-S_{wr}}\right)^4,$$

since then $\left(\frac{S_w-S_{wr}}{1-S_{wr}}\right)^{3+\frac{2}{\lambda}} - C(p_n, x)$ can be written as the product of S_w and a scalar $\alpha(S_w)$, that is:

$$\left(\frac{S_w-S_{wr}}{1-S_{wr}}\right)^4-\left(\frac{S_{wr}}{1-S_{wr}}\right)^4 = S_w\,\alpha(S_w).$$

3.3 General PDE Model

The partial upwind method not only works for CO_2 sequestration problem, but also works for other two-phase flow problems. Now, let us consider a more general PDE system with the CO_2 sequestration problem being a particular case. The general system of PDEs is presented in a way that the ambiguous terms are already reasonably separated according to the partial upwind method.

$$\tau_1(\mathbf{x},p)\frac{\partial S}{\partial t}+\theta_1(\mathbf{x},S,p)\frac{\partial p}{\partial t}-\nabla\cdot(a_1(\mathbf{x},S)\nabla S)$$
$$-\nabla\cdot(b_1(\mathbf{x},p)\nabla p)+\nabla\cdot\boldsymbol{\beta}_1(\mathbf{x},S,p,\nabla p)=q_1(t,\mathbf{x}), \tag{19}$$

$$\tau_2(\mathbf{x},p)\frac{\partial S}{\partial t}+\theta_2(\mathbf{x},S,p)\frac{\partial p}{\partial t}-\nabla\cdot(a_2(\mathbf{x},S,p)\nabla S)$$
$$-\nabla\cdot(b_2(\mathbf{x},p)\nabla p)+\nabla\cdot\boldsymbol{\beta}_2(\mathbf{x},S,p,\nabla p)=q_2(t,\mathbf{x}). \tag{20}$$

Even though in these equations, the term $\boldsymbol{\beta}_i(\mathbf{x},S,p,\nabla p)$ $(i=1,2)$ has the form of $\beta_i(\mathbf{x},S,p)\nabla p$, where β_i is a scalar function, we use for convenience the more general notation $\boldsymbol{\beta}_i(\mathbf{x},S,p,\nabla p)$. In this system, terms with a_i and b_i coefficients are treated as elliptic terms and the ones with $\boldsymbol{\beta}_i$ are treated as hyperbolic terms. $-\nabla\cdot(b_i\nabla p)$ and $\nabla\cdot\boldsymbol{\beta}_i$ are the separated terms using the partial upwind method.

For our CO_2 sequestration model, the coefficients are

$$\tau_1(\mathbf{x},p)=\phi\left(\rho_w-\rho_n\right), \tag{21}$$

$$\theta_1(\mathbf{x},S,p)=\phi(1-S)\rho_n'(p), \tag{22}$$

$$a_1(\mathbf{x},S)=\frac{\rho_w}{\mu_w}\frac{p_d}{\lambda}\frac{1}{1-S_{wr}}S_e^{2+\frac{1}{\lambda}}K, \tag{23}$$

$$b_1(\mathbf{x},p)=\frac{\rho_w}{\mu_w}C_{1,1}K+\frac{\rho_{co_2}}{\mu_n}C_{1,2}K, \tag{24}$$

$$\boldsymbol{\beta}_1(\mathbf{x},S,p,\nabla p)=-\frac{\rho_w}{\mu_w}\left(S_e^{3+\frac{2}{\lambda}}-C_{1,1}\right)K\nabla p$$
$$-\frac{\rho_n}{\mu_n}\left((1-S_e)^2(1-S_e^{1+\frac{2}{\lambda}})-C_{1,2}\right)K\nabla p, \tag{25}$$

$$q_1(t,\mathbf{x})=q^{CO_2}(t,\mathbf{x})+q^{H_2O}(t,\mathbf{x}), \tag{26}$$

$$\tau_2(\mathbf{x},p)=\phi\rho_w\left(1-X_w^{CO_2}\right), \tag{27}$$

$$\theta_2(\mathbf{x},S,p)=-\phi\rho_w S X_w^{CO_2\prime}(p), \tag{28}$$

$$a_2(\mathbf{x}, S, p) = \left(1 - X_w^{CO_2}\right) \frac{\rho_w}{\mu_w} \frac{p_d}{\lambda} \frac{1}{1-S_{wr}} S_e^{2+\frac{1}{\lambda}} K, \tag{29}$$

$$b_2(\mathbf{x}, p) = \left(1 - X_w^{CO_2}\right) \frac{\rho_w}{\mu_w} C_2 K, \tag{30}$$

$$\boldsymbol{\beta}_2(\mathbf{x}, S, p, \nabla p) = -\left(1 - X_w^{CO_2}\right) \frac{\rho_w}{\mu_w} \left(S_e^{3+\frac{2}{\lambda}} - C_2\right) K \nabla p, \tag{31}$$

$$q_2(t, \mathbf{x}) = q^{H_2O}(t, \mathbf{x}), \tag{32}$$

where $C_{1,1}$, $C_{1,2}$, and C_2 play the same role as the C in Eq. (18). The choice of $C_{1,1}$, $C_{1,2}$, and C_2 that we use are listed here:

$$C_{1,1} = \left(\frac{S_{wr}}{1-S_{wr}}\right)^4,$$

$$C_{1,2} = \frac{(1-2S_{wr})}{(1-S_{wr})^4},$$

$$C_2 = \left(\frac{S_{wr}}{1-S_{wr}}\right)^4.$$

3.4 Numerical Discretization

Suppose Ω is a polygonal domain. Let $\mathscr{E}^h$ be the mesh on Ω, comprised of elements denoted by E (intervals in 1D, triangles in 2D, and tetrahedra in 3D). Let γ denote the edge of the element and $\mathbf{n}_\gamma$ be a fixed normal direction for every γ. If γ is on the boundary, then $\mathbf{n}_\gamma$ is chosen to be the outward direction. Let Γ^h denote the collection of all the interior edges and $\Gamma^{h,\partial}$ the boundary edges. Let $\Gamma^{h,\partial+}$ denote the set of the outflow boundary and $\Gamma^{h,\partial-}$ the inflow boundary.

Define the finite element space as:

$$\mathbb{X}^h = \{v \in L^2(\Omega) : v \in \mathbb{P}^r(E), \forall E \in \mathscr{E}^h\}, \tag{33}$$

where r denotes the order of the polynomials and r is an integer bigger than or equal to 1. All functions in $\mathbb{X}^h$ have two different values on edge γ. Let us define the jump of a function on γ. Suppose γ is shared by two neighboring elements E_1 and E_2, and $\mathbf{n}_\gamma$ points from E_1 to E_2. For any function $v \in \mathbb{X}^h$, the jump on γ is defined to be

$$[v]|_\gamma = v|_{E_1} - v|_{E_2}. \tag{34}$$

If γ is on the boundary, then

$$[v]|_\gamma = v|_E. \tag{35}$$

The average is defined as:

$$\{v\}|_\gamma = \frac{1}{2}(v|_{E_1} + v|_{E_2}), \tag{36}$$

and if γ is on the boundary

$$\{v\}|_\gamma = v|_E. \tag{37}$$

We define the upwind as:

$$v^\uparrow|_\gamma = \begin{cases} v|_{E_1}, & \text{if } \{\nabla p_n \cdot \mathbf{n}_\gamma\} > 0, \\ v|_{E_2}, & \text{otherwise.} \end{cases} \tag{38}$$

If $\gamma \in \Gamma^{h,\partial+}$, and $\gamma \in E$, then

$$v^\uparrow|_\gamma = v|_E. \tag{39}$$

Let Δt denote the time step and N denote the number of time steps such that $T = N\Delta t$. Let $t^n = n\Delta t$ be successive discrete times. Define the variational forms $A_i(p_h^n, S_h^n, v)$ for $i = 1, 2$ and all $v \in \mathbb{X}^h$:

$$A_i(p_h^{n+1}, S_h^{n+1}, v) = \tag{40}$$

$$\sum_{E\in\mathscr{E}^h} \int_E a_i(\mathbf{x}, S_h^{n+1}, p_h^{n+1})\nabla S_h^{n+1} \cdot \nabla v - \sum_{\gamma\in\Gamma^h\cup\Gamma_1^{h,\partial}} \int_\gamma \{a_i(\mathbf{x}, S_h^{n+1})\nabla S_h^{n+1} \cdot \mathbf{n}_\gamma\}[v]$$

$$+\epsilon \sum_{\gamma\in\Gamma^h\cup\Gamma_1^{h,\partial}} \int_\gamma \{a_i(\mathbf{x}, S_h^{n+1})\nabla v \cdot \mathbf{n}_\gamma\}[S_h^{n+1}] + \sum_{E\in\mathscr{E}^h} \int_E b_i(\mathbf{x}, p_h^{n+1})\nabla p_h^{n+1} \cdot \nabla v$$

$$- \sum_{\gamma\in\Gamma^h\cup\Gamma_2^{h,\partial}} \int_\gamma \{b_i(\mathbf{x}, p_h^{n+1})\nabla p_h^{n+1} \cdot \mathbf{n}_\gamma\}[v] + \epsilon$$

$$\sum_{\gamma\in\Gamma^h\cup\Gamma_2^{h,\partial}} \int_\gamma \{b_i(\mathbf{x}, p_h^{n+1})\nabla v \cdot \mathbf{n}_\gamma\}[p_h^{n+1}]$$

$$- \sum_{E\in\mathscr{E}^h} \int_E \boldsymbol{\beta}_i(\mathbf{x}, S_h^{n+1}, p_h^{n+1}, \nabla p_h^{n+1}) \cdot \nabla v$$

$$+ \sum_{\gamma\in\Gamma^h} \int_\gamma \boldsymbol{\beta}_i^\uparrow(\mathbf{x}, S_h^{n+1}, p_h^{n+1}, \nabla p_h^{n+1}) \cdot \mathbf{n}_\gamma.$$

Define the linear forms $Q_i^n(v)$ for $i = 1, 2$:

$$Q_i^{n+1}(v) = \sum_{E \in \mathcal{E}^h} \int_E q_i(t^{n+1}, \mathbf{x}) v, \tag{41}$$

and the penalty forms for $i = 1, 2$:

$$\begin{aligned} J_i(p_h^{n+1}, S_h^{n+1}, v) = & \sum_{\gamma \in \Gamma^h \cup \Gamma_2^{h,\partial}} \frac{\sigma_\gamma^p}{h_\gamma} (\int_\gamma \{a_i(\mathbf{x}, S_h^n)\}) \int_\gamma [p_h^{n+1}][v] \\ & + \sum_{\gamma \in \Gamma^h \cup \Gamma_1^{h,\partial}} \frac{\sigma_\gamma^S}{h_\gamma} (\int_\gamma \{a_i(\mathbf{x}, S_h^n)\}) \int_\gamma [S_h^{n+1}][v]. \end{aligned} \tag{42}$$

Notice that the penalty term depends on the value of the elliptic coefficients. The boundary conditions are handled by the following forms, for $i = 1, 2$:

$$\begin{aligned} & B_i^{n+1}(S_h^{n+1}, p_h^{n+1}, \nabla p_h^{n+1}, v) = \\ & \epsilon \sum_{\gamma \in \Gamma_1^{h,\partial}} \int_\gamma a_i(\mathbf{x}, f_S(\mathbf{x}, t^{n+1})) \nabla v \cdot \mathbf{n}_\gamma f_S(\mathbf{x}, t^{n+1}) \\ & + \sum_{\gamma \in \Gamma_1^{h,\partial}} \frac{\sigma_S}{h_\gamma} \left(\int_\gamma a_i(\mathbf{x}, f_S(\mathbf{x}, t^n)) \right) \int_\gamma f_S(\mathbf{x}, t^{n+1}) v \\ & + \epsilon \sum_{\gamma \in \Gamma_2^{h,\partial}} \int_\gamma b_i(\mathbf{x}, f_P(\mathbf{x}, t^{n+1})) \nabla v \cdot \mathbf{n}_\gamma f_P(\mathbf{x}, t^{n+1}) \\ & + \sum_{\gamma \in \Gamma_2^{h,\partial}} \frac{\sigma_P}{h_\gamma} \left(\int_\gamma b_i(\mathbf{x}, f_P(\mathbf{x}, t^n)) \right) \int_\gamma f_P(\mathbf{x}, t^{n+1}) v \\ & + \sum_{\gamma \in \Gamma_1^{h,\partial}} \int_\gamma b_i(\mathbf{x}, p_h^{n+1}) g_P(\mathbf{x}, t^{n+1}) v \\ & - \sum_{\gamma \in \Gamma_2^{h,\partial}} \int_\gamma \boldsymbol{\beta}_i(\mathbf{x}, S_h^{n+1}, f_P(\mathbf{x}, t^{n+1}), \nabla p_h^{n+1}) \cdot \mathbf{n}_\gamma v \\ & - \sum_{\gamma \in \Gamma_1^{h,\partial}} \int_\gamma \beta_i(\mathbf{x}, f_S(\mathbf{x}, t^{n+1}), p_h^{n+1}) g_P(\mathbf{x}, t^{n+1}) v. \end{aligned}$$

We now defined the numerical scheme: find $(S_h^n)_{0\le n\le N-1} \subset \mathbb{X}^h$ and $(p_h^n)_{0\le n\le N-1} \subset \mathbb{X}^h$, satisfying for all $v \in \mathbb{X}_h$

$$\sum_{E\in\mathcal{E}^h}\int_E \tau_1(\mathbf{x}, p_h^{n+1})\frac{S_h^{n+1}-S_h^n}{\Delta t}v + \sum_{E\in\mathcal{E}^h}\int_E \theta_1(\mathbf{x}, S_h^{n+1}, p_h^{n+1})\frac{p_h^{n+1}-p_h^n}{\Delta t}v$$
$$+A_1(p_h^{n+1}, S_h^{n+1}, v) + J_1(p_h^{n+1}, S_h^{n+1}, v)$$
$$= Q_1(v) + B_1^{n+1}(S_h^{n+1}, p_h^{n+1}, \nabla p_h^{n+1}, v), \tag{43}$$

$$\sum_{E\in\mathcal{E}^h}\int_E \tau_2(\mathbf{x}, p_h^{n+1})\frac{S_h^{n+1}-S_h^n}{\Delta t}v + \sum_{E\in\mathcal{E}^h}\int_E \theta_2(\mathbf{x}, S_h^{n+1}, p_h^{n+1})\frac{p_h^{n+1}-p_h^n}{\Delta t}v$$
$$A_2(p_h^{n+1}, S_h^{n+1}, v) + J_2(p_h^{n+1}, S_h^{n+1}, v)$$
$$= Q_2(v) + B_2^{n+1}(S_h^{n+1}, p_h^{n+1}, \nabla p_h^{n+1}, v), \tag{44}$$

$$\sum_{E\in\mathcal{E}^h}\int_E S_h^0 v = \sum_{E\in\mathcal{E}^h}\int_E S_0 v, \tag{45}$$

$$\sum_{E\in\mathcal{E}^h}\int_E p_h^0 v = \sum_{E\in\mathcal{E}^h}\int_E p_0 v. \tag{46}$$

3.5 The Newton–Raphson Method for Linearization

We denote the basis of $\mathbb{X}^h$ by $(\phi_j)_{j=1}^J$ and expand the numerical approximations of saturation and pressure for $n = 0, \cdots, N$

$$S_h^n = \sum_{j=1}^J s_j^n \phi_j, \quad p_h^n = \sum_{j=1}^J p_j^n \phi_j.$$

We denote the vectors of degrees of freedom by $\mathbf{s}^n = (s_1^n, \cdots, s_J^n)$ and $\mathbf{p}^n = (p_1^n, \cdots, p_J^n)$. We can rewrite the discrete equations as a general nonlinear system of the form:

$$F_1(\mathbf{s}^{n+1}, \mathbf{p}^{n+1}) = 0,$$
$$F_2(\mathbf{s}^{n+1}, \mathbf{p}^{n+1}) = 0.$$

We use the Newton–Raphson method to solve for $\mathbf{s}^{n+1}$ and $\mathbf{p}^{n+1}$:

$$(\mathbf{s}_k, \mathbf{p}_k)^T = (\mathbf{s}_{k-1}, \mathbf{p}_{k-1})^T - \left(\frac{\partial(F_1, F_2)}{\partial(\mathbf{s}_{k-1}, \mathbf{p}_{k-1})}\right)^{-1}(F_1(\mathbf{s}_{k-1}, \mathbf{p}_{k-1}), F_2(\mathbf{s}_{k-1}, \mathbf{p}_{k-1}))^T, \tag{47}$$

where the subscript k denotes the kth iteration. The stopping criterion is

$$\|(\mathbf{s}_k, \mathbf{p}_k) - (\mathbf{s}_{k-1}, \mathbf{p}_{k-1})\|_2 \le tolerance\|(\mathbf{s}_k, \mathbf{p}_k)\|_2.$$

This algorithm involves calculating the Jacobian $\frac{\partial(F_1,F_2)}{\partial(\mathbf{s}_{k-1},\mathbf{p}_{k-1})}$, which is done analytically.

4 Numerical Results

In all simulations, the temperature is fixed at 50 °C.

4.1 CO_2 Injection Test on Smooth Solutions

We verify the scheme using the method of manufactured solutions. We obtain numerical convergence rates of the partial upwind DG method for the CO_2 sequestration model (Eqs. (11) and (12)) on smooth solutions. We describe below the functions of all the parameters and the exact solutions. The density, viscosity, and mass fraction are defined by:

$$\rho_n = 200 + 2 \cdot 10^{-6} p,$$
$$\mu_n = 1.6 \cdot 10^{-5} + 5 \cdot 10^{-13} p,$$
$$X_w^{CO_2} = 10^{-15} p^2.$$

The domain Ω is the unit interval and the final time is $T = 0.5$. The values of all other parameters are listed in Table 1. The exact smooth solutions are

$$p_n(x,t) = 10^5(x-1)^2 t + 8 \cdot 10^6, \quad S_w(x,t) = 0.75\sin(0.5\pi x)(1-t) + 0.25.$$

The source terms are calculated accordingly. Notice that the solutions are linearly dependent on time. Therefore, the backward Euler scheme for time discretization gives no consistency error. Hence, we root out the possibility that a very small time step is needed for the purpose of obtaining the expected convergence rate on space. We use $\epsilon = 1$ and set the penalties to be $\sigma_p = 10$, $\sigma_S = 10$. For the Newton's iteration, the tolerance is 10^{-10}. The starting point of the Newton iteration for each time step is the numerical solution from the previous time step. Tables 2 and 3 show the numerical errors and convergence rates of p_n and S_w for $r = 1, 2$, respectively. "P L^2-err" denotes the L^2 error for pressure and is defined to be

Table 1 Table for the parameters for verification example

Parameter	ϕ	K	p_d	λ	S_{wr}
Value	0.25	10^{-12}	5000	2	0.2

Table 2 Errors and convergence rates for smooth solutions for piecewise linears

Mesh size	P L^2-err	CR	P E-err	CR	S L^2-err	CR	S E-err	CR
0.125	9.60e−03		3.99e−01		3.82e−04		2.59e−02	
0.0625	2.61e−03	1.88	1.97e−01	1.02	9.93e−05	1.94	1.29e−02	1
0.03125	1.01e−03	1.37	9.76e−02	1.01	3.27e−05	1.6	9.54e−03	0.437
0.01562	5.04e−04	1	4.86e−02	1.01	1.45e−05	1.17	9.49e−03	0.00781
0.007812	2.63e−04	0.94	2.43e−02	1	5.21e−06	1.48	6.64e−03	0.516

Table 3 Errors and convergence rates for smooth solutions for piecewise quadratics

Mesh size	P L^2-err	CR	P E-err	CR	S L^2-err	CR	S E-err	CR
0.125	1.48e−05		6.20e−05		6.32e−06		1.04e−03	
0.0625	2.00e−06	2.88	8.13e−06	2.93	7.92e−07	3	2.59e−04	2
0.03125	2.39e−07	3.06	1.08e−06	2.92	1.01e−07	2.97	6.22e−05	2.06
0.01562	2.37e−08	3.34	2.06e−07	2.39	1.41e−08	2.84	1.40e−05	2.15
0.007812	2.74e−09	3.11	4.06e−08	2.34	2.05e−09	2.78	3.07e−06	2.19

$$\left(\sum_{E\in\mathcal{E}^h} \|p_n(\cdot, 0.5) - p_h^N(\cdot)\|^2_{L^2(E)} \right)^{\frac{1}{2}}.$$

"P E-err" denotes the energy error for pressure and is defined to be

$$\left(\sum_{E\in\mathcal{E}^h} \|p_n(\cdot, 0.5) - p_h^N(0.5)\|^2_{H^1(E)} + \sum_{\gamma\in\Gamma^h} \frac{\sigma_p}{h_\gamma} \|[p_n(\cdot, 0.5) - p_h^N(0.5)]\|^2_{L^2(\gamma)} \right)^{\frac{1}{2}}.$$

The errors " S L^2-err" and "S E-err" are defined similarly for the saturation S_w. "CR" denotes the convergence rate. For all these numerical simulations, it takes 3 or 4 Newton iterations to reach the stopping criterion.

Table 2 shows that we obtain first-order convergence rate for p_n in the energy norm, but not for S_w, because the energy error for the pressure dominates the total energy error. Since we solve pressure and saturation simultaneously, we are supposed to obtain first-order convergence rate for the total energy error, even though partial result (the saturation in this case) does not converge at the same rate. Table 3 shows that when $r = 2$, both p and S have the second-order convergence rates or more for the energy norm and third-order convergence rate for the L^2 norm. If we compare the results given by $r = 2$ with $r = 1$, we see that $r = 2$ gives much smaller errors, and thus gives us more accurate solutions. Therefore, we prefer $r = 2$ when we do numerical simulations for multiphase fluid flow problems.

In summary, we have obtained the expected convergence rates for smooth functions using the partial upwind method.

4.2 Two-Phase Incompressible Fluid Flow Problem

In this numerical example, we assume that there is no mass transfer between the two phases and the densities of the two phases are constant. Therefore, we obtain an incompressible two-phase fluid flow problem, which is a simpler problem than the CO_2 storage problem. Since there are many studies for this problem, we can compare the results obtained with the partial upwind method with other methods. In this work, we consider the injection of the non-wetting phase into the porous media filled with the wetting phase, because the purpose of this two-phase flow test is to pave the way for CO_2 sequestration simulation. We will also compare the partial upwind method with the usual DG method that uses the average numerical flux and show that the partial upwind method is superior for some cases. The two-phase incompressible fluid flow model can be written as:

$$-\phi\rho_n \frac{\partial}{\partial t} S_w - \nabla \cdot \left(\frac{k_{rn}}{\mu_n}\rho_n K \nabla p_n\right) = q^n, \tag{48}$$

$$\phi\rho_w \frac{\partial}{\partial t} S_w - \nabla \cdot \left(\frac{k_{rw}}{\mu_w}\rho_w K \nabla (p_n - p_c)\right) = q^w. \tag{49}$$

Example of Homogeneous Medium

The domain Ω is the unit interval. The initial pressure and saturation are

$$p_n(x, 0) = 2 \cdot 10^6, \quad x \in (0, 1),$$

$$S_w(x, 0) = \begin{cases} 0.3 + 2^5 \cdot x, & x \in (0, 2^{-6}), \\ 0.8, & x \in (2^{-6}, 1). \end{cases}$$

The values of the parameters are listed in Table 4. Figure 4 shows that the non-wetting phase front reaches almost 0.2 m at 30 s and almost 0.4 m at 60 s. Figure 5 shows the same numerical test using the average numerical flux. Comparing the two figures, we can see that both methods can solve this problem well and their results are almost identical. The Newton–Raphson method takes about 4 or 5 iterations to converge for both methods. We also point out that the saturation front for this problem is not very sharp. In the next example, we will change the parameters to obtain a sharper front and we will compare both the partial upwind and averaged flux methods.

We rerun the same example as before, except that the non-wetting phase viscosity μ_n is chosen to be 10^{-2} Pa·s, which is ten times larger than in the previous example. The resulting saturation front is sharper, thus more challenging to approximate numerically. First, we use the partial upwind method to solve the problem. The penalty values are chosen to be $\sigma_p = 1000$ and $\sigma_S = 0$. The numerical solutions are shown in Fig. 6. The pressure is shown in the left figure and we notice that the pressure gradient has an obvious change near 0.2 m, where also the saturation front

Table 4 Parameter values for incompressible two-phase flow in homogeneous medium

Parameter	Value
ϕ	0.2
K	10^{-12} (m^2)
p_d	5000 (Pa)
λ	2
S_{wr}	0.2
ρ_w	1000 (kg/m^3)
ρ_n	1000 (kg/m^3)
μ_w	10^{-3} (Pa · s)
μ_n	10^{-3} (Pa · s)
h	1/256
Δt	1 (s)
σ_p	10
σ_S	0

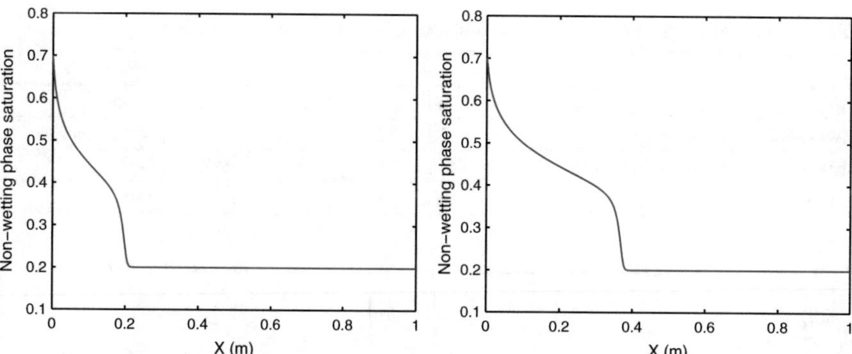

Fig. 4 Numerical results for S_n using partial upwind DG method. Left and right figures show the solutions at time $t = 30$ s and $t = 60$ s, respectively

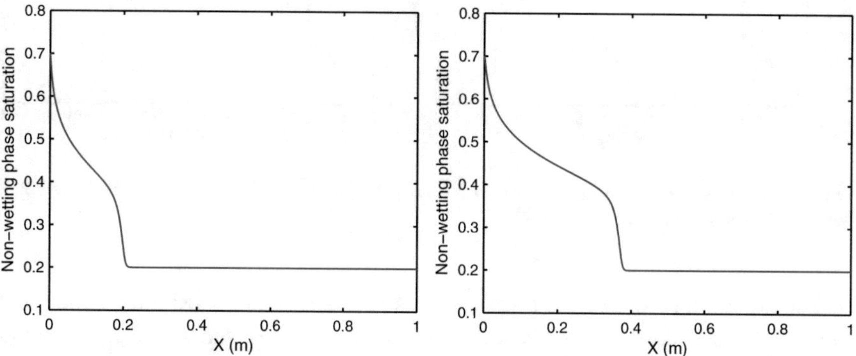

Fig. 5 Numerical results for S_n using the average numerical flux. Left and right figures show the solutions at time $t = 30$ s and $t = 60$ s, respectively

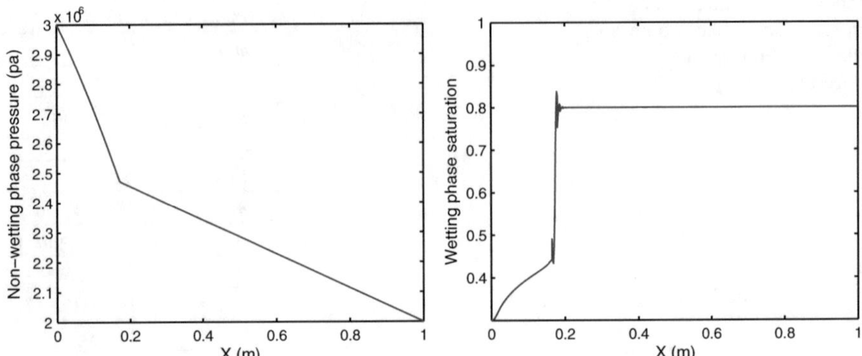

Fig. 6 Numerical results given by the partial upwind method for p (left) and S_w (right) at $t = 60$ s. $\sigma_p = 1000$ and $\sigma_S = 0$

Table 5 List of the parameters for the two-phase flow in homogeneous medium

Figures	Method	ϵ	h	Δt	σ_p	σ_S	Newton iter
Figure 6	Partial upwind	1	1/128	1 s	1000	0	Mostly 5–9
No solution	Average	1	1/128	1 s	1000	0	Not converge
Figure 7 (L)	Partial upwind	1	1/128	1 s	1000	1000	5–6
Figure 7 (R)	Average	1	1/128	1 s	1000	1000	5–6

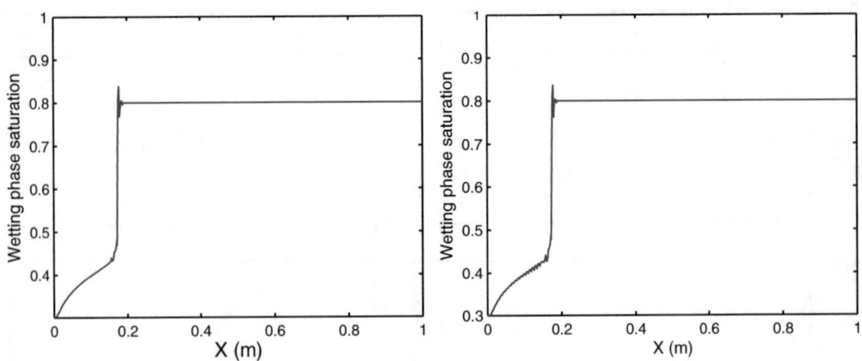

Fig. 7 Numerical results for S_w at $t = 60$ s given by the partial upwind flux (left) and the average flux (right). $\sigma_p = 1000$ and $\sigma_S = 1000$

is. The saturation front in the right figure is very sharp and it exhibits some local overshoot and undershoot. When we switch to the averaged numerical flux method using the same penalty values, the numerical solution blows up in the first time step. Table 5 summarizes the numerical parameters used for both methods.

In order to have a case where both methods are stable, we add more diffusion to the saturation by choosing a large value for σ_S. Figure 7 shows the saturation solutions after 60 s from both methods, when $\sigma_S = 1000$. The left figure uses the partial upwind flux, and the right one uses the average flux. Both methods capture the sharp front, but the right figure shows some wiggles on the interval $(0, 0.2)$. If we choose $\sigma_S = 100$, which means we add less diffusion, the average flux method blows up during the first time iteration, while we know from Fig. 6 that the partial upwind method works well even for $\sigma_S = 0$. From Table 5, we also see that both numerical methods use the same number of iterations to converge, if they do not blow up. Therefore, the partial upwind method is more robust than to the average flux method for this case.

Example of Heterogenous Medium

The domain is a heterogeneous porous medium in the sense that properties are different in the subdomain $(0.1562, 0.3125)$ than in the rest of the domain. The values of all the parameters are listed in Table 6. The pressure ranges from $8 \cdot 10^6$ to $8.05 \cdot 10^6$ Pa. The initial wetting phase saturation is 0.9. We run the test with $N = 256$ intervals and with $\Delta t = 1$ s. Figures 8 and 9 show the numerical results for $\sigma_S = 10$ and $\sigma_S = 0$, respectively. We show the curves of p_n and S_w at $t = 15$ and $t = 45$ s. At $t = 15$ s (figures in top row), the saturation front gradually passes the discontinuous point of the porous medium. More wetting phase is left in the high permeability region, resulting a saturation jump. At $t = 45$ s, we see that there is another saturation jump, where the porous medium property changes. We see from both figures that the one with $\sigma_S = 0$ seems to capture the saturation discontinuity slightly better than $\sigma_S = 10$, because there are less oscillations close to the discontinuities.

This numerical test shows the promising potential of the partial upwind DG method to solve multiphase multicomponent flows in heterogeneous media.

Table 6 Parameter values for incompressible two-phase flow in heterogenous porous medium

Parameter	Interval	Value
ϕ	$(0.1562, 0.3125)$	0.39
	$(0, 1)/(0.1562, 0.3125)$	0.4
K	$(0.1562, 0.3125)$	$5.26 \cdot 10^{-11}$ (m^2)
	$(0, 1)/(0.1562, 0.3125)$	$5.04 \cdot 10^{-10}$ (m^2)
p_d	$(0.1562, 0.3125)$	1324 (Pa)
	$(0, 1)/(0.1562, 0.3125)$	370 (Pa)
λ	$(0.1562, 0.3125)$	2.49
	$(0, 1)/(0.1562, 0.3125)$	3.86
S_{wr}	$(0.1562, 0.3125)$	0.1
	$(0, 1)/(0.1562, 0.3125)$	0.08
ρ_w	$(0, 1)$	1000 (kg/m^3)
ρ_n	$(0, 1)$	1000 (kg/m^3)
μ_w	$(0, 1)$	10^{-3} (Pa $\cdot$ s)
μ_n	$(0, 1)$	10^{-3} (Pa $\cdot$ s)

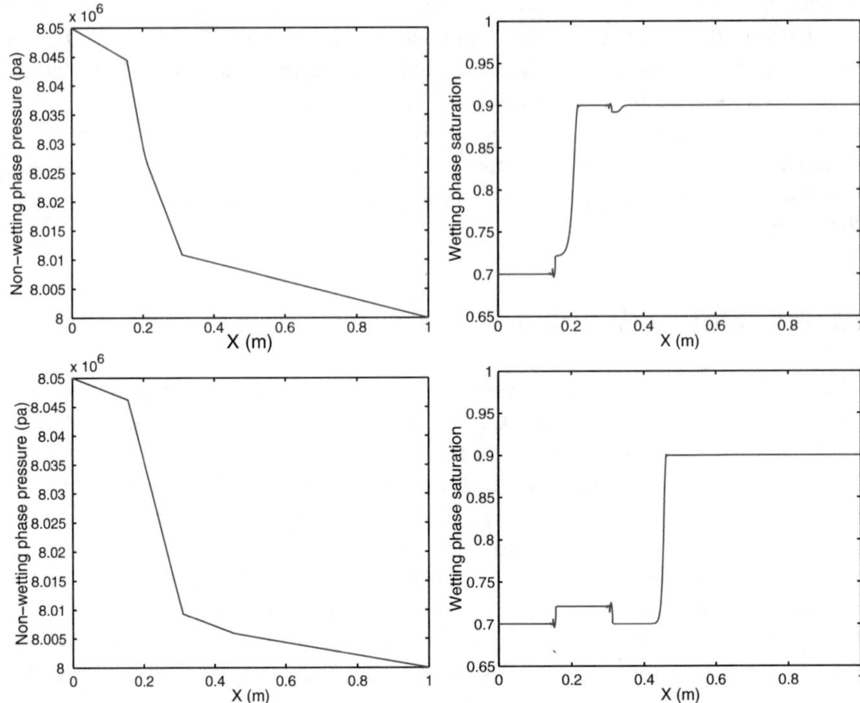

Fig. 8 Partial upwind numerical results for two-phase flow in heterogenous media. Figures show the numerical solutions p_n (left) and S_w (right) at time $t = 15$ s (top) and $t = 45$ s (bottom). Penalty values are $\sigma_p = 1000$ and $\sigma_S = 10$

4.3 Injection of CO_2 into Homogeneous Porous Medium

We now simulate the CO_2 sequestration problem on the domain $\Omega = (0, 1000)$. The initial non-wetting phase pressure is 250 bar and the initial wetting phase saturation is 0.95. Assume that CO_2 is injected at the rate of $\frac{\partial p}{\partial x} = -50$ for 3 years, at the endpoint $x = 0$. The parameters used in the simulations in this section are listed in Table 7. We first study the effect of varying the polynomial degrees and second the effect of varying the time steps.

CO_2 Injection Simulation for Different Orders of Approximation

We use $N = 256$ intervals for the mesh and $\Delta t = 5$ days for the time step. The numerical solutions are shown in Figs. 10 and 11 for $r = 1$ and $r = 2$, separately. We see that the CO_2-rich phase reaches approximately 270, 540, and 810 m after 1, 2, and 3 years, respectively. We also observe that the saturation of CO_2 gradually grows with time for a given point in space. Taking the point of 200 m, for example,

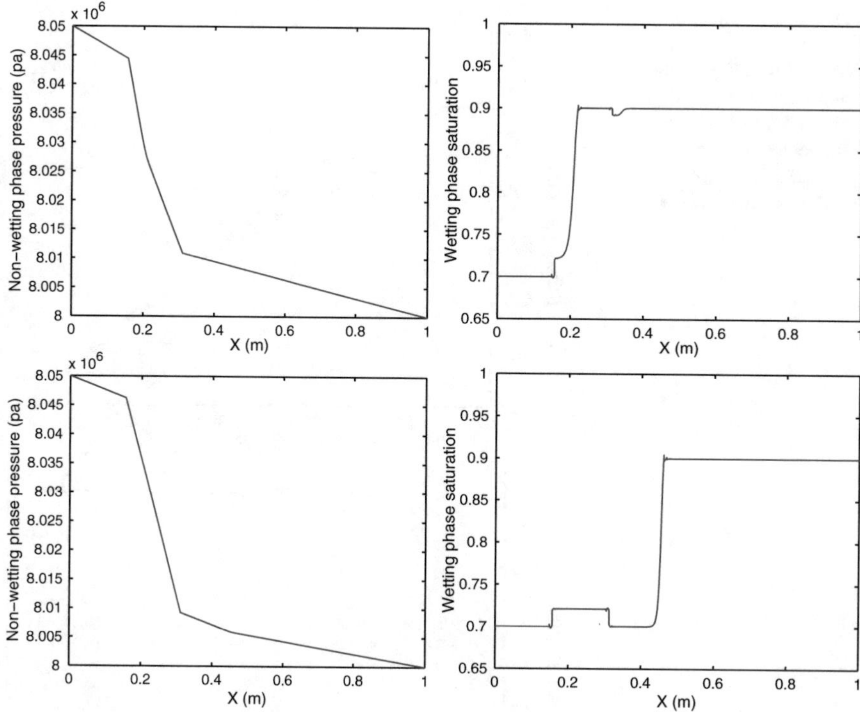

Fig. 9 Partial upwind numerical results for two-phase flow in heterogenous media. Figures show the numerical solutions p_n (left) and S_w (right) at time $t = 15\,\text{s}$ (top) and $t = 45\,\text{s}$ (bottom). Penalty values are $\sigma_p = 1000$ and $\sigma_S = 0$

Table 7 Parameter values for the simulation of CO_2 sequestration problem in homogeneous media

Parameter	Value
ϕ	0.25
K	10^{-12} (m^2)
p_d	5000 (Pa)
λ	2
S_{wr}	0.2
ρ_w	1000 (kg/m^3)
μ_w	10^{-3} (Pa · s)
ρ_n	Figure 1
μ_n	Figure 3
$X_w^{CO_2}$	Figure 2

CO_2 saturation is about 0.27 after the first year, 0.33 after the second year, and 0.36 after the third year. In addition, comparing Fig. 10 with Fig. 11, we observe that the solution obtained with $r = 2$ has less overshoot and has a sharper front than the solution obtained with $r = 1$. The effects of different values of σ_S are also studied. The case $\sigma_S = 10$ is shown in Fig. 11 and the case $\sigma_S = 0$ is shown in Fig. 12. We

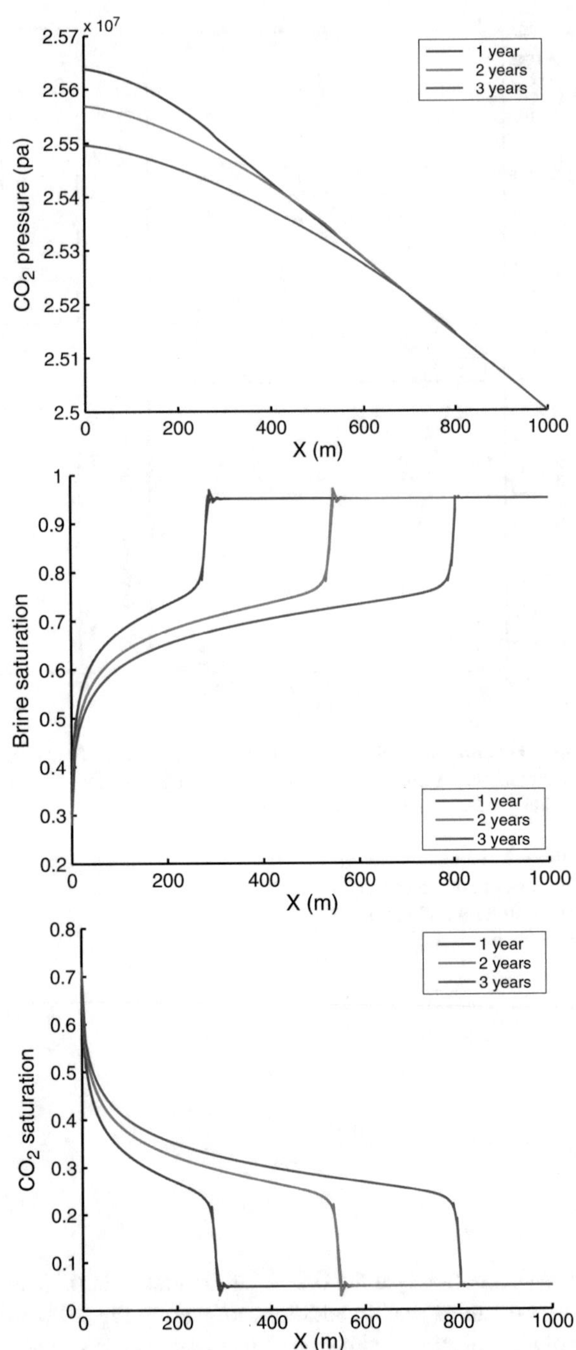

Fig. 10 CO_2 injection simulation in 1, 2, and 3 years. CO_2 pressure (top), brine saturation (middle), and CO_2 saturation (bottom). Parameters are: $r = 1$, $h = 1000/256\,\text{m}$, $\Delta t = 5$ days, $\sigma_p = 1000$, $\sigma_S = 10$

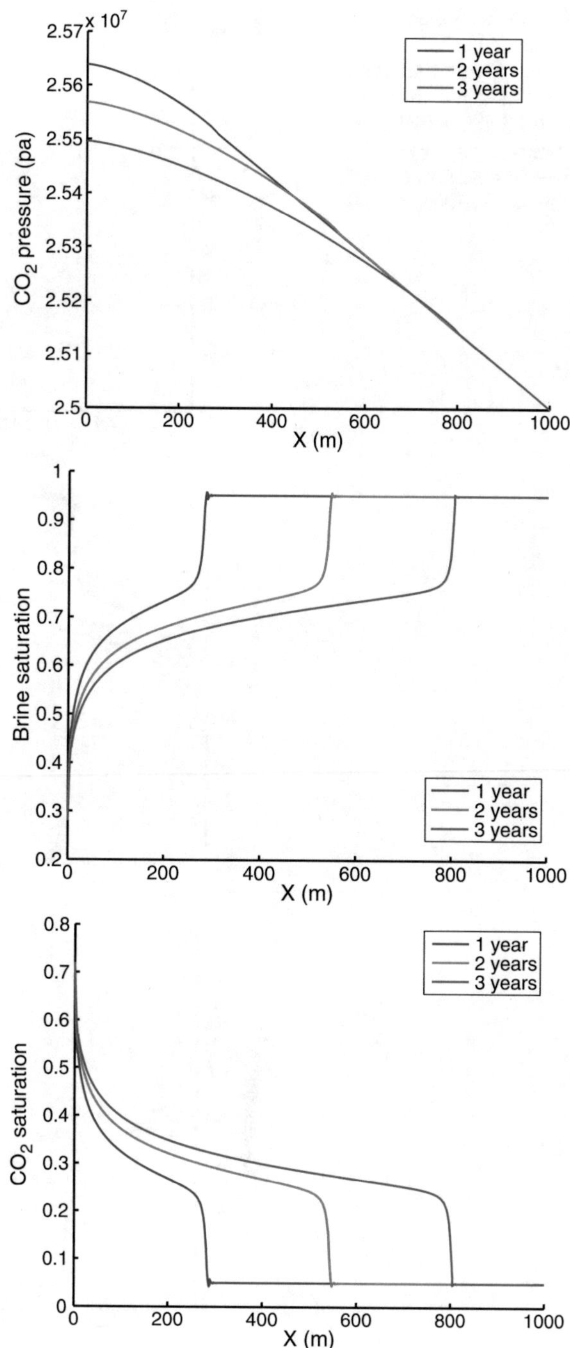

Fig. 11 CO_2 injection simulation in 1, 2, and 3 years. CO_2 pressure (top), brine saturation (middle), and CO_2 saturation (bottom). Parameters are: $r = 2$, $h = 1000/256\,\text{m}$, $\Delta t = 5$ days, $\sigma_p = 1000$, $\sigma_S = 10$

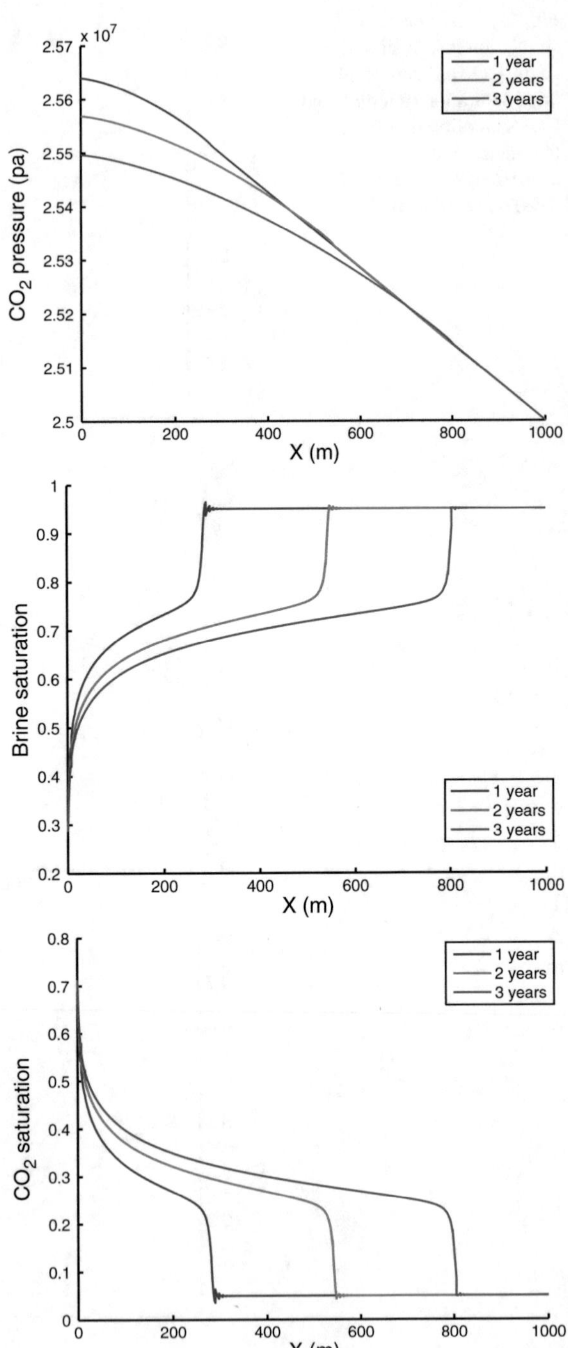

Fig. 12 CO_2 sequestration simulation in 1, 2, and 3 years. CO_2 pressure (top), brine saturation (middle), and CO_2 saturation (bottom). Parameters are: $r = 2$, $h = 1000/256\,\text{m}$, $\Delta t = 5$ days, $\sigma_p = 1000$, $\sigma_S = 0$

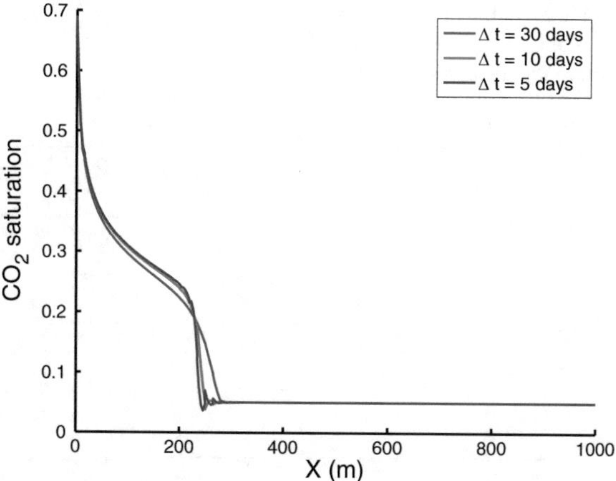

Fig. 13 Comparison of the numerical solutions for CO_2 saturation with different time steps. Parameters are: $r = 2$, $h = 1000/128$ m, $\sigma_p = 1000$, $\sigma_S = 0$

observe more oscillations near the saturation front in Fig. 12. It seems that the local oscillations are better controlled with larger penalty values for the term penalizing the saturation.

CO_2 Injection Simulation for Different Time Steps

In this section, we investigate how large of a time step we can choose when we use the backward Euler for the time discretization. We simulate the injection of CO_2 with $\Delta t = 5, 10, 30$ days, respectively, on 128 elements for a final time $T = 300$ days. Figure 13 shows the numerical solutions of the CO_2 saturation. We see that the solution with $\Delta t = 10$ days is only slightly diffusive than with $\Delta t = 5$ days. Therefore, $\Delta t = 10$ days gives an accurate enough solution. The solution with $\Delta t = 30$ days is more diffuse and fails to capture the sharp front. It is however more efficient and may be used when high accuracy is not the first priority. We also observe that the blue and the green curves have some small oscillations near the front. We note that since the problem is nonlinear, the time step does depend on the mesh size, but it is not clear how they are quantitatively related. We find that for a fixed mesh size, there is a limit to the maximum of the time step for the scheme to be stable. For instance, for this simulation, if we use 256 elements, the scheme blows up immediately with $\Delta t = 30$ days.

4.4 Injection of CO_2 into Heterogeneous Porous Medium

We consider a heterogeneous porous medium where the properties are different in the interval (0, 156.25) and the interval (156, 25, 1000). Table 8 lists the values for the parameters of the problem. The initial saturation for the wetting phase is 0.95.

Table 8 Parameter values for the CO_2 injection into the heterogenous porous medium

Parameter	Interval	Value
ϕ	(0, 1000)	0.25
K	(0, 156.25)	10^{-12} (m^2)
	(156.25, 1000)	10^{-13} (m^2)
p_d	(0, 156.25)	1000 (Pa)
	(156.25, 1000)	5000 (Pa)
λ	(0, 1000)	2
S_{wr}	(0, 156.25)	0.05
	(156.25, 1000)	0.1
ρ_w	(0, 1000)	1000 (kg/m^3)
μ_w	(0, 1000)	10^{-3} ($Pa \cdot s$)

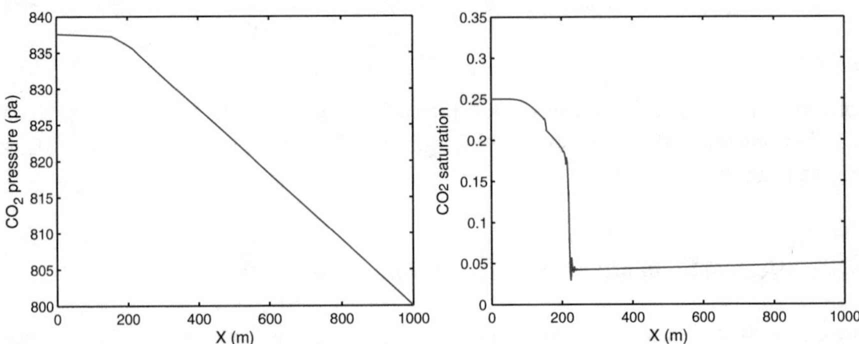

Fig. 14 CO_2 pressure and saturation after 10 years of injection into heterogenous porous media

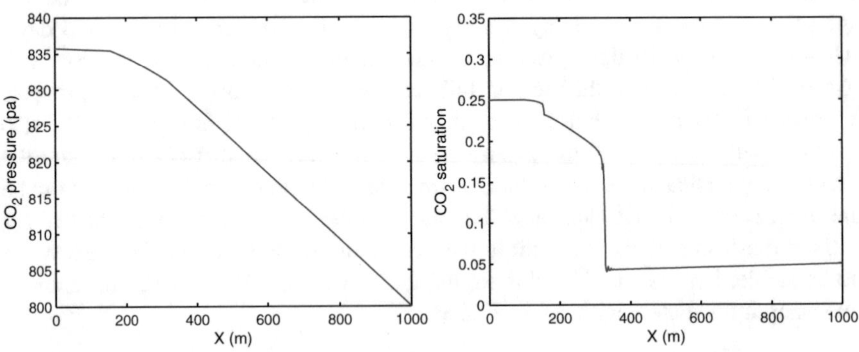

Fig. 15 CO_2 pressure and saturation after 15 years of injection into heterogenous porous media

The partial upwind method is used on a mesh with 512 elements and the time step is equal to 10 days. The penalties are chosen to be $\sigma_p = 1000$ and $\sigma_S = 10$. Figures 14 and 15 show the simulation for 10 years and 15 years, correspondingly. We can see that there is a jump for the saturation when the non-wetting phase goes from a high permeability medium to a low permeability medium. The Newton–Raphson method takes three steps to converge with a relative tolerance of 10^{-8}.

This test shows that the partial upwind method is a good candidate for simulating discontinuous solutions.

5 Conclusion

In this work, we propose the partial upwind method, which is a new version of the discontinuous Galerkin method that employs a carefully chosen decomposition of the elliptic and hyperbolic parts of the two-phase two-component model problem. The method is shown to be convergent, stable, and robust for several simulation test cases, including the case of incompressible two-phase flow and the case of injection of CO_2 in homogeneous and heterogeneous media in one dimension. Future work will study injection in higher-dimensional domains.

References

1. P. Bastian, A fully coupled discontinuous Galerkin method for two-phase flow in porous media with discontinuous capillary pressure. Comput. Geosci. **18**(5), 779–796 (2014)
2. P. Bastian, B. Riviere, Discontinuous Galerkin methods for two-phase flow in porous media. Technical Reports of the IWR (SFB 359) of the Universität Heidelberg, 2004
3. R.H. Brooks, A.T. Corey, *Hydraulic Properties of Porous Media*. Colorado State University Hydrology Papers, vol. 3 (Colorado State University, Fort Collins, 1964)
4. H. Class, A. Ebigbo, R. Helmig, H.K. Dahle, J.M. Nordbotten, M.A. Celia, P. Audigane, M. Darcis, J. Ennis-King, Y. Fan, et al., A benchmark study on problems related to CO2 storage in geologic formations. Comput. Geosci. **13**(4), 409–434 (2009)
5. Y. Epshteyn, B. Riviere, Fully implicit discontinuous finite element methods for two-phase flow. Appl. Numer. Math. **57**(4), 383–401 (2007)
6. Y. Epshteyn, B. Riviere, Convergence of high order methods for miscible displacement. Int. J. Numer. Anal. Model. **5**, 4763 (2008)
7. A. Ern, I. Mozolevski, Discontinuous Galerkin method for two-component liquid–gas porous media flows. Comput. Geosci. **16**(3), 677–690 (2012)
8. A. Ern, I. Mozolevski, L. Schuh, Discontinuous Galerkin approximation of two-phase flows in heterogeneous porous media with discontinuous capillary pressures. Comput. Methods Appl. Mech. Eng. **199**(23), 1491–1501 (2010)
9. O. Eslinger, Discontinuous Galerkin finite element methods applied to two-phase, air-water flow problems, PhD thesis, The University of Texas at Austin, 2005
10. V. Girault, J. Li, B. Riviere, Strong convergence of the discontinuous Galerkin scheme for the low regularity miscible displacement equations. Numer. Methods Partial Differential Equations **33**(2), 489513 (2017)
11. H. Huang, G. Scovazzi, A high-order, fully coupled, upwind, compact discontinuous Galerkin method for modeling of viscous fingering in compressible porous media. Comput. Methods Appl. Mech. Eng. **263**, 169–187 (2013)
12. W. Klieber, B. Rivière, Adaptive simulations of two-phase flow by discontinuous Galerkin methods. Comput. Methods Appl. Mech. Eng. **196**(1), 404–419 (2006)

13. J. Li, B. Riviere, High order discontinuous Galerkin method for simulating miscible flooding in porous media. Comput. Geosci. **19**(6), 12511268 (2015)
14. J. Li, B. Riviere, Numerical modeling of miscible viscous fingering instabilities by high-order methods. Transp. Porous Media **113**(3), 607628 (2016)
15. G. Scovazzi, A. Gerstenberger, S.S. Collis, A discontinuous Galerkin method for gravity-driven viscous fingering instabilities in porous media. J. Comput. Phys. **233**, 373–399 (2013)
16. G. Scovazzi, H. Huang, S.S. Collis, J. Yin, A fully-coupled upwind discontinuous Galerkin method for incompressible porous media flows: high-order computations of viscous fingering instabilities in complex geometry. J. Comput. Phys. **252**, 86–108 (2013)
17. G. Scovazzi, M.F. Wheeler, A. Mikelić, S. Lee, Analytical and variational numerical methods for unstable miscible displacement flows in porous media. J. Comput. Phy. **335**, 444–496 (2017)

Regularization Results for Inhomogeneous Ill-Posed Problems in Banach Space

Beth M. Campbell Hetrick

Abstract We prove continuous dependence on modeling for the inhomogeneous ill-posed Cauchy problem in Banach space X, then use these results to obtain a regularization result. The particular problem we consider is given by $\frac{\mathrm{d}u(t)}{\mathrm{d}t} = Au(t) + h(t), 0 \le t < T, u(0) = \chi$, where $-A$ generates a uniformly bounded holomorphic semigroup $\{e^{-zA} | Re(z) \ge 0\}$ and $h : [0, T) \to X$. In the approximate problem, the operator A is replaced by the operator $f_\beta(A)$, $\beta > 0$, which approximates A as β goes to 0. We use a logarithmic approximation introduced by Boussetila and Rebbani. Our results extend earlier work of the author together with Fury and Huddell on the homogeneous ill-posed problem.

1 Introduction

Ill-posed problems continue to be the focus of much research. As mathematical models are developed and used to describe natural phenomena, often the models exhibit instability: small changes in initial data may lead to large differences in corresponding solutions. In this case, we say that these problems are not continuously dependent on data and thus are ill-posed. We start with the ill-posed abstract Cauchy problem

$$\frac{\mathrm{d}u(t)}{\mathrm{d}t} = Au(t), 0 \le t < T,$$
$$u(0) = \chi, \tag{1}$$

where $-A$ is the infinitesimal generator of a uniformly bounded holomorphic semigroup $\{S(z) = e^{-zA} | \operatorname{Re}(z) \ge 0\}$ in a Banach space $(X, \|\cdot\|)$, $\chi \in X$, and

B. M. Campbell Hetrick (✉)
Gettysburg College, Gettysburg, PA, USA
e-mail: bcampbel@gettysburg.edu

A. Deines et al. (eds.), *Advances in the Mathematical Sciences*, Association for Women in Mathematics Series 15, https://doi.org/10.1007/978-3-319-98684-5_13

u is a function $u : [0, T] \to X$. Generalizing this, we consider the *inhomogeneous* ill-posed problem given by

$$\begin{aligned} \frac{\mathrm{d}u(t)}{\mathrm{d}t} &= Au(t) + h(t), \quad 0 \le t < T \\ u(0) &= \chi, \end{aligned} \tag{2}$$

with $h : [0, T) \to X$. We assume that h is differentiable on $(0, T)$ and that $h' \in L^1((0, T); X)$. We define an approximate well-posed problem

$$\begin{aligned} \frac{\mathrm{d}v(t)}{\mathrm{d}t} &= f_\beta(A)v(t) + h(t), \quad 0 \le t < T \\ v(0) &= \chi, \end{aligned} \tag{3}$$

where $\beta > 0$ and $f_\beta(A)$ approximates A. We prove continuous dependence on modeling; i.e., that if $f_\beta(A)$ is a suitable approximation of A, then a solution of the ill-posed problem (if it exists) is appropriately close to the solution of the well-posed model problem.

Previous results have been obtained using several different approximations $f_\beta(A)$. For example, $f_\beta(A) = A - \beta A^2$ is used in [1, 2, 4, 13, 15, 17, 20], and $f_\beta(A) = A(I + \beta A)^{-1}$ is used in [2, 4, 12, 19]. As explained in [10], both of these approximations result in an error of order $\mathrm{e}^{\frac{C}{\beta}}$, which leads to difficulty in extending results to the nonlinear case. In this paper we use a logarithmic approximation introduced by Boussetila and Rebbani [3],

$$f_\beta(A) = -\frac{1}{pT}\ln(\beta + \mathrm{e}^{-pTA}), \quad \beta > 0, \quad p \ge 1, \tag{4}$$

extending the results we obtained together with Fury and Huddell for the homogenous case in [10]. Boussetila and Rebbani use this approximation in Hilbert space, as do others who obtain more general results, including the nonlinear case [22] and the nonautonomous problem [8], where the operator A is time-dependent. Huang [11] extends the results to Banach space for the homogeneous case, and this paper presents results in Banach space for the inhomogeneous case. As in [10], our methodology differs from others in that we prove directly continuous dependence on the model. We show

$$\|u(t) - v_\beta(t)\| \le C(\sqrt{\beta})^{1-\frac{t}{T}} M^{\frac{t}{T}}, \quad 0 \le t < T \tag{5}$$

where $u(t)$ is an assumed solution of (2), $v_\beta(t)$ is the solution of (3) using the logarithmic function f_β in (4), and C and M are constants independent of β.

With observation error an inherent part of measurement and data collection, we are interested further in the approximate problem (3) with initial data slightly different than that in (2). Specifically, we consider the approximate problem (3) with initial data χ replaced by χ_δ, $\delta > 0$, where $\|\chi - \chi_\delta\| \leq \delta$. Using our above estimate (5), we obtain regularization results. Regularization results have been proved previously by authors including Melnikova [14]; Melnikova et al. [16]; Trong and Tuan [20, 21]; Huang and Zheng [12]; and Fury [9].

In Sect. 2, we present necessary background information. Section 3 contains our continuous dependence result, and in Sect. 4 we prove regularization. Throughout the paper, $B(X)$ denotes the space of bounded linear operators on X and $\varrho(A)$ the resolvent set of A.

2 Background

We follow [11] and use a functional calculus by deLaubenfels [5] to define $f_\beta(A)$. Here we review the development of $f_\beta(A)$ as outlined in [10]. Given that $-A$ generates $\{S(z) = \mathrm{e}^{-zA} \,|\, \mathrm{Re}(z) \geq 0\}$, define the function

$$G(s, A) = \frac{1}{\pi}\int_{-\infty}^{\infty} \frac{1-\cos(sr)}{r^2}\mathrm{e}^{\mathrm{i}rA}\mathrm{d}r\,, \quad s \geq 0\,.$$

Then $G(s, A)$ is a continuous function in s mapping into $B(X)$ and $\|G(s, A)\| \leq s\left(\sup_{\mathrm{Re}(z)\geq 0}\|\mathrm{e}^{-zA}\|\right)$. By switching within equivalent norms, we may assume without loss of generality that $\|G(s, A)\| \leq s$ for all $s \geq 0$ (cf. [11]). This leads to the functional calculus given by deLaubenfels:

$$f(A) = \left[\lim_{t\to\infty} f(t)\right]I + \int_0^\infty f''(s)G(s, A)ds \qquad (6)$$

for $f \in AC_r^1[0, \infty) := \{h \circ g \;:\; h \in AC^1[0, 1]\}$ where $g(t) = (1 + t)^{-1}$ and $AC^1[0, 1] := \{f \;:\; f' \text{ exists and is absolutely continuous on } [0, 1]\}$. For details on applying this functional calculus to obtain a formula for $f_\beta(A)$ as given by (4), see [11]. Using $f(s) = -\frac{1}{pT}\ln(\beta + \mathrm{e}^{-pTs})$ together with (6) yields

$$f_\beta(A) = -\frac{1}{pT}\ln\beta - \int_0^\infty \frac{\beta pT\mathrm{e}^{-pTs}}{(\beta + \mathrm{e}^{-pTs})^2}G(s, A)\mathrm{d}s\,.$$

Furthermore, Huang shows that $f_\beta(A)$ is a bounded operator on X satisfying $\|f_\beta(A)\| \leq -\frac{3}{pT}\ln\beta$ for $0 < \beta < (\sqrt{5} - 1)/2$.

With the functional calculus in place, we prove lemmas in [10] that are needed again in the inhomogeneous case.

Lemma 1 ([10, Lemma 1]) *Let $-A$ be the infinitesimal generator of a uniformly bounded holomorphic semigroup $\{S(z) = e^{-zA} : \mathrm{Re}(z) \geq 0\}$ on X and let $0 < \beta < (\sqrt{5} - 1)/2$. Then*

$$\| - Ax + f_\beta(A)x\| \leq \frac{8\beta}{pT}\|S^{-1}(2pT)x\|$$

for all $x \in Dom(S^{-1}(2pT)) = Ran(S(2pT))$.

For $x \in Dom(A)$, define the operator $g_\beta(A)$ in X by

$$g_\beta(A)x = -Ax + f_\beta(A)x \ .$$

Then $g_\beta(A)$ generates a C_0 semigroup $\{e^{tg_\beta(A)}\}_{t\geq 0}$ on X and we have the following bound:

Lemma 2 ([10, Lemma 2]) *For sufficiently small $\beta > 0$,*

$$\|e^{tg_\beta(A)}\| \leq \frac{2}{\sqrt{\beta}}\left(\sqrt{\frac{t}{pT}} + 1\right) \quad for \quad 0 \leq t \leq T \ .$$

3 Continuous Dependence on Modeling

A classical solution $u(t)$ of (2) is a function $u : [0, T] \to X$ such that $u(t) \in Dom(A)$ for $0 < t < T$, $u \in C[0, T] \cap C^1(0, T)$, and u satisfies (2) in X. A *strong* solution of (2) is a function $u(t)$ which is differentiable almost everywhere on $[0, T]$ such that $u'(t) \in L^1((0, T); X)$, $u(0) = \chi$, and $u'(t) = Au(t) + h(t)$ almost everywhere on $[0, T]$ [18]. We assume throughout that $u(t)$ is a strong solution of (2). The following theorem states conditions under which such a solution exists:

Theorem 1 *[18, Corollary 4.2.10] Let X be a Banach space and let A be the infinitesimal generator of a C_0 semigroup $U(t)$ on X. If $h : [0, T) \to X$ is differentiable almost everywhere on $[0, T]$ and $h' \in L^1((0, T); X)$, then for every $\chi \in Dom(A)$ the initial value problem (2) has a unique strong solution u on $[0, T]$ given by*

$$u(t) = U(t)\chi + \int_0^t U(t - s)h(s)ds. \tag{7}$$

We assume that h is differentiable on $(0, T)$ and that $h' \in L^1((0, T); X)$.

It is crucial to our work to know the circumstances under which the Cauchy problem is well-posed, and we review some essential results here. Following deLaubenfels, for $\epsilon > 0$ define a family of bounded operators

$$C_\epsilon = \frac{1}{2\pi \mathrm{i}} \int_{\Gamma_\phi} \mathrm{e}^{-\epsilon w^2} (w - A)^{-1} \mathrm{d}w$$

where Γ_ϕ is a complex contour contained within $\rho(A)$, running from $\infty \mathrm{e}^{\mathrm{i}\phi}$ to $\infty \mathrm{e}^{-\mathrm{i}\phi}$ with $0 < \phi < \frac{\pi}{2}$. Then $\{C_\varepsilon\}_{\varepsilon > 0}$ is a strongly continuous holomorphic semigroup on X generated by $-A^2$. In [7], deLaubenfels relates C-semigroups to the abstract Cauchy problem. A key theorem follows:

Theorem 2 ([7, cf. Corollary 4.2]) *Suppose A is the generator of a C-semigroup $\{W(t)\}_{t \geq 0}$. Then the abstract Cauchy problem* (1) *has a unique solution, for all $\chi \in C(Dom(A))$, given by $u(t, \chi) = W(t)C^{-1}\chi$. Problem* (1) *is then well-posed, in the following sense: when $\|C^{-1}(\chi_n - \chi)\|$ converges to 0, as n goes to infinity, then $u(t, \chi_n)$ converges to $u(t, \chi)$, uniformly on compact sets.*

Lemma 3 *The differential equation given by*

$$\frac{\mathrm{d}w(t)}{\mathrm{d}t} = Aw(t) + C_\epsilon h(t), \quad 0 \leq t < T$$
$$w(0) = C_\epsilon \chi,$$

where $-A$ is the infinitesimal generator of a uniformly bounded holomorphic semigroup $\{S(z) = \mathrm{e}^{-zA} \,|\, \mathrm{Re}(z) \geq 0\}$, has a unique strong solution.

Proof Since $-A$ is the infinitesimal generator of a uniformly bounded holomorphic semigroup, we have from deLaubenfels [6, Thm. 3.5] that A generates an entire C_ϵ group $\{W_\epsilon(T)\}_{t \geq 0}$ where $W_\epsilon(t) = C_\epsilon \mathrm{e}^{tA} = C_\epsilon S^{-1}(t)$. Then by Theorem 2 the abstract Cauchy problem has a unique solution for all initial data in $C_\epsilon(Dom(A))$. Note that for $\chi \in Dom(A)$, we have $C_\epsilon \chi \in C_\epsilon(Dom(A))$. Furthermore, since $h(t)$ is differentiable on $(0, T)$ and $h'(t) \in L^1((0, T); X)$, $C_\epsilon h(t)$ is differentiable on $(0, T)$ and $\frac{d}{dt}(C_\epsilon h(t)) = C_\epsilon h'(t) \in L^1((0, T); X)$ since C_ϵ is a bounded operator. Thus following Pazy (cf. Theorem 1), the differential equation given above has a unique strong solution. □

Lemma 4

$$C_\epsilon u(t) = W_\epsilon(t)\chi + \int_0^t W_\epsilon(t - s)h(s)ds.$$

Proof Consider the problem given by

$$\frac{dw}{dt} = Aw(t) + C_\epsilon h(t),$$
$$w(0) = C_\epsilon \chi. \tag{8}$$

By Theorem 2 and Lemma 3 above, (8) has a unique strong solution of the form

$$w(t, C_\epsilon \chi) = W_\epsilon(t)C_\epsilon^{-1}(C_\epsilon \chi) + \int_0^t W_\epsilon(t-s)C_\epsilon^{-1}(C_\epsilon h(s))ds$$
$$= W_\epsilon(t)\chi + \int_0^t W_\epsilon(t-s)h(s)ds.$$

We can show that $C_\epsilon u(t)$ is a solution of the differential equation given in (8). Further, $C_\epsilon u(0) = C_\epsilon \chi$. Thus $C_\epsilon u(t)$ is a solution of (8). Since this solution is unique, we have

$$C_\epsilon u(t) = w(t) = W_\epsilon(t)\chi + \int_0^t W_\epsilon(t-s)h(s)\mathrm{d}s.$$

□

Lemma 5

$$C_\epsilon v_\beta(t) = C_\epsilon \mathrm{e}^{tf_\beta(A)}\chi + \int_0^t C_\epsilon \mathrm{e}^{(t-s)f_\beta(A)}h(s)\mathrm{d}s.$$

Proof We have

$$C_\epsilon v_\beta(t) = C_\epsilon \left[\mathrm{e}^{tf_\beta(A)}\chi + \int_0^t \mathrm{e}^{(t-s)f_\beta(A)}h(s)ds\right]$$
$$= C_\epsilon \mathrm{e}^{tf_\beta(A)}\chi + \int_0^t C_\epsilon \mathrm{e}^{(t-s)f_\beta(A)}h(s)ds.$$

□

Note that C_ϵ commutes with A and thus with $S(t)$ and $\mathrm{e}^{(t-s)f_\beta(A)}$.

In the proof of our main result, we use the following lemma:

Lemma 6 (cf. [2, Lemma 5]) *For* $\epsilon > 0$,

$$\mathrm{e}^{tf_\beta(A)}C_\epsilon = \mathrm{e}^{tg_\beta(A)}W_\epsilon(t).$$

We now state our continuous dependence result.

Theorem 3 *Let* $-A$ *be the infinitesimal generator of a uniformly bounded holomorphic semigroup* $\{S(z) = \mathrm{e}^{-zA} | \mathrm{Re}(z) \geq 0\}$ *on* X *and let* $0 < \beta < \frac{\sqrt{5}-1}{2}$. *Let* $u(t)$ *and* $v_\beta(t)$ *be solutions of the ill-posed problem* (2) *and the approximate problem* (3), *respectively. Assume that* $u(t) \in \mathrm{Dom}(S^{-1}(2pT)) = \mathrm{Ran}(S(2pT))$ *and* $\|S^{-1}(2pT)u(t)\| \leq M'$ *for* $0 \leq t \leq T$. *Assume that h is differentiable on* $(0, T)$ *and that* $h' \in L^1((0, T); X)$. *Assume* $h(t) \in \mathrm{Dom}(S^{-1}(3pT)) = \mathrm{Ran}(S(3pT))$ *and* $\|S^{-1}(3pT)h(t)\| \leq N$ *for* $0 \leq t \leq T$.

Then there exist constants C and M, each independent of β, such that

$$\|u(t) - v_\beta(t)\| \leq C(\sqrt{\beta})^{1-\frac{t}{T}} M^{\frac{t}{T}} \quad \text{for} \quad 0 \leq t < T.$$

Recall from the proof of Lemma 3 that we may write $W_\epsilon(t) = C_\epsilon S^{-1}(t)$. It follows from our assumptions above that $\chi, h \in Dom(S^{-1}(t))$. Using this information together with Lemma 4 we can write

$$C_\epsilon u(t) = C_\epsilon S^{-1}(t)\chi + \int_0^t C_\epsilon S^{-1}(t-s)h(s)\mathrm{d}s.$$

Proof Using Lemmas 4 and 5 and the above remark, we have

$$\begin{aligned}
\|C_\epsilon(u(t) - v_\beta(t))\| &= \|C_\epsilon u(t) - C_\epsilon v_\beta(t)\| \\
&= \Big\| C_\epsilon S^{-1}(t)\chi + \int_0^t C_\epsilon S^{-1}(t-s)h(s)\mathrm{d}s \\
&\qquad - C_\epsilon \mathrm{e}^{tf_\beta(A)}\chi - \int_0^t C_\epsilon \mathrm{e}^{(t-s)f_\beta(A)}h(s)\mathrm{d}s \Big\| \\
&= \left\| C_\epsilon S^{-1}(t)\chi - C_\epsilon \mathrm{e}^{tf_\beta(A)}\chi \right\| \qquad (9) \\
&\quad + \left\| \int_0^t C_\epsilon S^{-1}(t-s)h(s)ds - \int_0^t C_\epsilon \mathrm{e}^{(t-s)f_\beta(A)}h(s)\mathrm{d}s \right\|. \qquad (10)
\end{aligned}$$

The term in (9) corresponds to work done in the homogeneous case. In [10] we show that

$$\left\| C_\epsilon \left(S^{-1}(t)\chi - \mathrm{e}^{tf_\beta(A)}\chi \right) \right\| \leq C \left(\sqrt{\beta}\right)^{1-\frac{t}{T}} M^{\frac{t}{T}}$$

where C and M are both constants independent of β and ϵ. We now focus on the term in (10). Using Lemma 6, we have

$$\begin{aligned}
&\left\| \int_0^t C_\epsilon S^{-1}(t-s)h(s)ds - \int_0^t C_\epsilon \mathrm{e}^{(t-s)f_\beta(A)}h(s)\mathrm{d}s \right\| \\
&\quad \leq \int_0^t \|C_\epsilon S^{-1}(t-s)h(s) - \mathrm{e}^{(t-s)f_\beta(A)}C_\epsilon h(s)\|\mathrm{d}s \\
&\quad \leq \int_0^t \|C_\epsilon S^{-1}(t-s)h(s) - \mathrm{e}^{(t-s)g_\beta(A)}W_\epsilon(t-s)h(s)\|\mathrm{d}s \\
&\quad \leq \int_0^t \left\| \left(I - \mathrm{e}^{(t-s)g_\beta(A)}\right) C_\epsilon S^{-1}(t-s)h(s) \right\| \mathrm{d}s.
\end{aligned}$$

Applying Lemmas 1 and 2 to the integrand above yields

$$
\begin{aligned}
&\left\|\left(I-\mathrm{e}^{(t-s)g_\beta(A)}\right)C_\epsilon S^{-1}(t-s)h(s)\right\|\\
&\qquad=\left\|-\int_0^{(t-s)}\frac{\mathrm{d}}{\mathrm{d}w}\mathrm{e}^{wg_\beta(A)}C_\epsilon S^{-1}(t-s)h(s)dw\right\|\\
&\qquad=\left\|-\int_0^{(t-s)}\mathrm{e}^{wg_\beta(A)}g_\beta(A)C_\epsilon S^{-1}(t-s)h(s)dw\right\|\\
&\qquad\le\int_0^{(t-s)}\left\|\mathrm{e}^{wg_\beta(A)}\right\|\left\|g_\beta(A)C_\epsilon S^{-1}(t-s)h(s)\right\|dw\\
&\qquad\le\int_0^{(t-s)}\frac{2}{\sqrt{\beta}}\left(\sqrt{\frac{w}{pT}}+1\right)\left\|g_\beta(A)C_\epsilon S^{-1}(t-s)h(s)\right\|dw\\
&\qquad\le\int_0^{(t-s)}\left[\frac{2}{\sqrt{\beta}}\left(\sqrt{\frac{w}{pT}}+1\right)\right]\frac{8\beta}{pT}\left\|S^{-1}(2pT)C_\epsilon S^{-1}(t-s)h(s)\right\|dw\\
&\qquad\le\int_0^{(t-s)}\frac{16\sqrt{\beta}}{pT}\left(\sqrt{\frac{w}{pT}}+1\right)\left\|S^{-1}(3pT)C_\epsilon h(s)\right\|dw,
\end{aligned}
$$

where $C_\epsilon S^{-1}(t-s)h(s)\in \mathrm{Dom}(S^{-1}(2pT))$.

Let $C=\sup_\epsilon\|C_\epsilon\|$. Together with the assumption that $\|S^{-1}(3pT)h(t)\|\le N$ for $0\le t\le T$, we have

$$
\begin{aligned}
&\int_0^{(t-s)}\frac{16\sqrt{\beta}}{pT}\left(\sqrt{\frac{w}{pT}}+1\right)\left\|S^{-1}(3pT)C_\epsilon h(s)\right\|dw\\
&\qquad\le\int_0^{(t-s)}\frac{16\sqrt{\beta}}{pT}\left(\sqrt{\frac{w}{pT}}+1\right)C\left\|S^{-1}(3pT)h(s)\right\|dw\\
&\qquad\le\int_0^{(t-s)}\frac{16\sqrt{\beta}}{pT}\left(\sqrt{\frac{w}{pT}}+1\right)CNdw\\
&\qquad\le\sqrt{\beta}C_2
\end{aligned}
$$

where C_2 is a constant independent of β and ϵ. Using $0<\beta<1$, we obtain

$$
\begin{aligned}
\|C_\epsilon(u(t)-v_\beta(t))\|&\le C\left(\sqrt{\beta}\right)^{1-\frac{t}{T}}M^{\frac{t}{T}}+T\sqrt{\beta}C_2\\
&\le C\left(\sqrt{\beta}\right)^{1-\frac{t}{T}}M^{\frac{t}{T}}+\left(\sqrt{\beta}\right)^{1-\frac{t}{T}}\left(\sqrt{\beta}\right)^{\frac{t}{T}}TC_2\\
&\le C\left(\sqrt{\beta}\right)^{1-\frac{t}{T}}M^{\frac{t}{T}}
\end{aligned}
$$

for a possibly different constant C, where again C is a constant independent of β. The bound on the right is independent of ϵ, so we let $\epsilon \to 0$ to get

$$\|u(t) - v_\beta(t)\| \leq C(\sqrt{\beta})^{1-\frac{t}{T}} M^{\frac{t}{T}}.$$

□

4 Regularization

We now consider the approximate problem (3) with perturbed initial data χ_δ and prove a regularization result. Following work on the homogeneous case (cf. [12] and [10]) and the extension to the inhomogeneous case in [9], we use the following definition:

Definition 1 ([9, Definition 4.1]) A family $\{R_\beta(t) | \beta > 0, t \in [0, T]\} \subseteq B(X)$ is called a *family of regularizing operators for the inhomogeneous Cauchy problem* (2) if for every solution $u(t)$ of (2) with initial data $\chi \in X$ and h differentiable on $(0, T)$ with $h' \in L^1((0, T); X)$, and for any $\delta > 0$, there exists $\beta(\delta) > 0$ such that

(i) $\beta(\delta) \to 0$ as $\delta \to 0$,
(ii) $\|u(t) - \left(R_\beta(t)\chi_\delta + \int_0^{t-s} R_\beta(t-s)h(s)\mathrm{d}s\right)\| \to 0$ as $\delta \to 0$ for each $t \in [0, T]$ whenever $\|\chi - \chi_\delta\| \leq \delta$.

Theorem 4 *Let $-A$ be the infinitesimal generator of a uniformly bounded holomorphic semigroup $\{S(z) = \mathrm{e}^{-zA} | \operatorname{Re}(z) \geq 0\}$ on X and let $f_\beta(A)$ be defined by* (4). *Then $\{R_\beta(t) := \mathrm{e}^{tf_\beta(A)} | 0 < \beta < (\sqrt{5}-1)/2, t \in [0, T]\}$ is a family of regularizing operators for problem* (2).

Note that due to the form of the solution of the approximate problems, the proof of this result for $t \in [0, T)$ follows immediately from the regularization result of Theorem 2 in [10]. We use the same ideas to prove the result below.

Proof Let $u(t)$ be a solution of (2) with $\|S^{-1}(2pT)u(t)\| \leq M'$ for all $t \in [0, T]$. Assume that h is differentiable on $(0, T)$ and that $h' \in L^1((0, T); X)$. Assume $h(t) \in \operatorname{Dom}(S^{-1}(3pT)) = \operatorname{Ran}(S(3pT))$ and $\|S^{-1}(3pT)h(t)\| \leq N$ for $0 \leq t \leq T$. Let $\delta > 0$ be given. Consider the approximate problem with perturbed data given by

$$\frac{dv_\beta(t)}{dt} = f_\beta(A)v_\beta(t) + h(t), \quad 0 \leq t < T$$

$$v_\beta(0) = \chi_\delta,$$

and assume $\|\chi - \chi_\delta\| \leq \delta$. This problem has solution

$$\begin{aligned} v_{\beta,\delta}(t) &= \mathrm{e}^{tf_\beta(A)}\chi_\delta + \int_0^{t-s} \mathrm{e}^{(t-s)f_\beta(A)}h(s)\mathrm{d}s \\ &= R_\beta(t)\chi_\delta + \int_0^{t-s} R_\beta(t-s)h(s)\mathrm{d}s. \end{aligned}$$

Since $f_\beta(A) \in B(X)$, we have

$$\|\mathrm{e}^{tf_\beta(A)}\| \leq \mathrm{e}^{t\|f_\beta(A)\|} \leq \mathrm{e}^{-\frac{3t}{pT}\ln\beta} = \beta^{-\frac{3t}{pT}}$$

for all $t \in [0, T]$ where $0 < \beta < (\sqrt{5}-1)/2$.

Let $t \in [0, T)$ and choose $\beta = \delta^{\frac{p}{6}}$. Then $\beta \to 0$ as $\delta \to 0$ and together with Theorem 3 we have

$$\begin{aligned} &\left\| u(t) - \left(R_\beta(t)\chi_\delta + \int_0^{t-s} R_\beta(t-s)h(s)\mathrm{d}s \right) \right\| \\ &= \|u(t) - v_{\beta,\delta}(t)\| \\ &\leq \|u(t) - v_\beta(t)\| + \|v_\beta(t) - v_{\beta,\delta}(t)\| \\ &\leq C(\sqrt{\beta})^{1-\frac{t}{T}} M^{\frac{t}{T}} \\ &\quad + \left\| \left(\mathrm{e}^{tf_\beta(A)}\chi + \int_0^{t-s} \mathrm{e}^{(t-s)f_\beta(A)}h(s) \right) \right. \\ &\quad\quad \left. - \left(\mathrm{e}^{tf_\beta(A)}\chi_\delta - \int_0^{t-s} \mathrm{e}^{(t-s)f_\beta(A)}h(s) \right) \right\| \\ &\leq C(\sqrt{\beta})^{1-\frac{t}{T}} M^{\frac{t}{T}} + \|\mathrm{e}^{tf_\beta(A)}(\chi - \chi_\delta)\| \\ &\leq C(\sqrt{\beta})^{1-\frac{t}{T}} M^{\frac{t}{T}} + \beta^{-\frac{3t}{pT}}\delta \\ &\leq C(\delta^{\frac{p}{12}})^{1-\frac{t}{T}} M^{\frac{t}{T}} + \delta^{1-\frac{t}{2T}}, \end{aligned}$$

which goes to 0 as $\delta \to 0$.

Now consider the case when $t = T$. It can be shown through an argument similar to that used in the proof of Theorem 1 in [10] that $\|u(T) - v_\beta(T)\| \leq K\sqrt{\beta}$, where K is a constant independent of β. Using $\beta = \delta^{\frac{p}{6}}$ still, we have

$$\begin{aligned} \left\| u(T) - \left(R_\beta(T)\chi_\delta + \int_0^{T-s} R_\beta(T-s)h(s)\mathrm{d}s \right) \right\| &\leq K\sqrt{\beta} + \beta^{-\frac{3}{p}}\delta \\ &= K\delta^{\frac{p}{12}} + \sqrt{\delta}, \end{aligned}$$

which goes to 0 as $\delta \to 0$. □

Acknowledgements The author would like to thank Rhonda J. Hughes for her inspiration and continued outstanding mentorship, along with Matthew Fury and Walter Huddell for their constant support and guidance.

References

1. K.A. Ames, On the comparison of solutions of related properly and improperly posed Cauchy problems for first order operator equations. SIAM J. Math. Anal. **13**, 594–606 (1982)
2. K.A. Ames, R.J. Hughes, Structural stability for ill-posed problems in Banach space. Semigroup Forum **70**, 127–145 (2005)
3. N. Boussetila, F. Rebbani, A modified quasi-reversibility method for a class of ill-posed Cauchy problems. Georgian Math. J. **14**, 627–642 (2007)
4. B.M. Campbell Hetrick, R.J. Hughes, Continuous dependence results for inhomogeneous ill-posed problems in Banach space. J. Math. Anal. Appl. **331**, 342–357 (2007)
5. R. deLaubenfels, Functional calculus for generators of uniformly bounded holomorphic semigroups. Semigroup Forum **38**, 91–103 (1989)
6. R. deLaubenfels, Entire solutions of the abstract Cauchy problem. Semigroup Forum **42**, 83–105 (1991)
7. R. deLaubenfels, C-Semigroups and the Cauchy problem. J. Funct. Anal. **111**, 44–61 (1993)
8. M. Fury, Modified quasi-reversibility method for nonautonomous semilinear problems. Electron. J. Differ. Equ. **Conf. 20**, 99–121 (2013); Ninth Mississippi State Conference on Differential Equations and Computational Simulations
9. M. Fury, Regularization for ill-posed inhomogeneous evolution problems in a Hilbert space. Discrete Contin. Dyn. Syst. **Suppl. 2013**, 259–272 (2013); 9th AIMS Conference on Dynamical Systems, Differential Equations and Applications
10. M. Fury, B. Campbell Hetrick, W. Huddell, Continuous dependence on modeling in Banach space using a logarithmic approximation, in *Mathematical and Computational Approaches in Advancing Modern Science and Engineering*, ed. by Bélair, J. et al. (Springer, Cham, 2016), pp. 653–663
11. Y. Huang, Modified quasi-reversibility method for final value problems in Banach spaces. J. Math. Anal. Appl. **340**, 757–769 (2008)
12. Y. Huang, Q. Zheng, Regularization for a class of ill-posed Cauchy problems. Proc. Am. Math. Soc. **133**, 3005–3012 (2005)
13. R. Lattes, J.L. Lions, The method of quasi-reversibility, in *Applications to Partial Differential Equations* (American Elsevier, New York, 1969)
14. I.V. Melnikova, General theory of the ill-posed Cauchy problem. J. Inverse Ill-Posed Probl. **3**, 149–171 (1995)
15. I.V. Melnikova, A. Filinkov, *Abstract Cauchy Problems: Three Approaches*. Chapman & Hall/CRC Monographs and Surveys in Pure and Applied Mathematics, vol. 120 (Chapman & Hall, Boca Raton, 2001)
16. I.V. Melnikova, Q. Zheng, J. Zhang, Regularization of weakly ill-posed Cauchy problems. J. Inverse Ill-Posed Probl. **10**, 503–511 (2002)
17. K. Miller, Stabilized quasi-reversibility and other nearly-best-possible methods for non-well-posed problems, in *Symposium on Non-Well-Posed Problems and Logarithmic Convexity*. Lecture Notes in Mathematics, vol. 316 (Springer, Berlin, 1973), pp. 161–176
18. A. Pazy, *Semigroups of Linear Operators and Applications to Partial Differential Equations* (Springer, New York, 1983)
19. R.E. Showalter, The final value problem for evolution equations. J. Math. Anal. Appl. **47**, 563–572 (1974)
20. D.D. Trong, N.H. Tuan, Regularization and error estimates for non homogeneous backward heat problems. Electron. J. Differ. Equ. **4**, 1–10 (2006)

21. D.D. Trong, N.H. Tuan, A nonhomogeneous backward heat problem: regularization and error estimates. Electron. J. Differ. Equ. **33**, 1–14 (2008)
22. D.D. Trong, N.H. Tuan, Stabilized quasi-reversibility method for a class of nonlinear ill- posed problems. Electron. J. Differ. Equ. **84**, 1–12 (2008)

Research in Collegiate Mathematics Education

Shandy Hauk, Chris Rasmussen, Nicole Engelke Infante, Elise Lockwood, Michelle Zandieh, Stacy Brown, Yvonne Lai, and Pao-sheng Hsu

Abstract The chapter sketches some of the landscape of current research in undergraduate mathematics education and offers useful information for present and future faculty members. Six research projects related to the teaching and learning of post-secondary mathematics are summarized. Approaches in the research reported here include individual interviews, classroom observations, national survey, and in-depth study of a particular instance or case of learning. The collegiate mathematics topics at the heart of the respective studies range from calculus, combinatorics, linear algebra, and foundations of proof to the application of mathematics to teaching in the development of future teachers.

Keywords Post-secondary mathematics education · Calculus · Combinatorics · Linear algebra · Proofs and proving · Teacher preparation

---

S. Hauk (✉)
WestEd, San Francisco, CA, USA
e-mail: shauk@wested.org; https://www.wested.org/personnel/shandy-hauk

C. Rasmussen
San Diego State University, San Diego, CA, USA

N. E. Infante
West Virginia University, Morgantown, WV, USA

E. Lockwood
Oregon State University, Corvallis, OR, USA

M. Zandieh
Arizona State University, Tempe, AZ, USA

S. Brown
California State Polytechnic University, Pomona, Pomona, CA, USA

Y. Lai
University of Nebraska–Lincoln, Lincoln, NE, USA

P.-s. Hsu
Independent Mathematician and Consultant, Columbia Falls, ME, USA

A. Deines et al. (eds.), *Advances in the Mathematical Sciences*, Association for Women in Mathematics Series 15, https://doi.org/10.1007/978-3-319-98684-5_14

1 Introduction

The work described in this chapter is research on the teaching and learning of undergraduate mathematics, written to be accessible to someone with advanced training in the mathematical sciences. The aim is to open and promote conversations between two communities of researchers: those in mathematics and those in post-secondary mathematics education. In this report, overviews of six different projects offer a view into the variety in topics and methods of research in collegiate mathematics education. The chapter arose from interactions among attendees at the 2017 AWM Symposium. For the first time, the symposium included a session on research in collegiate mathematics education. The six presentations in the session highlighted investigations into the teaching and learning of calculus, combinatorics, linear algebra, foundations of proof, and the application of mathematics in the college preparation of future school teachers. Research methods included individual and group interviews, classroom observations, national survey, and in-depth study of a particular instance or case. The research reported here relied on a variety of well-established perspectives on the nature of human cognition and knowledge structures.

Though research on the relationships amongst humans and ideas in education is far different from research rooted in relationships of ideas (e.g., in mathematics), the investigative processes share some characteristics. First, one must start with a few axioms and definitions. In education, the locally useful axioms form what is called a theoretical framework—a foundational perspective on the nature of thinking and human experience. In collegiate mathematics education research, necessary definitions include those for the methods of conducting a particular investigation (e.g., the ways that interview, observation, review of documents, and survey are used). No proofs exist in mathematics education research. Instead, like medical research where the participants are human, empirical evidence derived from capturing the nature of human experience is the basis of a compelling argument. In what follows, the authors touch on the theoretical and methodological foundations of their work and offer summary results and discussion of the implications of that work.

We start with two reports from the world of research on calculus. The first, by Rasmussen, is from a national research project into the organizational configurations and policies—institutional, departmental, and course-specific—that shape calculus instruction and lead to student persistence (or not) in mathematical course-taking. The second report about calculus, from Engelke Infante, is a deep dive into the classrooms of five instructors, examining how they used gestures in communicating about the second derivative test. Next, from the area of combinatorics, Lockwood reports on fascinating studies of what counts in counting; in particular, what are the ways we articulate the *multiplication principle*? It turns out the answer is nuanced and that an appropriate set of counting tasks can elicit important subtleties when students attempt to reinvent the statement of the principle. In a similar vein, the report from Zandieh on research and development in inquiry-driven linear algebra offers insight into the role of the instructor as a broker of mathematically rigorous meaning as students build their understanding of eigentheory and diagonalization. In the realm of mathematical discourse practices around proof, Brown reports that having students identify and explain the strengths and stumbles in a purported proof by writing the story of a conversation about it, a *proof script*, can be a valuable tool for instructors to identify potential expert blind spots about the nature of student struggles with proofs and proving. In the domain of applied mathematics, in the final section Lai gives a glimpse of what it might mean to apply mathematics to the work of teaching (e.g., as distinct from the ways mathematics is applied in engineering, or computer science, or physics) in teaching future teachers.

2 Calculus Program Structure

Calculus is typically the first mathematics course for science, technology, engineering, and mathematics (STEM) majors in the United States. Indeed, each fall approximately 300,000 college or university students, most of them in their first post-secondary year, take a course in differential calculus [10]. However, student retention in a STEM major and the role calculus plays in retention is a major problem [20, 40, 47]. This report provides an overview of characteristics of relatively more successful Calculus I programs across the country. Success was defined by a combination of student persistence in the calculus sequence; affective changes, including enjoyment of math, confidence in mathematical ability, interest to continue studying math; and passing rates.

2.1 Methods

Data for this report come from a 5-year, large empirical study funded by the National Science Foundation and run under the auspices of the Mathematical Association of America [13]. The project was conducted in two phases. In Phase 1 surveys were sent to a stratified random sample of students and their instructors at the beginning and the end of Calculus I courses designed to prepare students for the study of engineering or the mathematical or physical sciences. In Phase 2, the project team conducted explanatory case studies at 18 different post-secondary institutions, where the type of institution was determined by the highest degree offered in mathematics. In this report, the focus is the five case studies at doctoral degree granting institutions.

2.2 Results

The survey results from Phase 1 provided information on which institutions were more successful in terms of persistence, perceptions of math, and passing rates, as compared to other institutions of the same type. Surveys are valuable because they can gather information related to evidence of success. But, however well-crafted and implemented, surveys are limited in their ability to shed light on *why* and *how* institutions are producing students who are successful in calculus. The Phase 2 case studies addressed this shortcoming by identifying and contextualizing the teaching practices, training practices, and institutional support practices that contributed to student success in Calculus I.

Cross case analysis of the five doctoral degree granting institutions led to the identification of the following seven features that contribute to the success of a calculus program.

Coordination. Calculus I has a permanent course Coordinator. The Coordinator holds regular meetings where calculus instructors talk about course pacing and coverage, develop midterm and final exams, discuss teaching and student difficulties, etc. Exams and finals are common and in some cases the homework assignments are uniform.

Attending to Local Data. There was someone in the department who routinely collected and analyzed data in order to inform and assess program changes. Departments did this work themselves and did not rely on the university to do so. Data collected and analyzed included pass rates, grade distributions, persistence, placement accuracy, and success in Calculus II.

Graduate Teaching Assistant (GTA) Training. The more successful calculus programs had substantive and well-thought-out GTA training programs. These ranged from a weeklong training prior to the semester together with follow up work during the semester to a semester course taken prior to teaching. The course included a significant amount of mentoring, practice teaching, and observing classes. GTAs were mentored in how to use active learning strategies with students.

Active Learning. Calculus instructors were encouraged to use and experiment with active learning methods of teaching. In some cases, the department Chair sent out regular emails with links to articles or other information about teaching. One institution even had biweekly teaching seminars led by the math faculty or invited experts. Particular instructional approaches, however, were not prescribed or required for faculty at any of the institutions.

Rigorous Courses. The more successful calculus programs tended to challenge students mathematically. They used textbooks and selected problems that required students to delve into concepts, work on modeling-type problems, or even proof-type problems. Techniques and skills were still highly valued. In some cases, these were assessed separately and a satisfactory score on this assessment was a requirement for passing the course.

Learning Centers. Students were provided with out of class resources. Almost every institution had a well-run and well-utilized tutoring center. In some instances, this was a calculus only tutoring center and in others the tutoring center served linear algebra and differential equations. Tutoring labs had a director and tutors received training.

Placement. Programs tended to have more than one way to determine student readiness for calculus. This included: placement exams (which were monitored to see if they were doing the job intended), gateway tests 2 weeks into the semester and different calculus format (e.g., more time) for students with lower algebra skills.

2.3 Discussion

To interpret these findings, we have drawn on Tinto's academic and social integration perspective, which highlights the relationship between persistence and social and academic integration [31, 50]. From this perspective, student persistence is viewed as a function of the dynamic relationship between the student and other actors within the institutional environment, including the classroom environment. For example, almost without exception the students we talked with noted that they felt their calculus course was academically engaging and challenging (despite the fact that the vast majority had taken calculus in high school) and that there were a number of resources available to them to help them be successful. These resources included well-developed math help centers and availability of instructor and GTA office hours. Other factors that contributed to students' academic and social integration included student centered instruction, common space in the math department where students could gather to work on homework, dorms that provided them with opportunities to interact with like-minded students, and in some places a cohort system or strong student culture that provided cohesion among students. The fact that each of the more successful Calculus I programs regularly brought faculty together to develop shared resources and perspectives on teaching is also noteworthy.

In summary, the analysis of the five successful calculus programs at doctoral institutions highlighted a number of structural and programmatic features that other institutions would likely be interested in adapting. While there is no one size fits all, these common features offer a starting point for departments to initiate their own improvement efforts.

3 Gesture in Teaching Calculus

Every concept in mathematics can be represented in multiple ways. A function can be a graph, a formula, a set of points. A critical aspect of learning and understanding mathematical concepts is the ability to use and move between different representations of a common idea. Connections among varied representations and concepts form the foundations of advanced mathematics. Here, we explore how to assist students in making these connections and deepen their understanding. Understanding of concepts is shaped through bodily experiences, such as gesture. The theoretical framing for examination of human experience of mathematics from this perspective is called embodied cognition [32, 38]. Many researchers have indicated that gesture is a potentially powerful, but underutilized, means for making links during instruction [2, 8, 24, 45, 46].

Unfortunately, Calculus I has a high failure rate and is the reason many students leave their chosen field [19, 47]. Something different is needed in the classroom to help students succeed. Calculus is the study of motion, so the calculus classroom is a natural place to study how gesture is used in communicating about it. The purposeful use of gesture in calculus has great potential, especially as it is free to implement.

3.1 Methods

We took a careful and detailed look at how five instructors used gesture during naturally arising classroom activities that involved the second derivative test [21]. The method was what is called a qualitative case study [14]. Each lesson in which the second derivative test was introduced was transcribed and broken down into linking episodes—segments of classroom lecture or interaction in which links between key ideas are made [2]. These linking episodes were then analyzed to determine how many and what types of gestures were used in conjunction with the links being made.

The instructors used three primary forms of gesture: pointing, depictive, and writing. Pointing gestures are those used to index objects, locations, and inscriptions in the physical world. Depictive gestures are those made to represent the motion or shape of an object, such as a graph. Writing gestures are those in which writing or drawing is integrated with speech so that a writing instrument is used to indicate content of the accompanying speech (such as drawing a circle around a term in an equation while stating "this term"). Here, we summarize the results of three typical, distinct gesture use patterns. In the transcriptions, square brackets around an utterance indicates it was accompanied by the gesture described in the connected bracket. For example, a person pointing to a peak on a graph while saying "a maximum" in the sentence "We know we have a maximum" is transcribed as: "We know we have [a maximum] [pointing to peak]."

3.2 Results

Each instructor introduced the second derivative test while solving an optimization problem: either construct a fence that maximizes area for a given perimeter

(Instructors A, B, and C) or construct a fence of minimum perimeter for a fixed area (Instructors D and E). During this instruction, each instructor linked two key ideas: (1) the sign of the second derivative and the concavity of the function, and (2) the concavity of the function and the existence of a maximum or minimum.

Instructor A primarily used pointing gestures. At the beginning of her lesson, she drew an unlabeled graph with a maximum and a minimum value. As she pointed to the maximum value on her graph, she stated "we know we have [a maximum] [pointing to peak]," and then she pointed to the minimum on her graph and stated "we know we have [a minimum] [pointing to valley], but we don't know this starting out." She continued and asked the students "what does the second derivative tell us?" The students responded, "concavity." She proceeded to point to the maximum on the graph and asked "[what's the concavity right here?] [circular motion with her hand around the maximum on the graph]." This led to the conclusion that a concave down graph would have a maximum, which was pointed to again. Through a similar sequence of questions and points, the class arrived at the conclusion that a concave up graph would have a minimum. Hence, we see that she uses a sequence of points to draw attention to information on the graph and link that information to the existence of maxima and minima.

In contrast, Instructor C made very few references to what he had written on the board. Instead, he primarily used depictive gestures while facing the class. After drawing a small, downward curve on the board and quickly pointing to the maximum, he turned to face the students and moved to the center of the boards at the front of the classroom and exclaimed, "Watch! Watch! Watch!" and made sure that all eyes were on him before he proceeded (Fig. 1). He demonstrated that a concave down graph would have a maximum value (blue arching motion above the red line) while a concave up graph (blue arching motion below the red line) must have a minimum value. He related each of these ideas back to the idea of a horizontal tangent line (which he indicated with the horizontal red line motion).

Instructor B used a combination of the strategies used by Instructors A and C. She drew a representative graph on the board and took time to deliberately point to it (Fig. 2, left), and then she moved away from the board and reiterated the connection between concavity and the existence of a maximum (Fig. 2, right).

Fig. 1 Instructor C highlighting aspects of the second derivative test using gesture

Fig. 2 Above left: Instructor B pointing to the maximum on a concave down graph. Above right: Instructor B making a large, two-handed gesture for a concave down graph

3.3 Discussion

While each of the instructors used a similar example problem to introduce the second derivative test, there was significant variation in how they chose to link the key ideas of the concept. Even though there is noteworthy variation among the quantity and type of gestures used among instructors, more gesture is not necessarily indicative of a better lesson. A single, appropriately timed gesture of a particular type may be equally effective. For example, Instructor C made certain that his students would be paying attention to at least one sequence of gestures he made.

There is mounting evidence that instructors' use of gesture promotes student learning and that instructors can intentionally alter their gesture production during instruction to good effect [1, 3, 17, 27]. Gestures promote long-term memory [15–17]. Hence, instructors are encouraged to spend a little time during lesson preparation thinking about how to purposefully incorporate gesture to facilitate students' understanding.

4 Combinatorics

Combinatorics is a rich and accessible topic, but counting problems are difficult for students to learn and for teachers to teach (e.g., see [5, 18]). One approach for mitigating such difficulties is to study foundational concepts pertaining to combinatorial enumeration. The Multiplication Principle (MP), which is sometimes referred to as The Fundamental Principle of Counting (e.g., as in [43]), is a one such foundational aspect of counting. This principle is fundamental to combinatorics, underpinning many standard formulas and counting strategies. Generally, the MP says that for independent stages in a counting process, the number of options at each stage can be multiplied together to yield the total number of outcomes of the entire process (specific statements are provided below). It is central to (and can justify) many of the counting formulas to which students are introduced.

4.1 Methods

Because it is such an important topic that is central to combinatorial enumeration, my colleagues and I conducted two related studies about the MP. First, we conducted a textbook analysis of statements of the MP, reported in [36] and second, we conducted a reinvention study with a pair of undergraduate student novice counters, reported further in [35]. Together, these studies contribute to our field's understanding of both the MP as a mathematical construct and how students reason about the MP. I briefly highlight results from each of these studies and report some potential implications of the work.

4.2 Results

In the textbook analysis, we examined 64 textbooks that were used at universities across the country, and because some books had more than one statement, this yielded 73 statements in total. This textbook analysis facilitated an in-depth conceptual analysis of the MP, and it yielded two major results. First, we identified and reported three different statement types, structural, operational, and bridge statements. Structural statements characterize the MP as involving counting structural objects (such as lists or k-tuples). Operational statements characterize the MP as determining the number of ways of completing a counting process. Bridge statements simultaneously characterizes the MP as counting structural objects and specifies a process by which those objects are counted. Statements by Bona [11] in Figure 1, Rosen [44] in Figure 2, and Tucker [51] in Figure 3 offer examples of these three respective statement types.

These statement types highlight three different ways of presenting the MP, particularly in terms of how counting processes and sets of outcomes are framed [34]. They also help to characterize a surprising amount of variation we observed among the statement types—among the 73 statements, we identified 22 structural statements, 33 operational statements, and 18 bridge statements (Figs. 3, 4, and 5).

> ***Generalized Product Principle:*** *Let $X_1, X_2, \ldots, X_k$ be finite sets. Then the number of k-tuples (x_1, $x_2, \ldots, x_k$) satisfying $x_i \in X_i$ is $|X_1| \times |X_2| \times \ldots \times |X_k|$.*

Fig. 3 Bona's [11] statement of the MP (Structural)

> ***The Product Rule:*** *Suppose that a procedure can be broken down into tasks. If there are n_1 ways to do the first task and n_2 ways to do the second task after the first task has been done, then there are n_1n_2 ways to do the procedure.*

Fig. 4 One of Rosen's [44] statements of the MP (Operational)

> ***The Multiplication Principle**: Suppose a procedure can be broken down into m successive (ordered) stages, with r_1 different outcomes in the first stage, r_2 different outcomes in the second stage, …, and r_m different outcomes in the mth stage. If the number of outcomes at each stage is independent of the choices in the previous stages, and if the composite outcomes are all distinct, then the total procedure has $r_1 \times r_2 \times \ldots \times r_m$ different composite outcomes.*

Fig. 5 Tucker's [51] statement of the MP (Bridge)

For the second major result of the textbook analysis, we identified three key mathematical ideas that are central to the MP and, we conjectured, would be important aspects of student reasoning about the MP. These key mathematical ideas are that a statement of the MP should: (a) require independence of a number of options for stages in a counting process, (b) allow for dependence of option sets, and (c) require composite outcomes to be distinct. If a statement does not attend to these issues, then it is possible to apply the MP in a way that could yield an incorrect answer. We elaborate certain problems that demonstrate the importance of these key mathematical ideas both in [36] and [35].

While the textbook analysis was theoretically interesting to us, we also wanted to examine students' reasoning about the MP, including the extent to which these key mathematical ideas arose in their conceptions of the MP. Given our findings from the textbook analysis, we followed up with another study in which we had two undergraduate students reinvent a statement of the MP.

These students were undergraduates who were enrolled in a vector calculus class; had not taken a discrete mathematics course in college. We interviewed them in 8 h-long sessions during which they collaboratively solved problems and iteratively refined their statements. The students first engaged in counting activity that involved multiplication, and then we asked them to articulate a statement of when they would use multiplication to solve a counting problem.

For the sake of space, we juxtapose their first and last statements, providing a brief description of their progression. Their first statement said:

> Use multiplication in counting problems when there is a certain statement shown to exist and what follows has to be true as well.

We interpret that the students were trying to account for multiple constraints, and this yielded a reasonable but far from a rigorous initial statement of the MP. Throughout the sessions we gave students certain problems that were designed to target the key mathematical ideas mentioned above, and we used certain problems to elicit conversations that led students to make small adjustments to their statements.

Ultimately, the students refined their statement of the MP and came up with the final statement:

> If for every selection towards a specific outcome, there is no difference in the number outcome, regardless of the previous selections, then you multiply the number of all the options in each selection together to get the total number of possible unique outcomes.

This final statement attends to each of the key mathematical ideas, and through this process we gained insight into students' reasoning about the MP.

4.3 Discussion

The results from both the textbook analysis and the interviews with students emphasize that the MP is a nuanced idea with subtle and important mathematical features. One conclusion from these studies is that if students do not grapple these subtleties, they may apply the MP without understanding potential issues that may arise. For researchers and teachers, we would encourage them to take seriously the MP, even if it appears to deal with a familiar and straightforward operation. For students who are learning counting, we suggest that they should be thoughtful and careful about how they apply the MP when solving counting problems.

5 Linear Algebra

Mounting evidence exists regarding the positive impacts of active learning on student success and attitudes in undergraduate science, technology, engineering, and mathematics (STEM) courses (e.g., [22, 29]). In addition, linear algebra is an important content course because of its wide applicability in many STEM fields. Our team has been conducting research and curriculum development in linear algebra for over 10 years. The work has included creating the NSF-funded *Inquiry-Oriented Linear Algebra* (IOLA) curriculum materials (http://iola.math.vt.edu/; [4, 53]).

Inquiry-oriented instruction is a form of active learning in which students contribute to the reinvention of important mathematical ideas; this is in contrast to forms of active learning in which students' activity is marked by practicing or applying principles that have been previously explained or demonstrated. We draw on Rasmussen and Kwon's [41] two-part definition of inquiry: (1) students inquire into the mathematics and (2) instructors inquire into student thinking. In particular, instructors engage students in problem solving tasks that elicit rich mathematical discussions. Within those discussions, the instructor points out connections between the mathematics that students are generating and the established mathematical definitions and symbolisms that the instructor wants students to learn. In this way, the instructor serves as a *broker* between the classroom community and the larger mathematical community [42].

5.1 Methods: IOLA Materials

This task sequence, which we refer to as "The Blue to Black Unit," supports students' reinvention of change of basis, eigentheory, and how they are related through diagonalization. Task 1 builds from students' experience with linear transformations in $\mathbb{R}^2$ to introduce them to the idea of stretch factors and stretch directions and how these create a non-standard coordinate system for $\mathbb{R}^2$. In Task 2, students create matrices that convert between standard and non-standard coordinate systems and relate these to the stretching transformation of Task 1 to reinvent the equation $A\mathbf{x} = PDP^{-1}\mathbf{x}$. In Task 3, students build from their experience with stretch factors and directions to create for themselves ways to determine eigenvalues and eigenvectors given various pieces of information about transformations. In Task 4, students work with examples in $\mathbb{R}^3$ to develop the characteristic equation as a solution technique, as well as connect ideas about eigentheory to their earlier work with change of basis through the idea of diagonalization.

For this short report, we provide an overview of student activity in Task 1 and then illustrate in more detail one example of student and teacher inquiry from Task 1. More detail regarding student work and instructor facilitation on this task, as well as Task 2 of this unit, can be found in [55]. Task 3 is discussed in [39].

5.2 Results: Illustrative Example

Student inquiry in Task 1 (see Fig. 6) involves exploring a graphical situation, but also finding ways to symbolize graphical activity using vectors and matrices. The inquiry involves creating or choosing appropriate ways to symbolize mathematical processes and relationships. One student who we will call Donald explained his work to the class (see Fig. 7). He said that the points along $y = x$ stay fixed so the upper right and lower left corners of the box stay fixed. Note his circles around these two points on the right graph in Fig. 7. To determine what happened to the upper left corner point, $(-2, 2)$, he noted that it is along a line parallel to $y = -3x$ that starts at the point $(-1, 1)$. On the left graph in Fig. 7 we can see that he has drawn a number of line segments parallel to $y = -3x$. One of these has dots marked by Donald at $(-1, 1)$, $(-2, 2)$, and $(-3, 5)$. Donald indicated that the point $(-2, 2)$ would move (be transformed) to the point $(-3, 5)$ which he drew in on the right in Fig. 7 as the new upper left corner of the stretched parallelogram.

Donald's inquiry into the mathematics led him to note geometrical relationships within the task setting. In response, the instructor talked the class through her board work (Fig. 8, next page), which she connected to what Donald had explained. She pointed out the vector $(-2, 2)$ can be written as $(-1, -1) + (-1, 3)$, and that when the linear transformation is applied to this equation, we know that $(-1, -1)$ stays fixed and $(-1, 3)$ is doubled resulting in $(-1, -1) + (-1, 3) = (-3, 5)$.

The instructor's explanation provided a bridge between Donald's geometric argument and an argument based on the ideas and notation of vector equations and linear transformations. This is an example of the instructor serving as a *broker* between the classroom community and the larger mathematical community.

Donald was not the only student or group of students who had drawn lines on their paper parallel to $y = -3x$ or parallel to $y = x$. Later in the class the instructor recalled this student gridding activity and used it to introduce the graphic in Fig. 9 (next page). The instructor's discussion of the solution process using the

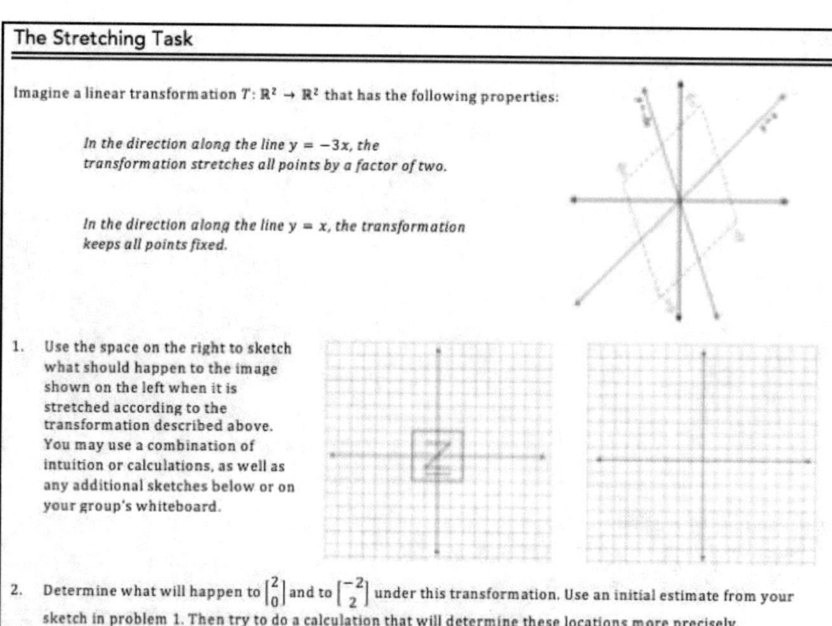

The Stretching Task

Imagine a linear transformation $T: \mathbb{R}^2 \rightarrow \mathbb{R}^2$ that has the following properties:

In the direction along the line $y = -3x$, the transformation stretches all points by a factor of two.

In the direction along the line $y = x$, the transformation keeps all points fixed.

1. Use the space on the right to sketch what should happen to the image shown on the left when it is stretched according to the transformation described above. You may use a combination of intuition or calculations, as well as any additional sketches below or on your group's whiteboard.
2. Determine what will happen to $\begin{bmatrix} 2 \\ 0 \end{bmatrix}$ and to $\begin{bmatrix} -2 \\ 2 \end{bmatrix}$ under this transformation. Use an initial estimate from your sketch in problem 1. Then try to do a calculation that will determine these locations more precisely.
3. Determine a matrix that allows you to calculate what happens under the transformation to any point on the plane. Use it to check your sketch or improve its accuracy.

Fig. 6 Task 1: The stretching task

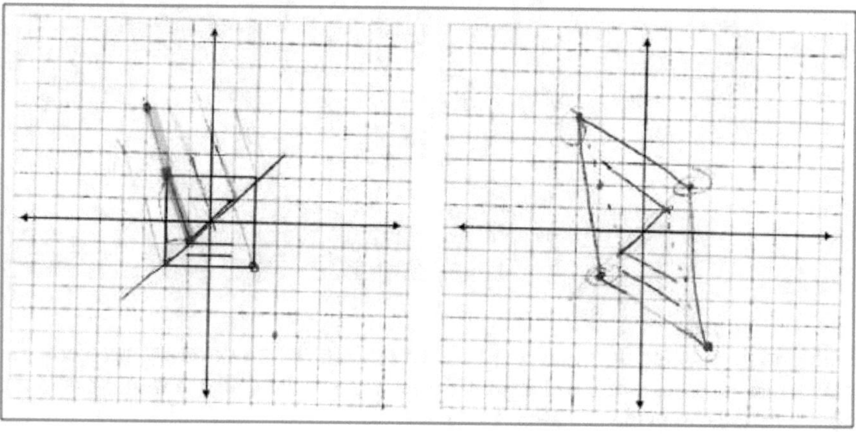

Fig. 7 Donald's work

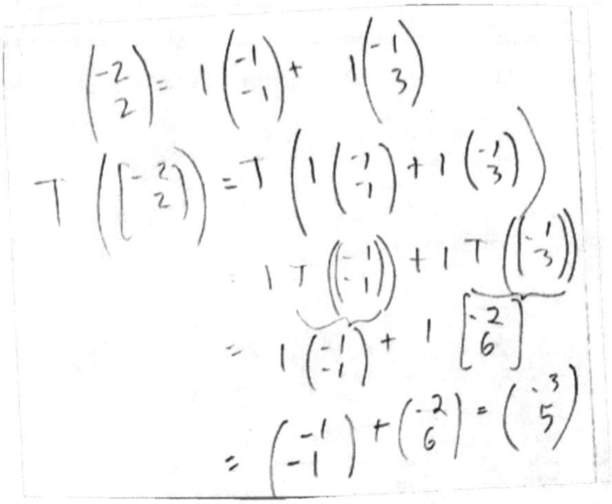

Fig. 8 In recapping Donald's work, his instructor wrote out the work shown

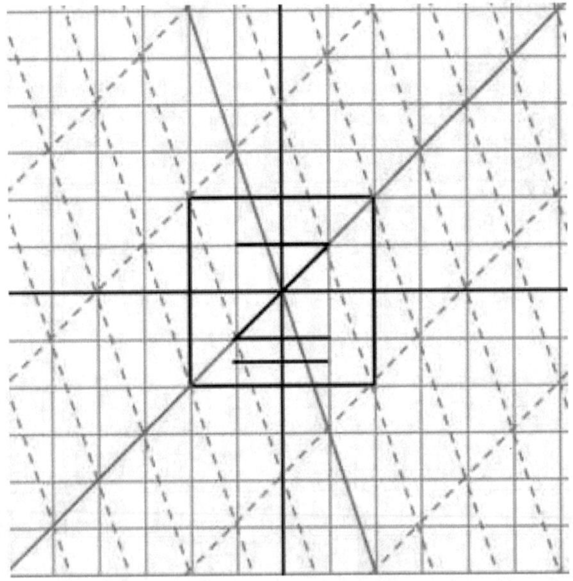

Fig. 9 A unifying graphic that illustrates stretch directions and foreshadows $A = PDP^{-1}$

blue gridding served as a unifying summary of student work on the problem, but also a foreshadowing of the change of basis idea that the students would use in the next task to develop the equation $A = PDP^{-1}$.

5.3 Discussion

This brief vignette serves as an example of both student inquiry into mathematical ideas and an instructor's bridging from the students' mathematics to additional mathematical ideas that she wanted the students to learn. For this kind of classroom inquiry to occur, one needs both (1) tasks that allow for and (2) instructor interactions that encourage these explorations and discussions.

6 Unearthing Problematics Through Student Proof Scripts

The aim of the reported study was to explore students' ways of reasoning about contradictions through the examination of students' responses to the proof script task shown in Fig. 10. The proof employed in this task was selected for two reasons. First, the proof provided an avenue to examine students' reasoning about contradictions outside of the context of a proof by contradiction. Indeed, the reader will notice the proof does not demonstrate that the negation of the theorem leads to a contradiction but shows that the theorem's hypotheses lead to its conclusion.

Second, while research has shown that proofs by contradiction present specific difficulties for students [6, 26, 33], research has yet to identify those difficulties that are salient to students. Moreover, little is known about students' ways of reasoning about those difficulties. For example, little is known about students' reasoning about logically degenerate cases (i.e., cases within a proof, whose hypotheses lead to a contradiction either with other given hypotheses, their consequences, or statements in the broader mathematical theory) even though proofs of this form are common in introductory topology, number theory, and real analysis courses (e.g., consider how one proves: For any real numbers x and y, if $x \leq y$ and $y \leq x$, then $x = y$.).

6.1 Methods

To bridge this gap in the research literature and identify the difficulties that are salient to students, the proof script methodology was employed. This method draws on Commognitive Theory in which it is argued, "thinking is a form of communication and ... learning mathematics is tantamount to modifying and extending one's discourse" [48, p. 567]. Furthermore, proof scripts were appropriate for the study since proof scripts afford a window into students' ways of reasoning that are salient to students by unseen by experts; that is, they provide a means to avoid expert blind spots [30, 37, 56, 57].

As part of a larger study, 20 students enrolled in Introduction to Proof courses were randomly assigned the task shown in Fig. 10. Using a thematic analysis [12], common student difficulties were identified. The following problematic aspects were anticipated: (1) the variable m is used but not defined; (2) there are only two cases, shouldn't there be four? and (3) does the contradiction in Case 2 imply the proof is flawed?

Instructions: Create a 1-2 page dialog that introduces and explains the theorem and its proof. Highlight the problematic points in the proof with questions and answers. In your submission:

- The dialog should occur between two students, you and a mathematics student named Gamma, which you can denote with either G or g.
- Start by reading the proof and identifying what you believe are the *problematic points* for a learner when attempting to understand the theorem/statement or its proof. List these *problematic points* in a bulleted list.
- **Write a dialog** between you and Gamma in which you explain the theorem and the proof to Gamma, paying special attention to the *problematic points* you identified and listed in your bulleted list. In the dialog you should both pose questions to Gamma and answer any questions Gamma might ask. (THIS IS THE MAIN PART OF THE ASSIGNMENT)

You may add comments at the end or within the dialog using [ ].

Theorem 1: For every integer m, if $3\nmid((m^2-1)$ then $3|m$

1) *Proof*: Assume $3\nmid(m^2-1)$.
2) Since $(m^2-1)=(m+1)(m-1)$, we can say $3\nmid(m+1)(m-1)$.
3) By a previous theorem we know,
4) if $3\nmid(m+1)(m-1)$ then $3\nmid(m+1)$ and $3\nmid(m-1)$.
5) Since $3\nmid(m+1)$, it follows from the Division Algorithm, that
6) $(m+1)=3k+1$ or $(m+1)=3k+2$ for some integer k.
7) *Case 1*. Let $(m+1)=3k+1$
8) Then $m+1-1=3k+1-1$
9) $m=3k$
10) Since k is an integer, $3|m$.
11) This is our desired result.
12) *Case 2*. Let $(m+1)=3k+2$
13) Then $m+1-1=3k+2-1$
14) $m=3k+1$
15) $m-1=3k+1-1$
16) $m-1=3k$
17) Since k is an integer, $3|(m-1)$,
18) This is a contradiction.
19) It follows from *Case 1* and *Case 2*, that $3|m$. □

Fig. 10 Proof script assignment

6.2 Results

Analyses indicated that all of the problematic aspects anticipated prior to data collection were observed in the students' scripts. Specifically, 6 out of 20 students noted the statement "let m be an integer" should be included; 7 out of 20 posited (incorrectly) that two additional cases (i.e., a total of four cases) were required; and 4 out of 20 argued the contradiction in Case 2 implied either the theorem was false or the proof was flawed.

While it was anticipated students would find the contradiction in Case 2 problematic, two ways of reasoning were observed which were common but unanticipated. First, 4 out of 20 students addressed confusion over the role of the contradiction by arguing an alternative "better" proof—a proof by contraposition—should be used (see Examples 1 and 2).

Example 1

Student 1: [...] As for case 2, the contradiction cannot prove the case, because since you stated that 3 does not divide $m + 1$ then 3 divides m should be true. Same as case 1.

Gamma: How will I be able to prove the theorem then?

Student 1: Approach the proof with a different direction. Negate the statement or prove it by contraposition.

Gamma: Okay, thanks. [*End-of-dialog.*]

Example 2

Student 2: The theorem suggests that for every integer m, if 3 does not divide $m^2 - 1$, then 3 divides m.

Gamma: I proved this directly using the division algorithm

Student 2: But it would be less problematic to prove this theorem by the contrapositive.

This finding is surprising since research indicates university students experience difficulties accepting proofs by contraposition [7, 23]. Indeed, Goetting found students were "wary of the validity of the 'backwards' arguments" [23, p. 124].

Second, 6 out of 20 students argued Case 2 as unnecessary since the Division Algorithm produces a disjunctive result as in Example 3.

Example 3

Gamma: Case 2 has proven that $3|(m - 1)$ which is the contradiction of $3|m$ then does that mean the theorem is false?

Student 3: From line 6, it is stated that the remainders are 1 or 2, that means only one case is needed to be true for the 3|m to be true. Moreover, case 2 is not quite necessarily needed when case 1 is already proven to be true to make the theory true. [*End-of-dialog.*]

These remarks suggest the following reasoning:

(a) $m + 1 = 3k + 1$ or $m + 1 = 3k + 2$;
(b) $m + 1 = 3k + 1$ implies $3|m$;
(c) Hence, $m + 1 = 3k + 1$ or $m + 1 = 3k + 2$ implies $3|m$.

In other words, given $A \vee B$ it is sufficient to prove $(A \Rightarrow Q)$ to conclude Q. The issue here is that such reasoning is flawed. $(A \vee B) \Rightarrow Q$ is logically equivalent to $(A \Rightarrow Q) \wedge (B \Rightarrow Q)$. Thus, $(A \Rightarrow Q)$ is necessary but not sufficient to conclude $(A \vee B) \Rightarrow Q$. What can account for the commonality of this response? Given that students argued, "if one of those are true ... then the statement is true," it appears students held a disjunctive interpretation of the cases. In other words, they (incorrectly) reasoned: $[(A \vee B) \Rightarrow Q] \equiv [(A \Rightarrow Q) \wedge (B \Rightarrow Q)]$; hence, $A \Rightarrow Q$ is sufficient and $B \Rightarrow \sim Q$ can be ignored. This finding is of interest, for it suggests ways of reasoning about proof-by-cases that may not impact one's interpretations when each case produces the desired conclusion, but might impact one's reasoning about logical dilemmas (i.e., arguments of the form $A \vee B$, $A \Rightarrow C$, $B \Rightarrow D$, $C \vee D$), which are common to non-constructive existence proofs.

6.3 Discussion

Many would argue a key characteristic of effective instruction is the inclusion of curricular activities that address common student difficulties. Due to a large body of research indicating experts may be blind to novices' difficulties, educational researchers have developed methods for identifying difficulties experienced by students but unseen by experts. Findings from this study indicate that, even though prior research has shown students experience difficulties with proof by contraposition, there are contexts in which students may gravitate towards this form of indirect proof. The findings also indicate further research is needed on students' ways of reasoning about proof-by-cases. Indeed, most Introduction to Proof instructors have asked students to prove "For any integer n, $(n^2 - n)$ is even" only to have students turn in:

> *Proof*: Let n be an integer. By the division algorithm, $n = 2k$ or $n = 2k + 1$ for some integer k. If $n = 2k$ then $(n^2 - n) = 4k2 - 2k = 2(2k2 - k)$. Since $(2^2 - k)$ is an integer, $(n^2 - n)$ is even. Q.E.D.

As instructors, we often respond, "You forgot the case $n = 2k + 1$" believing the student was being absent-minded. However, the results of the scripting task provide an alternative interpretation. Namely, some students are reasoning, "Case 2 is not quite necessarily needed when Case 1 is already proven to be true."

7 Mathematics Applied to Teaching

Consider the following two tasks. Each was designed to be a set of discussion items in a mathematics class for high school mathematics teacher candidates, that is, undergraduates or graduates who seek certification to teach high school.

Task 1

(a) Derive the identity $x^0 = 1$, for all x, assuming that the additive law for exponents must extend from positive integers to 0, and that $a^1 = a$, for all a.

(b) (i) State the definitions of function, inverse of a function, and partial inverse of a function.

(ii) Using the definition of function and inverse of a function, and function and partial inverse of a function, explain how the following functions are examples or not of these ideas:

$$\sin(x),\ \arcsin(x)$$

$$\cos(x),\ \arccos(x)$$

$$x^2,\ \pm\sqrt{x}$$

$$x^3,\ \sqrt[3]{x}.$$

Task 2

(a) A student in your algebra class says, "Why is $x^0 = 1$? It seems like it should be 0, because anything times 0 is 0." How would you respond?

(b) (i) State the definitions of function, inverse of a function, and partial inverse of a function. Then interpret these definitions using different representations of functions, such as tables and graphs.

(ii) Construct as many partial inverses as you can, of:

$$f(x) = \sin(x)$$
$$f(x) = \cos(x)$$
$$f(x) = x^2$$
$$f(x) = x^3.$$

Although each pair of tasks draws on similar mathematics, there is a qualitative difference between how Tasks 1 and 2 draw on that mathematics. Learning to do mathematics for one's own sake differs from learning to do mathematics so as to help someone else learn it. Learning facts of how high school mathematics fits into higher mathematics is different from constructing instances of higher mathematics in the context of high school mathematical ideas. Below, I summarize several investigations related to these ideas. After providing motivation for such investigations, I offer a few details of methods and results for two research projects on which I have worked related to the application of mathematics to teaching.

7.1 Methods: Motivation

As Wu [54] argued, the "Intellectual Trickle-Down Theory" does not work. This is the approach that assumes prospective teachers will learn mathematical theory and can be expected to then independently derive the underlying structures of high school mathematics. Future and current high school teachers find the content and experience of their undergraduate mathematics courses to be irrelevant to their teaching, including the norms and skills for communication (see, e.g., [25, 49, 52]). It seems unlikely that teachers would deliberately use knowledge that they perceive to be irrelevant, even if that knowledge is in fact relevant.

One strategy that has been advocated for teacher education is to use tasks where mathematics is "applied" to teaching [9], such as in Task 2. While this approach is theoretically promising, it is an open question to what extent these tasks would fulfill their anticipated function. Understanding the issue of how exactly these tasks elicit knowledge and how experiences with these tasks may transfer are critical for improving teacher education using the approach. I report on some observations to date from ongoing projects focused on these issues. These investigations used task-based interview as the primary method of data collection and several rounds of qualitative categorization and coding of the interview data.

7.2 Results: Situating Teachers in Context

Heather Howell, Geoffrey Phelps, and I [28] analyzed a set of tasks where mathematics is applied to teaching (e.g., as in Task 2). The study was based on the logic that if knowledge for teaching includes mathematics but is more than mathematics, then having incorrect mathematical knowledge should lead to an incorrect answer on tasks, and that other factors related to teaching should also lead to incorrect answers. We examined 23 high school teachers' responses to 11 tasks (from a pool of 55 tasks). We found that mathematical errors contributed to only half of the incorrect responses. The other half arose, instead, from teaching factors such as alternate expectations about student error, or not understanding students' responses, or not grasping a stated teaching objective. Moreover, some interviewees' responses to pure mathematics questions asked in the interview showed understanding of mathematics yet those interviewees had applied the mathematics incorrectly to teaching situations. We have used our analysis to argue that the way that these tasks elicit mathematical knowledge differs from how a pure mathematical task would, because the tasks situate the mathematics and the person using it in the large and complex context of teaching.

In ongoing work, Erin Baldinger and I interviewed teachers on pairs of tasks where the context switches between asking the interviewee to answer the question *as a teacher in high school* and *as a student in a university mathematics course*. Our data contain multiple instances of interviewees who assert the same mathematical argument is valid when in the role of a teacher of a high school student who produced that argument and invalid when asked to judge the proof as a student in a university mathematics course—and vice versa. These data suggest that context may impact the inferences drawn from mathematical analysis of a problem, even when the task is otherwise exactly the same.

7.3 Discussion

It may be useful for mathematics departments to view mathematical knowledge for teaching as mathematics applied to teaching. As with other forms of applied mathematics, learning mathematics in isolation may be less useful than learning in context. For instance, differential equations are not useful to a mathematical biologist who cannot use differential equations to model natural phenomena. Contextualizing mathematics in teaching has the potential to improve teacher education.

8 Conclusion

Though extensive detail was beyond the scope of this chapter, each of the projects reported on here was grounded in the essentials of high quality educational research. Before any data were collected, each effort had articulated a well-developed theoretical framework, a careful selection of methods for gathering data that included informed consent from participating students and instructors, techniques for storage and analysis of data, and plans for peer-reviewed dissemination of the results. While each example discussed teaching, the work of the included projects went beyond documentation of reflective instruction. Also, each did more than report on the scholarship of teaching and learning (i.e., where the goal is to examine what may be happening in a particular instructional situation, explain, and share it with thoughtful peers). What the research projects reported on here have in common is: rigorous, systematic investigation to accumulate a preponderance of evidence about a result. The purpose of such educational research in its full reporting (e.g., in a peer-reviewed journal) is to provide compelling warrant and sufficient detail to support others in well-informed efforts to investigate the generalization and/or transfer of ideas to other contexts.

As noted at the outset, the aim here was a sketch of the landscape in collegiate mathematics education research. For a fuller picture, we encourage the interested reader to browse the listed references. Also, there are accessible reports of research in the materials available through the Mathematical Association of America's *Special Interest Group on Research in Undergraduate Mathematics Education* website http://sigmaa.maa.org/rume (e.g., short research papers are available for free in the group's conference proceedings, click the Proceedings link).

Acknowledgements Support for this work was funded by the National Science Foundation under grant numbers 0910240, 1504551, and 165215. The opinions expressed do not necessarily reflect the views of the foundation. The Calculus Case Study team members for the doctoral degree granting institutions included Jessica Ellis, Chris Rasmussen, and Dov Zazkis.

References

1. M.W. Alibali, M.J. Nathan, R.B. Church, M.S. Wolfgram, S. Kim, E. Knuth, Teachers' gestures and speech in mathematics lessons: forging common ground by resolving trouble spots. ZDM Math. Educ **45**, 425–440 (2013). https://doi.org/10.1007/s11858-012-0476-0
2. M.W. Alibali, M.J. Nathan, M.S. Wolfgram, R.B. Church, S.A. Jacobs, C.J. Martinez, E. Knuth, How teachers link ideas in mathematics instruction using speech and gesture: a corpus analysis. Cogn. Instr. **32**(1), 65–100 (2014). https://doi.org/10.1080/07370008.2013.858161
3. M.W. Alibali, A.G. Young, N.M. Crooks, A. Yeo, M.S. Wolfgram, I.M. Ledesma, et al., Students learn more when their teacher has learned to gesture effectively. Gesture, **13**(2), 210–233 (2013). https://doi.org/10.1075/gest.13.2.05ali
4. C. Andrews-Larson, M. Wawro, M. Zandieh, A hypothetical learning trajectory for conceptualizing matrices as linear transformations. Int. J. Math. Educ. Sci. Technol. **48**(6), 809–829 (2017). https://doi.org/10.1080/0020739X.2016.1276225

5. S.A. Annin, K.S. Lai. Common errors in counting problems. Math. Teach. **103**(6), 402–409 (2010)
6. S. Antonini, M.A. Mariotti, Reasoning in an absurd world: difficulties with proof by contradiction, in *Proceedings of the 30th PME Conference*, Prague, vol. 2 (2006), pp. 65–72
7. S. Antonini, M.A. Mariotti, Indirect proof: what is specific to this way of proving? ZDM Math. Educ. **40**, 401–412 (2008)
8. F. Arzarello, D. Paoloa, O. Robutti, C. Sabena, Gestures as semiotic resources in the mathematics classroom. Educ. Stud. Math. **70**, 97–109 (2009). https://doi.org/10.1007/s10649-008-9163-z
9. H. Bass, Mathematics, mathematicians, and mathematics education. Bull. Am. Math. Soc. **42**(4), 417–430 (2005)
10. R. Blair, E.E. Kirkman, J.W. Maxwell, *Statistical Abstract of Undergraduate Programs in the Mathematical Sciences in the United States. Conference Board of the Mathematical Sciences* (American Mathematical Society, Providence, 2012)
11. M. Bona, *Introduction to Enumerative Combinatorics* (McGraw Hill, New York, 2007)
12. V. Braun, V. Clarke, Using thematic analysis in psychology. Qual. Res. Psychol. **3**(2), 77–101 (2006)
13. D. Bressoud, V. Mesa, C. Rasmussen (eds.), *Insights and Recommendations from the MAA National Study of College Calculus* (Mathematical Association of America, Washington, DC, 2015)
14. L. Cohen, L. Manion, K. Morrison, *Research Methods in Education* (Routledge, London, 2011)
15. S.W. Cook, Z. Mitchell, S. Goldin-Meadow, Gesturing makes learning last. Cognition **106**, 1047–1058 (2008). https://doi.org/10.1016/j.cognition.2007.04.010
16. S.W. Cook, T.K. Yip, S. Goldin-Meadow, Gesturing makes memories that last. J. Mem. Lang. **63**, 465–475 (2010). https://doi.org/10.1016/j.jml.2010.07.002
17. S.W. Cook, R.G. Duffy, K.M. Fenn, Consolidation and transfer of learning after observing hand gesture. Child Dev. **84**(6), 1863–1871 (2013). https://doi.org/10.1111/cdev.12097
18. M.M. Eizenberg, O. Zaslavsky, Students' verification strategies for combinatorial problems. Math. Think. Learn. **6**(1), 15–36 (2004)
19. J. Ellis, M.L. Kelton, C. Rasmussen, Student perceptions of pedagogy and associated persistence in calculus. ZDM Math. Educ. **46**, 661–673 (2014). https://doi.org/10.1007/s11858-014-0577
20. J. Ellis, B.K. Fosdick, C. Rasmussen, Women 1.5 times more likely to leave STEM pipeline after calculus compared to men: lack of mathematical confidence a potential culprit. PLoS One **11**(7), e0157447 (2016). https://doi.org/10.1371/journal.pone.0157447
21. N. Engelke Infante, The second derivative test: a case study of instructor gesture use. Paper presented at the 38th annual conference of the North American chapter of the international group for the psychology of mathematics education, Tucson, AZ, 2016
22. S. Freeman, S.L. Eddy, M. McDonough, M.K. Smith, N. Okoroafor, H. Jordt, M.P. Wenderoth, Active learning increases student performance in science, engineering, and mathematics.Proc. Natl. Acad. Sci. **111**(23), 8410–8415 (2014)
23. M. Goetting, The college students' understanding of mathematical proof, Doctoral dissertation, University of Maryland (1995)
24. S. Goldin-Meadow, Beyond words: the importance of gesture to researchers and learners. Child Dev. **71**(1), 231–239 (2000)
25. M. Goulding, G. Hatch, M. Rodd, Undergraduate mathematics experience: its significance in secondary mathematics teacher preparation. J. Math. Teach. Educ. **6**(4), 361–393 (2003)
26. G. Harel, L. Sowder, Students' proof schemes: results from exploratory studies, in *Research on Collegiate Mathematics Education. III*, ed. by A. Schoenfeld, J. Kaput, E. Dubinsky (American Mathematical Society, Providence, 1998), pp. 234–283
27. A.B. Hostetter, K. Bieda, M.W. Alibali, M.J. Nathan, E. Knuth, Don't just tell them, show them! Teachers can intentionally alter their instructional gestures. Paper presented at the 28th annual conference of the cognitive science society, Vancouver, BC, 2006

28. H. Howell, Y. Lai, G. Phelps, Assessing mathematical knowledge for teaching beyond conventional content knowledge: do elementary models extend? Paper presented at the annual meeting of the american educational research association, Washington, DC, 2016
29. M. Kogan, S.L. Laursen, Assessing long-term effects of inquiry-based learning: a case study from college mathematics. Innov. High. Educ. **39**(3), 183–199 (2014)
30. B. Koichu, R. Zazkis, Decoding a proof of Fermat's Little Theorem via script writing. J. Math. Behav. **32**, 364–376 (2013)
31. G. Kuh, T. Cruce, R. Shoup, J. Kinzie, R. Gonyea, Unmasking the effects of student engagement on first-year college grades and persistence. J. High. Educ. **79**(5), 540–563 (2008)
32. G. Lakoff, R. Nunez, *Where Mathematics Comes From: How the Embodied Mind Brings Mathematics into Being* (Basic Books, New York, 2000)
33. U. Leron, A direct approach to indirect proofs. Educ. Stud. Math. **16**(3), 321–325 (1985)
34. E. Lockwood, A model of students' combinatorial thinking. J. Math. Behav. **32**, 251–265 (2013). https://doi.org/10.1016/j.jmathb.2013.02.008
35. E. Lockwood, B. Schaub, An unexpected outcome: students' focus on order in the multiplication principle, in *Proceedings of the 20th Conference on Research on Undergraduate Mathematics Education*, San Diego, CA (2017)
36. E. Lockwood, Z. Reed, J.S. Caughman, An analysis of statements of the multiplication principle in combinatorics, discrete, and finite mathematics textbooks. Int. J. Res. Undergrad. Math. Educ. **3**(3), 381–416 (2017). https://doi.org/10.1007/s40753-016-0045-y
37. M.J. Nathan, K.R. Koedinger, M.W. Alibali, Expert blind spot: when content knowledge eclipses pedagogical content knowledge, in *Proceedings of the Third International Conference on Cognitive Science*, ed. by L. Chen et al. (USTC Press, Beijing, 2001), pp. 644–648
38. R. Nemirovsky, F. Ferrara, Mathematical imagination and embodied cognition. Educ. Stud. Math. **70**, 159–174 (2009). https://doi.org/10.1007/s10649-008-9150-4
39. D. Plaxco, M. Zandieh, M. Wawro, Stretch Directions and Stretch Factors: A Sequence Intended to Support Guided Reinvention of Eigenvector and Eigenvalue. *Challenges and Strategies in Teaching Linear Algebra*, ed. by S. Stewart, C. Andrews-Larson, A. Berman, M. Zandieh (Berlin: Springer, 2018), pp. 172–192.
40. President's Council of Advisors on Science and Technology (PCAST), *Engage to Excel: Producing One Million Additional College Graduates with Degrees in Science, Technology, Engineering, and Mathematics* (The White House, Washington, DC, 2012)
41. C. Rasmussen, O.N. Kwon, An inquiry-oriented approach to undergraduate mathematics. J. Math. Behav. **26**, 189–194 (2007)
42. C. Rasmussen, M. Zandieh, M. Wawro, How do you know which way the arrows go? The emergence and brokering of a classroom mathematics practice, in *Mathematical Representations at the Interface of the Body and Culture*, ed. by W.-M. Roth (Information Age, Charlotte, 2009), pp. 171–218
43. B. Richmond, T. Richmond, *A Discrete Transition to Advanced Mathematics* (American Mathematical Society, Providence, 2009)
44. K.H. Rosen, *Discrete Mathematics and Its Applications*, 6th edn. (McGraw Hill, New York, 2007)
45. W.M. Roth, Gestures: their role in teaching and learning. Rev. Educ. Res. **71**(3), 365–392 (2001)
46. W.M. Roth, D. Lawless, Scientific investigations, metaphorical gestures, and the emergence of abstract scientific concepts. Learn. Instr. **12**, 285–304 (2002)
47. E. Seymour, N.M. Hewitt, *Talking About Leaving: Why Undergraduates Leave the Sciences* (Westview, Boulder, 1997)
48. A. Sfard, When the rules of discourse change, but nobody tells you: making sense of mathematics learning from a commognitive standpoint. J. Learn. Sci. **16**(4), 567–615 (2007)
49. C.S. Ticknor, Situated learning in an abstract algebra classroom. Educ. Stud. Math. **81**(3), 307–323 (2012)
50. V. Tinto, Linking learning and leaving, in *Reworking the Student Departure Puzzle*, ed. by J.M. Braxton (Vanderbilt University, Nashville, 2004)

51. A. Tucker, *Applied Combinatorics*, 4th edn. (Wiley, New York, 2002)
52. N. Wasserman, M. Villanueva, J.-P. Mejia-Ramos, K. Weber, Secondary mathematics teachers' perceptions of real analysis in relation to their teaching practice, in *Proceedings of the 18th Conference on Research in Undergraduate Mathematics Education*, Pittsburgh, PA (2015)
53. M. Wawro, C. Rasmussen, M. Zandieh, C. Larson, Design research within undergraduate mathematics education: an example from introductory linear algebra, in *Educational Design Research-Part B: Illustrative Cases*, ed. by T. Plomp, N. Nieveen (SLO, Enschede, 2013), pp. 905–925
54. H. Wu, The mis-education of mathematics teachers. Not. Am. Math. Soc. **58**(3), 372–384 (2011)
55. M. Zandieh, M. Wawro, C. Rasmussen, An example of inquiry in linear algebra: the roles of symbolizing and brokering. PRIMUS: Probl. Resour. Issues Math. Undergrad. Stud. **27**(1), 96–124 (2017). https://doi.org/10.1080/10511970.2016.1199618
56. D. Zazkis, J.P. Cook, Interjecting scripting studies into a mathematics education research program. Scripting Approaches in Mathematics Education. *Advances in Mathematics Education*, 205–228 (2018). https://doi.org/10.1007/978-3-319-62692-5_10
57. R. Zazkis, D. Zazkis, Script writing in the mathematics classroom: imaginary conversations on the structure of numbers. J. Res. Math. Educ. **16**(1), 54–70 (2014)

Index

A. Deines et al. (eds.), *Advances in the Mathematical Sciences*, Association for Women in Mathematics Series 15, https://doi.org/10.1007/978-3-319-98684-5

Printed in the United States
By Bookmasters